Siting Energy Facilities

Siting Energy Facilities

RALPH L. KEENEY

Woodward-Clyde Consultants
San Francisco, California

 1980

ACADEMIC PRESS

A Subsidiary of Harcourt Brace Jovanovich, Publishers
New York   London   Toronto   Sydney   San Francisco

ACADEMIC PRESS, INC.
111 Fifth Avenue, New York, New York 10003

United Kingdom Edition published by
ACADEMIC PRESS, INC. (LONDON) LTD.
24/28 Oval Road, London NW1 7DX

Library of Congress Cataloging in Publication Data

Keeney, Ralph L., Date
    Siting energy facilities.

    Bibliography: p.
    1. Energy facilities--Location. 2. Decision-
making. I. Title.
TJ163.2.K43   621.042   80–764
ISBN 0–12–403080–7

PRINTED IN THE UNITED STATES OF AMERICA

80 81 82 83   9 8 7 6 5 4 3 2 1

To my mother
Anna Murel Keeney

CHAPTER TITLES

CONTENTS

PREFACE

Energy problems will be among those most crucial to our society for the remainder of the century. Many of these problems take on national and international proportions: How much oil will be available and from where? Should liquid natural gas be imported? Is the breeder reactor necessary? How much of our energy can be supplied from solar power? and so on. As part of the answers to these questions, a large number of major energy facilities will need to be built in the ensuing years. This book describes a tool for making the process of finding sites for these facilities more efficient and more responsive to the concerns of our society. The result should be better sites and a siting process that is understandable and defensible.

Major energy facilities include power plants, dams, refineries, import terminals, storage facilities, waste disposal sites, mines, pipelines, and transmission lines. The days are past when such facilities could be sited by engineering and economic criteria alone, yet most existing siting procedures are based on these criteria. In contrast, the decision analysis siting approach described in this book is a tool for comprehensively addressing contemporary siting problems. A major focus of the approach is the systematic search for and identification of suitable candidate sites for the proposed facility. The evaluation of the candidate sites explicitly includes environmental impacts, health and safety, socioeconomic effects, and public attitudes, in addition to engineering and economic criteria. The procedure allows the inclusion of the uncertainties and value judgments that are a significant part of all energy siting problems.

The decision analysis siting approach is prescriptive. It provides substantial help to decision makers who must select a site for an energy facility, and it documents the reasoning behind the site selection for regulatory agencies and public appraisal. The decision makers include companies interested in building the facility, communities in the vicinity of the proposed facility, state and federal agencies with a responsibility related to the facility, and interest groups and intervenors with a stake in the proceedings. The procedure described in this book can be used to help any of these decision makers. In fact, it would be desirable if the different decision makers would use the format and substance of the siting analysis as an aid for structuring their discussion processes and for focusing on substantial disagreements.

Every siting procedure necessarily makes several assumptions. These include structural assumptions for the analysis and assumptions about the possible alternative sites, the collection and treatment of data and uncertainty, the value judgments necessary for evaluating the sites, and the interpretation of the results of the analysis. These assumptions are usually implicit, often unnecessarily restrictive, and sometimes inappropriate. As opposed to other siting approaches, the fundamental assumptions on which the decision analysis siting approach is based are explicitly stated (see the Appendix). For appraisal of the appropriateness of any siting approach, its assumptions should be examined by decision makers responsible for the energy siting decision.

Energy facility siting is complex. It is difficult to do a good analysis of such a problem, and yet it is important to try because of the opportunity for gaining significant insights about the best alternative. The aim of the decision analysis siting approach is to address the complexity head-on. Complexity is part of the problem, so it should not be left out of the analysis because it seems too difficult to include. Its inclusion provides at least a basis for appraising the importance of the fact that certain aspects of a siting problem are not well addressed. The spirit is not to ask what *could* be done in the analysis and then do it, but rather to ask what *should* be done and then do our best to do that. Our standard is the ideal, rather than the level that is common practice. Our goal is to raise the state of practice of siting to be much closer to the state of the art. Then siting analyses should provide more insight and help in significantly improving the siting of major energy facilities. The potential benefits to our society seem worthy of this pursuit.

Outline of the Book

The material in this book can be categorized into three sections: problem definition, the methodological and procedural aspects of the decision analysis siting approach, and illustrations of its use. The first two chapters define what is meant by an energy facility siting problem and indicate the approach and motivation for the decision analysis siting procedure. The nature and complexity of the problem are detailed in Chapter 1. Chapter 2 summarizes approaches currently used on energy siting problems and indicates their weaknesses. The decision analysis approach is then outlined and contrasted with these.

As a practical introduction to the decision analysis approach, a case study is presented in Chapter 3. This example concerns the identification of potential nuclear power plant sites in the northwestern part of the United States for the Washington Public Power Supply System. This case study is presented early to motivate the methodological and procedural details of the approach; these follow in Chapters 4–8. Another case study, concerning the selection of a site for a pumped storage power plant, is discussed in detail in Chapter 9.

Each of the Chapters 4–8 describes one step of the decision analysis siting approach: identifying candidate sites, specifying objectives and attributes of the siting study, describing possible site impacts, evaluating site impacts, and analyzing and comparing candidate sites. The presentations include basic models and concepts, procedures for implementing these, and numerous examples illustrating their use. An understanding and working knowledge of this material are necessary to conduct decision analysis siting studies and to thoroughly appraise any siting study.

Chapter 10 discusses the relevance of siting studies to related energy problems. The Appendix summarizes and appraises the fundamental assumptions on which the decision analysis siting procedure is founded.

Audience

The intended audience for this book includes individuals interested in energy problems, siting problems, and decision analysis, as well as those interested in their interrelationships.

There are three main groups of individuals specifically interested in energy siting problems: those who build and operate energy facilities, those who regulate or influence them, and those who conduct analyses of siting problems. Builders include individuals in architectural and engineering firms, design and construction firms, consulting firms, and some government agencies. Operators include utility companies and the petroleum industry. Official regulators are usually state or federal agencies, but public interest groups and intervenors also fall unofficially into the regulatory category. Systems analysts, engineers, operations researchers, and management scientists are the main segments of the analysis community with a potential interest in the subject matter of this book.

However, because input from a large number of disciplines—economics, ecology, biology, meteorology, geology, seismology, sociology, psychology, geography, and so on—is necessary for a siting analysis of an energy facility, some individuals from these disciplines will have an interest in selected parts of the book.

Individuals interested in energy problems other than siting or in siting problems not related to energy may also find this book useful. As discussed in Chapter 10, the approach outlined and illustrated is relevant to many energy problems, including the choice of a fuel source, the capacity of a facility, the timing of facility construction, and standard setting for facilities. The book is also appropriate for siting of other major facilities as diverse as airports, harbors, canals, military installations, and ski areas, to name a few. Decision analysts will also find aspects of the book of interest. In particular, the decision analysis screening models for identifying good alternatives should be useful in other contexts.

Readers with different goals may wish to read only parts of this book. The basic siting problem and decision analysis approach are discussed in Chapters 1 and 2. A reader especially interested in applications may first read the case studies reported in Chapters 3 and 9 and the last two sections of Chapter 6. Numerous other actual cases are summarized throughout Chapters 4–8. The details of the approach, stressing implementation, are in those chapters.

The only prerequisite for the book is an appreciation of rudimentary probability theory. However, an understanding of the fundamentals of decision analysis will give the reader a head start. This can be obtained in the first five chapters of "Decision Analysis" by Howard Raiffa (Addison-Wesley, Reading, Massachusetts, 1968), or the first ten chapters of "Fundamentals of Decision Analysis" by Alvin W. Drake and Ralph L. Keeney (Center for Advanced Engineering Study, Massachusetts Institute of Technology, 1978).

ACKNOWLEDGMENTS

In the early 1970s, Howard Raiffa of Harvard University interested me in examining the usefulness of conducting decision analyses of major siting problems. In a sense this book is an elaboration of siting efforts described in our book "Decisions with Multiple Objectives" (Wiley, New York, 1976). Then, in early 1975, through the efforts and with the collaboration of Keshavan Nair of Woodward-Clyde Consultants, we conducted an analysis of sites for a proposed nuclear power plant for the Washington Public Power Supply System (see Chapter 3). This led to other studies of potential sites for energy facilities and eventually to this book.

Several others also made significant contributions to the final product. David E. Bell of Harvard University, Helmut Jungermann of the Technical University of Berlin, and Craig W. Kirkwood of Woodward-Clyde Consultants provided many interesting discussions and helpful comments on an earlier draft. I have benefited from numerous other discussions that have influenced my thinking on aspects of siting problems. Participants in these discussions included William Buehring of Argonne National Laboratory, Richard de Neufville of the Massachusetts Institute of Technology, Wes Foell of the University of Wisconsin, Sarah Lichtenstein of Decision Research, Jan Norris of the U.S. Nuclear Regulatory Commission, Richard Richels and Ronald Wyzga of the Electric Power Research Institute, Rakesh Sarin of the University of California at Los Angeles, David Tillson of the Washington Public Power Supply System, Detlof von Winterfeldt of the University of Southern California, and many colleagues at Woodward-Clyde Consultants including Gail Boyd, Patricia Fleischauer, Steve James, Ram Kulkarni, Craig Kirkwood, K. T. Mao, Keshavan Nair, Ashok Patwardhan, Gordon Robilliard, and Alan Sicherman.

Many excellent suggestions for presentation were made and the actual drafting of the figures was done by Fumiko Goss and Sadako McInerney of Woodward-Clyde Consultants. Lynn Schwartz competently assisted in proofreading the entire volume.

Partial support for this effort was provided by the Office of Naval Research with contract N00014-78-C-0688, monitored by Randy Simpson. Woodward-Clyde Consultants also provided funds to complete the manuscript. To all of the individuals and organizations providing help in this undertaking, I am gratefully indebted. Last, but certainly not least, I thank Janet Beach for her interest, enthusiasm, and companionship during the entire project of writing this book.

CHAPTER 1

THE SITING PROBLEM

During the next ten years, hundreds of corporations, utilities, and government agencies will construct a large number of major energy facilities. These include power plants, dams, refineries, import terminals, storage facilities, waste disposal facilities, mines, pipelines, and transmission lines, all of which are strategic to the nation and critical to our way of life. Each case poses a complex siting problem worthy of careful examination. Appropriate sites must be found for these facilities. However, there is no effective siting policy for selecting such sites in either the United States or Europe (Meier [1975], Carter [1978], Organization for Economic Co-operation and Development [1979]).

The magnitude of the overall problem is tremendous. Energy demand in the United States is growing, and many expect this growth to continue (Reichle [1977]). Even with no growth in total energy consumption, the sources of our energy will be altered because of new technological knowledge and changing public priorities, as well as the need for replacement of facilities. In either situation, new facilities need to be constructed. Between 1975 to 1985, the U.S. Federal Energy Administration [1976] has estimated that energy investment in the U.S. will be approximately 580 billions (in 1975 dollars). A major share will be spent on new energy facilities. This state of affairs gives rise to many crucial questions:

What amount of energy should be supplied?
Which sources should be used to supply it?
How many facilities of what capacity are needed or wanted?
When should they be built?
Where should they be located?

1

This book addresses the last question: Where should specific energy facilities be located? Although this question may be the least important of the five, it is still critically important. To a large degree it can be addressed independently of the other four. In fact, the questions can be examined in the order stated, recognizing, of course, that iteration is sometimes required.

The answers to the "what amount" and "which sources" questions are mainly dependent on national energy policy, which is meant to reflect the public's needs, attitudes, and social conscience as well as the availability and implications of using a particular material. The "how many" and "when" questions are strategic decisions for the organization which will operate the facilities and they involve timing to match the supply to the demand. Regardless of the resolution of these four questions, there will be a need to evaluate sites for particular energy facilities.

Our point of view is the following. Because the stakes and potential consequences of any particular siting decision are large, it is worthwhile to conduct a formal analysis to aid the professional intuition of the decision makers. Without excessive costs or effort, an analysis can be very responsive to the needs of these decision makers. It should provide insight about the best site† as well as necessary information for regulatory processes. Historically, most analyses of specific siting problems have not provided this. What has been lacking is not information, but a framework to integrate and incorporate it with the values of the decision makers to permit examination the overall implications of each alternative site. Such a framework is discussed and illustrated in this book.

Outline of the Chapter

In this chapter, our purpose is to define carefully what we mean by "the siting problem." What exactly is it, and what is it not? Section 1.1 bounds the siting problem by identifying features which it is not meant to include. Since we are proposing to conduct analyses of siting problems, one might ask for whom. The decision makers, described in Section 1.2, are the individuals or groups whose siting problem is being analyzed. Section 1.3 discusses the five general concerns that delineate the categories of impact relevant to evaluating energy sites. They also provide a basis for identi-

† Our use of the term "best site" is not defined exclusively by technological, economic, and environmental considerations. Suppose an organization had identified a site which they felt was very good for a particular facility. Furthermore, suppose they felt that a better site with regard to technological, economic, and environmental considerations could be found with a very exhaustive (time consuming and expensive) search. If the organization did not feel the possible improvement in site characteristics was worth the time and effort, the "very good" site would be, by our definition, the best site from the point of view of that organization.

fying the objectives of any particular siting study. The features which create the complexity of siting problems are defined in Section 1.4. Collectively, these first four sections define our siting problem. In Section 1.5, we argue that a formal analysis of such problems is necessary to make responsible decisions and justify them when required or desirable. Section 1.6 outlines the remainder of the book.

1.1 Bounding the Siting Problem

In this book, we intend to address a large class of important problems. The approach developed for this purpose is generally applicable to siting. To put the approach into practice requires the consideration of many aspects specific to the particular study being analyzed. Both the general method and procedures for applying it to specific problems are covered in detail. A wide variety of case studies illustrates the breadth and relevance of the approach.

However, it is clearly important to limit the considerations addressed in the siting problems discussed in this book. This focusing is done in this section. It is also worthwhile to point out assumptions which do not need to be made in order to utilize the approach beneficially. The general assumptions made by, and considerations addressed with, our procedures are discussed in the rest of this chapter.

1.1.1 LIMITING THE PROBLEM SCOPE

The general rule for deciding whether or not a consideration should be included in evaluating alternative sites is whether or not that inclusion could have a significant impact on the site evaluations. As an example, if each location identified as a potential site for a nuclear power plant was in the same type of seismic region, then seismic considerations could be eliminated from the comparative analysis of alternative sites. However, such a consideration may be relevant in deciding whether or not to use any site. Similarly, if the cost of acquiring land for all candidate sites was equivalent, this could be omitted in a comparative consideration. This general rule allows us to eliminate many considerations from siting studies.

Benefits of the Energy

For many siting problems, it is reasonable to assume that the benefits of the proposed project will be essentially the same for each of the alternatives under consideration. If one is considering various sites for a

1000 MWe power plant, the generated electricity will probably have the same benefits regardless of where the plant is located within a region. This is not precisely true since line losses may be different for the various sites. However, if such differences are felt to be important, the losses could be considered as one of the costs† of a particular location.

For some energy facilities, the benefits may depend on the specific facility. This is most likely the case when there are benefits in addition to the product (e.g., electricity or natural gas) of the proposed facility. For example, in evaluating hydroelectric power plants, the recreational benefits resulting from the proposed dams may very well be site dependent and such considerations need to be included in any analysis of the situation. However, the benefits from the electricity produced may be excluded from the comparative analysis.

If the proposed projects are not intended to provide the same amount of the product, then, of course, the benefits must be included in the problem. For most of this book, we shall exclude this feature since it concerns the capacity question raised at the beginning of this chapter. However, as discussed in Section 10.4, the method provided for siting is appropriate for addressing the capacity problem. Furthermore, siting studies of facilities with equal capacity can provide critical information for such a study.

National Issues

There are many decisions made at a national level that are extremely important to energy companies, yet almost irrelevant to siting decisions. As an example, the decision about how much capacity to supply clearly affects how many facilities will be built. But in the decision to build a specific facility, the evaluation of which site is best does not depend strongly on the number of facilities being built elsewhere.

Although the choice of whether a coal or nuclear base-load power plant is better for a particular situation could depend on whether or not the United States develops a breeder reactor, the choice of a site for a proposed nuclear power plant would not depend on the existence of the breeder. In the latter case, each proposed site would be affected in almost the same way with or without the breeder.

Other national (or international) issues which seem to have little effect on siting are: the deregulation of natural gas prices, an oil embargo, or possible legislation curtailing operations of fossil fuel power plants because of the "greenhouse effect." These issues would, of course, have a

---

† Costs, as used in this book, include both economic and noneconomic costs. We shall refer to economic costs explicitly as such.

large influence on any decision concerning the first four questions raised in the beginning of the chapter.

Fuel Cycle Implications

In many situations, aspects of the fuel cycle do not need to be considered in comparing sites. For example, if one is considering using coal from West Virginia as fuel for a coal-fired power plant in North Carolina, the impacts of mining the coal will probably be identical for each of the possible sites. It may be that transportation of the fuel in certain cases is "almost" site independent. By this, it is meant that differences are small enough to neglect. This may be the case for nuclear power plant siting or liquid natural gas import terminals if they are located within a few hundred miles of each other.

To the extent that wastes from facilities will be handled in a similar manner independent of a site, these considerations can also be eliminated from a comparative evaluation. The most obvious case of this is the storage of nuclear waste material. Likewise, if the wastes generated from offshore drilling operations were to be disposed of in the same way for each of several proposed sites, disposal might be neglected in an evaluation of sites.

Economic Costs of Operation and Maintenance

Many of the operation and maintenance economic costs do not depend on the site. If the facilities are essentially identical and if maintenance is an activity which occurs mainly indoors, such an assumption seems reasonable. This would be true to a large degree with facilities such as liquefaction plants, fuel storage centers, and many base-load power plants, such as nuclear, coal, or solar. For hydroelectric or geothermal power plants, offshore facilities, or pipelines and power lines, the appropriateness of eliminating the economic costs of operation and maintenance probably needs to be considered on a case-specific basis.

1.1.2 ASSUMPTIONS NOT REQUIRED FOR SITING

Given the boundaries of the siting problem discussed above, it may be possible to conclude that our general problem is rather limited in scope and relevant only after all the policy decisions have been made. That is, only after one has decided to build a facility, when it should be built, what capacity it should have, and what fuel to use, does one get to the siting problem. This view is not true. In fact, none of these questions need to be decided before a siting study is conducted.

Because of the interdependent nature of all the questions raised above, it may be best in many circumstances to first conduct a siting study to provide crucial information for the other questions. For instance, if one is choosing between a 1000 MWe coal or nuclear base-load facility, one would clearly want to compare the best coal site with the best nuclear site. These sites could be identified by separate siting studies. Then, of course, the coal/nuclear choice would need to include aspects such as national issues, fuel cycle implications, and economic costs of operations and maintenance because these could have different implications for the different facilities.

A decision about the best capacity for a facility may depend strongly on the best available sites. If there is an excellent site available for an 800 MWe pumped storage unit, but the best possible site for a 1000 MWe unit is not as good, this may have a significant impact on the choice of the plant capacity. To choose the better of the 800 MWe and the 1000 MWe sites requires siting studies. As mentioned above, some of the considerations eliminated from siting studies in Section 1.1.1 would then need to be considered in comparing facilities of different capacities. In particular, some recognition must be given to the differential benefits of the product.

The decision to build or not is analogous to the capacity decision. The distinction is that one of the options has a capacity of zero. However, the decision to build either an 800 or 1000 MWe facility may be easier to make than the choice between no facility and a 1000 MWe facility. Assumptions which may be appropriate for the former problem are definitely inappropriate for the latter. For example, the same transmission corridors may be used for both the 800 and 1000 MWe plants. Hence, this aspect may be eliminated from the comparison. This is clearly not the case in comparing the 1000 MWe facility to no plant at all.

If a siting study conducted now implies that site A is better than site B, it does not follow that this same order will be true in three years. There are many illustrative examples. Site A may be better than site B now, even though the transmission line needed at site A costs $100 million and that at site B only $60 million. However, if it happens that transmission line costs double in three years, site B with transmission line costs of $120 million may then be preferred to site A with transmission line costs of $200 million. Of course, other factors would also be changing, but the idea should be clear. Consequently, in deciding when to build a facility, one would like to know the best facility in each of the time periods of interest. These would come from separate siting studies. The timing decision would incorporate this with projected demand data and forecasted benefits to determine the best option.

The siting decision is a fundamental component in many policy decisions regarding energy. To the extent that one feels that sites of about "equal value" are available for anything one decides to build, these policy decisions can be addressed before finding a site. In such a policy analysis, one assumes that a site will be found which has a high enough value to make the policy decision legitimate. However, as public priorities have changed in the last few decades, the number of potentially adequate sites for large-scale facilities has been significantly reduced. This phenomenon is mainly a result of increasing emphasis on environmental quality. The assumption that good sites can always be found is rapidly becoming less appropriate. This means that more emphasis will be placed on siting studies done prior to policy decisions.

Before proceeding, it should be noted that restricting our attention to finding a site for a single facility is actually a weak restriction. If one is concerned with siting multiple facilities, such as offshore drilling platforms, this can be treated as a series of siting problems for single facilities. Alternatively, a facility may be defined as composed of multiple units (i.e., drilling platforms) and the problem again becomes one of finding a site for a "single" facility. The overall problem would be more involved in both cases, but the methodological approach would still be appropriate.

1.2 The Decision Maker for Siting Problems

The main purpose of this book is to help a decision maker select a site for a proposed facility. This is very different from describing the procedure by which a site will be chosen for that same proposed facility. The former concerns what is referred to as prescriptive decision analysis — helping to decide what should be done. The latter concerns descriptive decision studies — trying to describe what will be done. This prescriptive orientation has major implications for the role of, and in fact the existence of, a "decision maker" in the siting problem.

Our term "decision maker" can refer to an individual or to a group of individuals acting in concert. There may be more than one decision maker in any siting problem. The characteristic which distinguishes decision makers from others is that they be in a position to exercise some control over the decision to be made. As we shall see, this definition implies that the identity of the decision makers in any particular problem will depend on our perspective.

1.2.1 A DESCRIPTIVE VIEW OF THE SITING PROBLEM

A large number of individuals and groups interact in a complex manner in the overall process which eventually leads to the existence of an energy facility. Descriptively, there are many decision makers. Let us simply describe some of the decision makers involved in the building of a base-load nuclear power plant of 1000 MWe capacity. First, the utility company concerned feels they may need the additional electricity in 12 to 15 years time. They may try to interest other utilities in combining with them in this venture. This consortium must then have the proposed facility designed by an engineering firm. Design decisions will be made in this phase. A second firm may do the siting study for the consortium. Many professional judgments about earthquake faults, water availability, and so on will be made. After a site is chosen, it is proposed to the Nuclear Regulatory Commission for licensing. Several other federal agencies, such as the Department of the Interior, Bureau of Reclamation, and Army Corps of Engineers, may each have separate decisions to make regarding the appropriateness of the proposed facility. Several state agencies such as a facilities siting council and public utilities commission will also have decisions to make regarding the project. The licensing process provides for intervenors to be designated to participate. These may include environmentalist groups, local interest groups, concerned scientists, or almost any other group concerned with the proposed facility. Each of these groups will have to decide how and when to participate in influencing the overall decision in its favor; furthermore, each interested party can, via a suit, require that part or all of the process be brought into the courts. This introduces another set of legal decision makers.

Eventually the facility proposed for a particular site is either accepted or rejected. The entire process leading to this result may have required a few iterations, several years of time, and the involvement of scores of decision makers (Winter and Conner [1978]). To try to describe the entire process would be a Herculean task. It may be important to do this, and it certainly would be interesting, but it is not our task in this book to provide a procedure for doing it.

1.2.2   A Prescriptive View of the Siting Problem

Clearly, there are many decision makers in major energy siting problems. The purpose of this book is to prescribe how one of these decision makers, or perhaps more than one, *should* act—which decisions they should make and why—if they wish to behave responsibly and consistently in such a complex decision environment. To define our terms, a client will refer to any decision maker for whom an analysis is conducted in order to aid the client's decision making process. Clients may be companies, governmental agencies, or other interested parties. It is not necessary that the client provide funds for the analysis. Each client has a decision to make which may or may not depend on what the other decision makers in the process do. For instance, the selection by a utility company of a site for a coal power plant may depend on perceived reactions (decisions) of environmental and local interest groups with respect to the alternative sites.

Any client's decision should be based on two categories of information:

1. the possible impacts of choosing each of the candidate sites and
2. the values of the client used in evaluating these impacts.

The first category includes the information about environmental and economic impacts, consequences to others affected by the proposed facilities, possible delays in licensing, values of other decision makers involved, and so on. This information is gathered by searches of existing data, site-specific studies, models describing possible impacts (e.g., economic models), and professional judgments. The values of the client incorporate his or her attitudes toward risk, value tradeoffs among impacts, relative concern for others impacted, and so on. Such values are elicited from the client to help evaluate the alternatives and select, from the client's viewpoint, the best site.

In any siting study, there is a question about how much of the impacts and values to include in the formal analysis. Some features are easier to include than others. In all cases, the time, cost, and effort necessary to conduct the overall analysis increase with the amount of information required. On the other hand, the insight from a siting study also increases as the time, cost, and effort are increased. Choosing the proper amount of each is an art. However, because the cost and time for a thorough siting study are small compared with these requirements for the entire siting process from conception to existence of a facility, and because many laws (e.g., National Environmental Policy Act) require justifying the siting process in order to obtain the licenses necessary for the facility, it is probably fiscally wise, if not legally necessary, to conduct a thorough analysis.

The fact is that the possible impacts and values are the heart of the siting problem. Each must be considered either formally or informally, and there are significant advantages to handling them formally.

The Single-Client Problem

From a prescriptive perspective, the most common siting problem is that of a single client. Our major concern is with the client who is searching for the best site, or at least a very good one, to locate a major facility. Examples are a utility company finding a site for a power plant, an oil company finding a site for a refinery, and a segment of the government finding a site for the storage of oil or for nuclear waste disposal.

Such a client has the responsibility for the major siting decision in choosing among the many possible sites available. Other decision makers only decide whether the client's chosen site is satisfactory.

For the single-client problem, both the judgments of impacts and evaluation of the impacts will be provided by the client or other designated parties. The impact data will necessarily come from a host of disciplines, including economics, meteorology, seismology, water resources, environmental sciences, and medical fields. One might expect that a good analysis would uncover the true information about the likely impacts at each of the various sites and, hence, there would be substantial agreement among various experts about the potential impacts. On the other hand, there are no true values. There could be a large disparity between the client's values and others in society. In fact, the reason for regulatory processes is to insure that the actions taken by those being regulated do not conflict with general values of society. Thus, the client would normally take the perceived values of society into account when establishing his or her values for siting. For instance, because of the increased environmental consciousness of society, utility companies, oil companies, and the government each places a higher value on environmental impacts than they did a decade ago.

The Multiple-Client Problem

Sometimes the major siting decision from the prescriptive viewpoint is the responsibility of a consortium of clients. Examples include a regional power pool (i.e., group of utility companies) selecting a power plant site, two government agencies selecting a hydroelectric dam site for multiple purposes, and the government and several oil companies selecting a site for a deep-water port for oil imports.

The multiple clients may use the same groups of individuals to collect and process the information about impacts at each site. This aspect of the problem would not be so different from the single-client version. Given this information, all the clients may be in reasonable agreement about the possible impacts. However, the clients may not be in agreement about the appropriate value structure for evaluating these impacts. If, indeed, the different value structures lead to identifying different sites as being better than others, the conflict must be resolved before proceeding. The unique feature of the multiple-client problem, which makes it different from the single-client problem, is the requirement for resolving such conflicts. Value judgments are again needed to decide how to resolve conflicts which persist after careful scrutiny. We discuss this critical aspect of the multiple-client problem in detail in Section 7.7.

1.3   General Concerns in Siting

What is it that makes one site better than another for a particular energy facility? Is the siting problem really important enough to justify the attention that it receives? The answer is yes. The desirability of one site relative to another depends on a multitude of factors. The differences in desirability among various candidate sites can be great.

For any particular siting problem, it is necessary to consider carefully a large number of factors to see which are relevant to the problem at hand. However, each of these factors is indicative of a more general concern. The five general concerns, which are fundamental to all siting problems, are the following:

the environment,
economics,
socioeconomics,
health and safety, and
public attitudes.

Collectively they characterize our interest in the siting problem. It is not necessary that their domains be mutually exclusive. However, we do want them to be collectively exhaustive so that any possible siting impact of interest is included in at least one of these concerns. The fact that different sites can lead to vastly different implications in terms of these general concerns, motivates us to carefully examine the alternatives. In siting problems, there is a need to assess the implications of each of the candidate sites with respect to these concerns. However, some of the impacts may be equivalent for each of the proposed facilities. In this case, as discussed in Section 1.1.1, these factors will have no influence on the choice of which site is better, and need not be included in the analysis. Let us clearly define what we mean to include in each general concern.

1.3.1   THE ENVIRONMENT

Environmental impact refers to the impact on the ecosystem. The elimination or disruption of members of various flora and fauna species are of particular interest. With any electrical generation facility there will be environmental impacts related to mining and processing the fuel, and transporting it to the power plant site. The construction of the plant and the transmission line will have environmental effects. Additional impacts of operations occur via air, water, and land pollution and possible radiation. Other energy facilities, such as refineries and pipelines, will also have various environmental effects associated with construction and operation. In all facilities, the disposal of wastes such as spent nuclear material, sludge, liquid wastes, flyash, and by-products also produces environmental effects.

1.3.2   ECONOMICS

The most important aspect of many siting problems is and should be economics. Basically, the goal is to build and operate the facility at the lowest possible cost, but many factors make this difficult to evaluate. For instance, there are land acquisition costs, capital construction costs, operation costs, maintenance costs, and transmission or transportation costs for the product. It is important to note that these costs may include components due to various legal requirements, such as meeting environmental and safety standards, and that these may occur in different time periods over the lifetime of the facility.

Major uncertainties affect the total costs of almost any major energy facility. In recent years, the estimated total construction costs for such facilities have invariably been significantly lower than the resulting actual costs, partly due to the uncertainties. These uncertainties include the availability and costs of future fuel and water supplies, the reliability of the system, natural phenomena such as earthquakes and storms, and the likelihood of possible government actions. Such factors can have a major impact on cost and should be considered when examining the economic impacts of the various alternatives. Another economic concern is the ability to raise funds during construction. This can be exacerbated by construction delays, which escalate construction costs. For many problems, the dividends paid to shareholders and the price charged to customers for the product are important. Both of these may depend on pricing decisions of regulatory authorities (e.g., the Federal Power Commission or a state Public Utility Commission). It may be that such decisions would not depend on the siting option per se, but on the cost of the option. In such cases, the pricing of the product can be excluded from the site comparison.

1.3.3   SOCIOECONOMICS

Socioeconomic impact means the impact on individuals living near a proposed facility site, exclusive of health and safety. Socioeconomic impacts are felt in several ways. A major effect arises from the taxes which any large energy facility would pay. This may have a significant impact on property taxes of residents. Usually state laws indicate how the economic benefits via taxes from the facility will be spread among the citizens living in the area in which the power plant is located.

Another major socioeconomic effect is the boom–bust cycle associated with the rapid increase and then decrease of people and activity resulting from construction and completion of the facility. During the construction phase of a project, a large number of workers either move to or commute to the site area. This affects the social institutions and economic vitality of the towns near the site. Since individuals living in nearby towns would probably feel the boom–bust impacts most severely, it may be reasonable to categorize socioeconomic impacts into those affecting individuals in the immediate vicinity of the proposed plant and those affecting other citizens in the general area.

Another effect, which we categorize here as socioeconomic, is the aesthetic impact of the facilities and their operations. This includes the impact of the plant itself, cooling towers, transmission facilities, pollutants, and noise. These concerns might have been categorized as environmental except that we define environmental impact to mean impact on the ecosystem, and socioeconomic impact to mean impact on humans other than health and safety and public attitudes.

In some cases, it might be possible that an alternative site would have some archaeological value. In such cases, any potential damage would probably have to be mitigated because of regulatory requirements. This eliminates the need to include archaeological aspects in the site comparison except as an additional economic cost due to the mitigation. However, if it is not possible to mitigate all the archaeological disturbance, then such an aspect should be included.

1.3.4 HEALTH AND SAFETY

The health and safety concerns include mortality, morbidity, and injuries due to normal operations or accidents. All power facilities have risks associated with them which may result in fatalities, sickness, or injury (see, for example, Inhaber [1978, 1979]). With any energy facility, fatalities may result from construction of the facility, acquisition of the fuel, transportation and storage of the fuel and resulting product to, from, and at the facility, pollution or radiation from the facility, and waste disposal.

Parenthetically, one may ask why build these energy facilities at all if they are so unsafe as to possibly result in some fatalities. Can we not construct completely safe facilities? The answer is clearly no. It is indeed important to minimize individual risks. However, the issue is: are the risks worth the benefits which are derived from having a facility? It should be evident that no energy facilities are free from potential fatalities and that the absence of energy facilities would also cause some fatalities because of loss or lack of needed end-use energy.

Returning to the basic problem, there are several factors which add to the complexity of the health and safety concerns. Considering only fatalities, there will be different impacts if the individuals are workers or uninvolved citizens; if the accidents result from natural causes (e.g., earthquakes) or human error; if several individuals die at once or the same number die in a series of minor incidents; if the fatalities result from a well-understood cause (e.g., flood due to a hydroelectric dam break) or a less-understood cause (e.g., radiation). The list could be continued, but this will be deferred until Section 7.5.5 where the risk of potential fatalities is discussed.

1.3.5 PUBLIC ATTITUDES

In addition to local residents, the public is meant to include other interest groups such as environmental groups, business groups, and consumers of the product of the proposed energy facility. Many of the impacts on the public are accounted for within the socioeconomic, environmental, economic, and health and safety concerns. However, even if all of the site options have identical impacts in terms of these four concerns, the public may prefer one option to the other, partly because of the fact that different sites affect different people (particularly local groups) and these groups may have different attitudes. What is important is the public's perception about the degree to which their attitudes and feelings matter (see Section 6.6).

The manner in which a client uses the viewpoints (and perhaps even participation) of the public in evaluating sites for a particular facility might have major implications for future decisions in the regulatory processes affecting the proposed site. These implications, which could be very important, should be carefully considered in deciding and designing the appropriate public input.

1.3.6. A PERSPECTIVE

The five general concerns categorize domains of possible end consequences to individuals who will be impacted by the siting of an energy facility. In trying to ascertain what these consequences might be for each of the particular sites, investigations must be conducted within a wide range of disciplines. These disciplines are mainly concerned with the means by which the end consequences eventually occur.

Specifically, most siting studies will require information about the meteorology, geology, seismology, hydrology, ecology, and geography of a region. They should also be concerned with the sociology and psychology of the people in that region. All of this information is important not in itself, but as a means to the final impacts. To include the means as separate concerns would be tantamount to double counting, and, hence, they are not included as such.

1.4 Features Complicating the Siting Problem

No one who has really thought about it feels that the siting problem is simple. On the other hand, with the five general concerns in mind, perhaps one should just look for a site which measures up as much as possible in terms of each of them, and then select it. That does not sound too difficult. However, that is essentially what we do propose, and it is extremely difficult, as we shall see. In this section, we shall discuss a host of features which complicate the siting problem. These features render it very difficult to determine how well a particular site does, in fact, measure up to what is desired.

1.4.1 NUMEROUS POSSIBLE SITES

The purpose of a siting study is to identify a particular location for a particular energy facility. However, at the origin of the study, the client often feels that the site may be located anywhere in a loosely defined, relatively large region of interest. For example, a utility company wanting to site a 1200 MWe fossil or nuclear power plant may have the entire state in mind as a region of interest. An oil company or LNG company may have 500 miles of coast as a region of interest for an import terminal and conversion facility.

Within any region of interest, there are often literally thousands of potential sites for energy facilities. This can be recognized just by comparing the area needed for the facility, usually on the order of a square mile, to the area of the region of interest. Many—in fact, most—of these potential sites are inappropriate for a variety of reasons. Some can easily be eliminated from consideration. Often however, after the easy cases are eliminated, there are still too many potential sites to evaluate thoroughly. One must reduce the number of potential sites to a manageable group of candidate sites, which will then be carefully compared with each other. This process requires the balancing of the time and effort required to care-

fully carry out the task against the likelihood that more cursory procedures may inadvertently eliminate some or all of the best sites. It is not an
easy balancing act to do.

1.4.2   MULTIPLE OBJECTIVES

In every siting problem, we should expect that there would be multiple
objectives. For each of the general concerns, it is easy to generate several
specific objectives which might influence the desirability of a site. The
overall desirability depends on the possible impact with respect to each of
these.

However, for a specific problem it is not always necessary to have an
objective for each of the general concerns. The reason for such an omission, as always, is that the sites cannot be distinguished in terms of this
concern. For instance, in trying to site a geothermal power plant in a
restricted region of interest, it may be that the socioeconomic impact
would be expected to be identical for all of the proposed sites. This is not
to say that the socioeconomic impact can be precisely forecasted, but that
whatever it is will be independent of the specific site chosen. In a very
restricted case, it may be that only economics matter. However, for most
siting problems, we would expect that economics and environmental impacts, at least, would be different for the various sites.

1.4.3   SEVERAL INTEREST GROUPS

For many siting problems, the client may be concerned with the impacts of the proposed facility on several groups of individuals. External to
the client organization, these groups may include the consumers of the
product, business interests, citizens in communities which may be impacted by the proposed sites, organizations of environmentalists or sports
enthusiasts, and heritage committees. Within the client company, there
may be groups with differing opinions, for example, company management, shareholders, and employees who will work at the proposed facility.

In some cases, even within the local area of a proposed facility, it is
necessary to differentiate the impacts on the various groups. Section 6.8
illustrates a risk analysis conducted for a proposed LNG terminal and
vaporization facility in Texas. This analysis was conducted because of the
small possibility of accidents which could result in public fatalities. In
order to determine if any group of individuals was at much greater risk

than others, the risks were separately calculated for the following: residents of each of two nearby towns, visitors who frequently used the area, commercial fishermen, and recreational boaters. Such information would not only be relevant to decisions about the appropriateness of a particular site, but it might suggest measures which could be used to improve a particular site.

1.4.4 INTANGIBLES

Many of the objectives in a siting problem will involve intangible factors. By this we mean factors for which there is no obvious way to measure the impact. An example is the aesthetic impact of a transmission line that obscures the view of an otherwise natural mountain scene. Another is the social disruption that individuals in a community may feel as a result of the rapid influx of construction workers. Psychological and moral considerations such as the fear of living downstream from a dam or the question of whether a company should be able to build facilities on Indian lands are also representative of the intangible aspects of siting problems. In many problems, the morale of the workers at a proposed facility is an intangible factor which requires careful consideration in the siting decision.

1.4.5 DEGREE OF IMPACT

One may look at the previous three features as follows. For any particular siting problem, there will be a number of objectives. Some of these will involve concerns of the client, such as economics, which can be measured in terms of dollars; other objectives will describe the possible impacts on each of the interest groups which the client wishes to consider; and some of the objectives may address intangible factors. The last two features particularly compound the difficulty created by the degree of impact feature.

For each proposed site and each objective, it is not enough to state that there will or will not be some impact. It probably will be necessary, as well as informative, to indicate the degree of impact that might be expected. For most problems, we would expect that there would be some environmental impact at each of the proposed sites. For example, it is difficult to clear a corridor for a pipeline or transmission line without having some environmental impact. Thus the issue is not if there is an impact but rather how much impact and how important it is.

Generalizing from this, it is easy to see that one needs a scale for measuring each of the objectives. Because of the importance of this requirement, Chapter 5 is devoted to the concomitant problems of specifying objectives and associated measures for siting problems. It is evident that it will be particularly difficult to identify measures for intangible factors. For most cases, these measures must be constructed as discussed in Section 5.4.

1.4.6 LONG TIME HORIZONS

The impacts of interest in most siting studies do not all occur at the same time. Furthermore, many continue over the lifetime of the project. Among these are the economic costs of operation and maintenance and other impacts of operation, such as pollution. In some cases, impacts are even felt after the lifetime of the project. The aesthetic impact of a decommissioned power plant may be greater than when it was in use. Because of the public focus on the waste storage necessary for a nuclear power plant, it is perhaps worthwhile to recall from Section 1.1.1 that this concern is excluded from siting of nuclear power facilities since it is independent of the power plant site. However, the selection of a site for nuclear waste storage is a separate siting problem and the procedure outlined in this book is appropriate for the task.

The chief motivation for an interest in the time horizon is that the value of a particular impact may depend strongly on when it occurs as well as what it is. This phenomenon is well understood with economic concerns. It is more or less universally agreed that a cost of 5 million dollars tomorrow is less desirable than a cost of 5 million in two years. One method which is consistent with such a time preference for money is discounting at a fixed rate. However, discounting implicitly makes many other assumptions which may not be appropriate for examining energy investments. The specific objections to the approach are presented in Section 7.3.

The timing of other impacts may also be critical. Most people would agree that a population influx into a small community of 1000 per year for five years is preferred to a single influx of 5000, even though the total numbers are the same. An accidental discharge of liquid waste into a nearby stream resulting in the death of 2000 salmon may be of greater concern if it occurs in the first year of operation than in the 25th year. And although it is not pleasant to think about, would an accident at an energy facility resulting in five fatalities be less preferred if it happened in the near future than the more distant future?

1.4.7 UNCERTAINTIES ABOUT IMPACTS

The uncertainty feature greatly increases the difficulty of responsibly dealing with each of the previous features. Unfortunately, it is simply impossible to forecast all of the impacts for any proposed energy facility at any site. There may be large uncertainties about the possible environmental impacts, all future costs, the likelihood of accidents, and the impacts of such accidents.

The major reasons for the existence and persistence of these uncertainties are

1. little or no data can be gathered for some events,
2. some data are very expensive or time consuming to obtain,
3. natural phenomena such as earthquakes and droughts affect impacts,
4. people move, affecting future impacts,
5. priorities, and hence, perceived impacts change over time, and
6. it is impossible to forecast the future completely.

Examples of each may be given:

Nuclear power is one of the potential sources for expanding our electricity production. Consequently, the likelihood of plant accidents causing releases of radiation which might result in public fatalities is of interest. However, the accident at Three Mile Island in March, 1979 is the only accident of this sort in a commercial power plant. In its analysis of this accident, the President's Commission (Kemeny *et al.* [1979]) concluded either that there would be no cases of cancer due to the release of radiation or that the number of cases would be so small that it would not be possible to detect them. Thus, it is not possible to make any direct estimate of the likelihood of such accidents or the possible resulting fatalities. Even if there had been several such accidents in the past decade, there would be disagreement on the interpretation of this data for estimating the likelihood of future accidents. In such cases, the estimates of the likelihood of events or of specific impacts will necessarily be uncertain.

In appraising corridors for transmission lines or pipelines, it may be very expensive and almost impossible to determine the geological structure everywhere along the route. One would clearly want to have general information about the route, but route selection would probably have to be made with only this information. This would imply the existence of uncertainties which could affect the cost of constructing the proposed facility.

The most obvious influence of natural phenomena is on the likelihood of accidents. And since, for example, we cannot predict exactly when and where an earthquake will occur, or its magnitude, there will be uncertainties about dam failures or facility damage. Natural events also affect the impact of normal operations. A drought may result in less power and irrigation water being supplied from a reservoir, a severe snowstorm may hinder the repair of a transmission line, and tornadoes may render a refinery temporarily out of order. Since prediction of natural phenomena is imperfect, especially over the long time horizons included, uncertainties about impacts remain.

To the extent that there are hazards to individuals in the locality of an energy facility, the movements of people in and out of the area will influence health and safety impacts. The inherent uncertainties about future population shifts cause uncertainties about impacts.

The inability to perceive all the changes in priorities and tastes which occur over time is closely related to the uncertainties about impacts. Impacts which are felt to be critical today may diminish in importance with time and be replaced by other concerns. Roughly speaking, the importance of economic costs relative to environmental and health factors was much greater 30 years ago than it is today. Currently there seems to be increased importance placed on the aesthetics of energy facilities and on considering the viewpoints of groups directly impacted by the proposed facilities.

The fact that it is impossible to predict the future needs no elaboration. However, for all the frustrations this circumstance causes us in general, it is interesting to note how dull life might be if we could predict the future—all of the future.

There is no field investigation or research program which can completely eliminate all of the uncertainties, and yet their resolution may be critical to the final implications of the various siting options. It is, therefore, prudent to acknowledge and explicitly consider the uncertainties in appraising the site options.

1.4.8 TIME DELAYS IN LICENSING AND CONSTRUCTION

There are several possible causes of delay in getting an energy facility into operation. Such delays, which can occur in either the licensing or construction phase of a proposed facility, can have a significant impact on the economic viability of the project. Some of these delays are site independent and need not be included in the site comparison. An example would be delays in the delivery of a standard component, such as the reactor, which would be the same for each of the sites. Other delays, such as one due to a court suit over the license to cross a particular piece of land with transmission facilities, may be site dependent and necessary for the site comparison.

1.4.9 OPERATING RELIABILITY

Many types of energy facilities have scheduled maintenance which requires that the operations of the facility be curtailed or stopped for a certain time period. Because of the planning for these, one expects the time involved to be essentially the same for similar facilities. However, the circumstances which may force unscheduled shutdown or curtailment of the facilities' operations may depend strongly on the site. The main causes for such differences would be natural phenomena, such as severe storms, and accessibility to allow continuing operations or correcting any problems.

Hydroelectric power plants obviously must cut their electricity production during severe droughts. To the extent that one site is more protected from such occurrences than another, it may be more reliable. Storm waves, tsunamis, and high winds may limit access to ports (e.g., importing oil to refineries or LNG to regasification facilities) and break pipelines linking offshore facilities to land. Such occurrences reduce operating time, hence reducing the amount of product produced, and result in the need for costly repairs. Even in calm weather, the inaccessibility of certain transmission line corridors or the naturally treacherous seas around some offshore facilities render repair work more costly and time consuming than at other sites.

1.4.10 VALUE TRADEOFFS

As indicated in Section 1.4.2, most siting problems involve multiple objectives. If it were possible, the client would clearly prefer to optimize with respect to each of these objectives simultaneously. It would be ideal to minimize environmental impact and economic costs at the same time that health and safety are promoted as much as possible, much to the satisfaction of the public. Realistically this simply cannot be done. There may be some dominated sites† that need not be further considered for the proposed facility. However, almost invariably, there is a set of nondominated sites for which better achievements in terms of one objective can be obtained only at the expense of worse achievement of some other objective. It is at this point that value tradeoffs are required.

---

† For a particular siting study, a site is dominated if another site exists that is better in terms of at least one objective and no worse in terms of *each* of the other objectives.

Value tradeoffs must be elicited from the client and represent the value judgments of the client. They indicate the level of achievement of an objective that one is willing to accept in order to improve by a specified amount the achievement of any other objective. For instance, one value tradeoff might indicate that the client's value structure is such that it is worth increased capital construction costs of 10 million dollars to reduce the miles of mature forest through which a transmission line would pass from 40 to 30. Such a change in impacts could result from a different route for the transmission line. By presenting this example, we are not saying that the value tradeoffs must be formally made or publicly disclosed. We are saying that value tradeoffs are inevitably part of siting problems. One can address this feature either formally or informally. However, the possibility of ignoring value tradeoffs and hoping they will go away just does not exist.

1.4.11 EQUITY

It would be nice to satisfy as much as possible each of the groups interested in a particular siting problem. However, value tradeoffs will have to be made because the lot of some groups can only be improved at the expense of others, once dominated alternatives are eliminated from further consideration. However, there is a special consideration necessary when determining the value tradeoffs between groups, that is, equity. It is important to be as fair as possible to each group. Determining what is fair is often a very difficult task involving complex value judgments.

1.4.12 RISK ATTITUDES

Because of the uncertainties (i.e., risks), the attitudes of the clients toward risk is important. To illustrate with a simple example, a facility which will be sure to cost 2 billion dollars may be preferred to one which has a one-half chance of costing either 2.4 or 1.4 billion, even though the expected cost in the latter case is 1.9 billion. The uncertainty may cause many problems in addition to construction costs *per se,* such as difficulties in dealing with rate regulation agencies and in financing the project. Value judgments of the client are required to specify his or her attitude toward risk. In Chapter 9, the assessments of a utility company executive for costs of a proposed pumped storage facility indicate how important attitudes toward risk can be.

To avoid the misconception that attitudes toward risk are only relevant to economic concerns, consider this example from the nuclear power siting study discussed in Chapter 3. One important measure in the problem represented the percentage of salmon which might die in a stream because of a possible accident during construction or operation of the facility. Everyone concerned with the study felt that it was easily preferable to have a sure 50% loss of salmon rather than a one-half chance of either no loss or a 100% loss. In both cases the expected loss is 50%, but everyone preferred to avoid the risk of a 100% loss.

1.4.13 UNCERTAINTIES IN GOVERNMENT DECISIONS

The actions of the federal and state governments can have a large influence on the relative desirability over time of various sites for a proposed energy facility. For example, a state government decision to eliminate the operation of wells in a certain area because of a drastic drop in the water table could result in soaring costs for a pumped storage unit which replenished its water supply from such wells. A future federal government decision requiring the installation of either pollution or safety equipment on all facilities of a certain type meeting an external criterion (e.g., the air pollution level in the area of a fossil fuel plant, the storm frequency at offshore facilities) could have a tremendous differential impact on the candidate sites being considered now. However, siting decisions must be made now, and uncertainties about future government actions will always be present. Thus, it may be necessary to consider the possibilities of the various government actions in evaluating current siting decisions.

1.5 The Argument for Formal Analyses of Siting Problems

If it were not the case before reading to this point, it should now be evident that energy siting problems are both important and complex. This should provide a motivation for further reading. Our purpose is to demonstrate that careful analysis of these problems is worthwhile.

1.5.1 MOTIVATION FOR A FORMAL ANALYSIS

There are a number of characteristics which render it worthwhile to conduct an analysis of a decision problem. Siting problems have several of these, any one of which would justify a formal study. The four main characteristics are the following:

High Stakes. The difference in desirability between sites can be enormous: for example, hundreds of millions of dollars or severe local environmental damage.

Complicated Structure. The features discussed in Section 1.4 clearly indicate the complexity inherent in most siting problems. This makes it very difficult to appraise informally each of the alternative sites in any particular case.

No Overall Experts. Because of the breadth of concerns in siting problems, there are no overall experts in siting. Different individuals are, however, experts in relevant disciplines such as economics, engineering, and the various branches of science.

Need to Justify Decisions. In order to insure that the public interest is accounted for in siting decisions, petitioners who wish to build energy facilities must justify their proposed actions to obtain authorization from regulatory agencies.

Collectively, these characteristics describe a very involved problem. Yet, there is no doubt that siting decisions will continue to be made. One might argue that the problems are so involved that formal analysis cannot be expected to *solve* the problem. I agree. Formal analysis will not *solve* the problem, nor is it intended to do that. It is intended to *help* the client make responsible decisions. It is precisely the complexity which implies that a formal analysis should be done.

1.5.2 THE GOALS OF FORMAL ANALYSIS

There are two basic goals of a formal analysis of siting problems. The first is to reduce the likelihood of a poor outcome (or increase the likelihood of a good outcome) or, more precisely, to improve the quality of the decision taken. This goal is worthwhile because of the high stakes and is difficult to accomplish because of the complicated structure.

By improving the quality of the decision, we mean to include several things. The analysis should stimulate constructive discussion and provide a framework for identifying and resolving conflicts. It should produce insights which may not be obvious because of the complex nature of the problem. An explicit analysis makes it easier to completely and logically keep track of all that is relevant. Because there are no overall experts, one of the most important functions of analysis is to provide a framework for integrating the information from the diverse disciplines working on different aspects of the siting problem. In some sense, the model plays the role of the nonexistent overall expert, and yet it is just a useful tool of the client, analysts, regulators, and various experts who are interested in the siting problem.

The second goal of a formal siting study is to provide a rationale and documentation for supporting the siting decision to regulatory authorities, as well as sometimes to shareholders and the public. It should not only indicate what information was collected from where, but how it was used and why it implied the proposed site was the best of the alternatives. The model is also a device which can answer "what if" questions often posed by regulatory bodies.

1.5.3 THE CHALLENGE TO SITING ANALYSES

In order to achieve the two stated goals, the analysis of a siting problem must be thorough and competent. Thorough means that all of the relevant features of the siting problems outlined in Section 1.4 need to be considered. Competent refers to both the depth and completeness with which these features are analyzed and the logic which is used to integrate the parts of the analysis. The analysis needs to be theoretically sound and practical. The method of analysis for siting problems proposed in Chapter 2 does meet these requirements.

Before proceeding, it is necessary to note that some siting studies are conducted only because of the second goal—justifying the chosen site. Such analyses, which are usually conducted after a site has been chosen by an informal procedure, are likely to be inappropriate for two reasons. First, the chosen site may not be that good. Second, the time pressure of producing a siting study for justification of a prior decision sometimes results in a less than thorough analysis. There are some exemplary cases of this type of analysis which have not been sufficient to convince regulatory authorities of the appropriateness of a site. Of course, it is difficult to ascertain whether it was the site or the analysis or both which led to rejection.

For these reasons as well as the fact that a good analysis can be of significant help in selecting the best site, we feel it is very desirable to conduct the siting study prior to selecting a site. Existing laws, most notably the National Environmental Policy Act,† also support this contention by requiring analysis prior to site selection.

1.6   Outline of the Book

This book is intended to do two things: (1) to demonstrate that the concepts, methodology, and procedures of decision analysis can capture the issues and substance of the siting problem much better than any alternative methodology and (2) to illustrate, with procedural sections and many case studies, how to implement this approach.

Chapter 2 begins with a discussion of the requirements desired of a siting methodology. This sets the standard by which such methodologies should be compared. Existing approaches are summarized, and their strengths and weaknesses with respect to the desired requirements are examined. The decision analysis approach to siting is then presented in a compact form. The manner in which decision analysis addresses the weaknesses of other approaches is discussed.

A complete siting study using decision analysis is presented in Chapter 3. The problem involved identifying a site for a 3000 MWe nuclear power plant facility in the Pacific Northwest. The client for this study was the Washington Public Power Supply System. The chapter is organized to illustrate each of the steps in a decision analysis of siting. This is intended to provide the reader who is interested in the results and concepts of a siting procedure with an easy way to grasp them. Technical and procedural details are omitted as much as possible.

Chapters 4–8 provide the details of the decision analysis procedure for siting energy facilities. The procedure is divided into five parts, one per chapter:

1. identifying candidate sites,
2. specifying objectives and attributes of the siting study,
3. describing possible site impacts,
4. evaluating site impacts, and
5. analyzing and comparing candidate sites.

In each of these chapters, the methodological procedures necessary to implement the approach are presented and illustrated by actual cases. The rationale for assumptions which are necessarily made in selecting the de-

---

† Public Law 91-190, 42 U.S. Congress 4321-4347, January 1, 1970, amended by Public Law 94-52, July 3, 1975, and Public Law 94-83, August 9, 1975. See the Council on Environmental Quality [1978].

cision analysis approach is discussed and the implications of such assumptions indicated. The potential client is then in a position to appraise the appropriateness of decision analysis for any particular siting problem.

Some sections in Chapters 4–8 contain methodological results. These results provide tools for explicitly addressing some of the complex features of siting problems summarized in Section 1.4. However, the text is written to allow the reader to skip these sections and still get a complete picture of how to conduct a decision analysis of a siting problem. In addition, aspects of several siting studies are frequently used to illustrate particular points of the approach.

Chapter 9 presents a study for locating a 600 MWe pumped storage unit in a southwestern state. The purpose is to illustrate the components of a decision analysis siting study after presenting all the procedures. Having seen the approach in detail, the reader is better able to appraise the potential role of decision analysis in siting. One can compare decision analysis both to what is desired of a siting methodology and to what other siting methodologies do in fact provide. As you might guess, this book would not have been written if I did not believe decision analysis was much better than the alternatives.

Chapter 10 discusses the relationship of the siting problem to other energy problems. These include licensing, designing an energy facility, selecting an energy source for a power plant, determining the optimum capacity for a facility, choosing the date when a facility should begin operations, sequencing a series of power plant facilities to come into operation, and setting energy standards. As will be indicated, the siting problem as defined in this book is an important component of each of these significant energy problems.

CHAPTER 2

METHODOLOGIES FOR SITING
ENERGY FACILITIES

The dual purposes of a particular siting study, as stated in Section 1.5, are to help find the best available site and to provide a rationale and documentation for that decision. These are obviously different from the purposes of a siting methodology. A siting methodology should provide a framework for achieving the purposes of siting. Hence, the methodology is in some sense a means to aid in achieving the end purposes of particular siting studies.

In Section 2.1, the requirements, i.e., desired properties, for such a methodology are discussed. This is intended to clarify the meaning of "provide a framework for achieving the purposes of siting." Current siting methodologies and procedures are presented in Section 2.2. The purpose is to outline the conceptual approach used in these methodologies and to provide a procedural overview of their implementation. Their shortcomings with respect to the requirements of Section 2.1 are summarized. Section 2.3 presents an overview of the decision analysis siting procedure. The manner in which it addresses the features of siting problems that are discussed in Section 1.4 and measures up to the requirements of Section 2.1 are indicated. The final Section 2.4 presents a perspective on so-called objective, value-free analyses.

2.1   Requirements of a Siting Methodology†

If the purpose of a specific siting study were only to find the best available site, we would not need to be so concerned with details of the method by which this is accomplished. However, the additional purpose of providing a rationale and documentation for the siting decision emphasizes the fact that the methodology itself is important.

The requirements of a siting methodology can be discussed without any specific knowledge of siting. There are three broad general requirements of siting methodologies:

1. to promote quality analysis,
2. to be practical, and
3. to improve perception of the analysis.

In simple terms, one could say that these requirements mean that it must be theoretically possible to do a good siting study with the methodology, practically reasonable to carry it out, and understood by all that the first two requirements have been met.

2.1.1   QUALITY ANALYSIS

In order to be more specific about the meaning of quality analysis, it is necessary to refer briefly to the general conduct of siting studies. Most siting studies can be broken down into two components, identifying possible sites and selecting one of them. Both of these steps are important and each must be conducted carefully and responsibly in order to have a quality analysis.

In identifying possible sites, this means that all of the criteria used in the process should be clearly stated and justified. In the selection stage, quality analysis means that each of the features of siting problems should be addressed for each site, and the manner in which all of this information is integrated into the overall evaluation of each site must be made clear. Because of the inherent uncertainties in any siting study, the methodology should provide for useful procedures to conduct sensitivity analyses of all inputs.

A quality analysis should clarify the professional judgments, value judgments, and data which were used in the study. It should indicate how and why this information was used. The manner in which the siting features are addressed by this information should also be indicated. But

---

† Sections 2.1 and 2.2 draw heavily on material in Ford, Keeney, and Kirkwood [1979].

perhaps even more important than detailing all that was done, a good analysis should specifically mention what was not done. No analysis addresses everything relevant to a decision. Such an attempt would be impractical, if not impossible. With a clear understanding of what is omitted, it is much easier to interpret the study results in the light of any relevant omitted factors.

In summary, in order for a methodology to promote quality analysis, it should be logically sound, defensible, and useful for decision making.

2.1.2 PRACTICALITY

We have specified some necessary properties for quality analysis. If it does not already exist, clearly a methodology could be designed, in theory, to provide those properties. But what is important is whether this can be done in practice. The methodology should be practical, that is, it should be possible to conduct siting studies in the real environment with the methods and procedures provided.

Basically this means two things. There must be a pool of expertise available to implement the methodology and the cost, time needed, or effort required should not be excessive. One must be careful here in defining "excessive." Since the time needed can be reduced with more intensive efforts, which implies a greater cost, "excessive" might be roughly thought of in terms relative to total facility costs. Since major energy facilities can cost hundreds of millions, or even billions, of dollars, a substantial analysis could be conducted for a fraction of 1% of the total cost.

2.1.3 PERCEPTION OF THE ANALYSIS

Because of the political and social climate in which most siting studies are conducted, many bodies such as regulating agencies, customers, shareholders, and interest groups get involved in the siting process. Thus it is very important that the perception by these groups of any analysis be good. This means the methodology should responsibly and demonstrably address concerns of these bodies. Of course, it would be nice to please each of these bodies to the maximum extent, but this is not possible, since interests of some groups are diametrically opposed to those of others.

Three main characteristics would appear to affect perception the most. The first is understandability. It is desired to present the analysis and its implications in a manner such that an interested and enlightened individual can understand it. This does not imply that there cannot be some so-

phisticated analysis. However, the motivation for such analysis and the
basic concepts used should be understandable. The second characteristic
is whether the impacts to and concerns of the various groups are accounted for in the analysis. For example, citizens of a town near a proposed major facility may like to have their opinions and values accounted
for in a decision which could significantly affect their lives. Moral concerns constitute the third characteristic. Was the process fair, legal, and
rational, or was it just part of a smokescreen to obscure the real reasons
for which an action was taken? To the extent that a methodology makes
the value judgments clear and the evaluation process transparent, it is
much easier both to see through a smokescreen and to demonstrate a
responsible analysis.

2.2   Currently Used Siting Procedures

We do not intend to describe a number of methodologies which have
been used in siting studies, such as cost–benefit analysis, and then comment on their strengths and weaknesses for addressing considerations of
the siting problem.† Instead, we will focus on the components of siting
studies and indicate general approaches common to many procedures
used on these components. There are several reasons for this.

It is not easy to ascertain, and hence describe, exactly what comprises
a particular methodology. Different sources will ascribe different features
to the same methodology. Second, most methodologies are described in
theory and the manner in which that theory is applied varies greatly.
Sometimes the applications contain flaws not inherent in the methodology. Finally, there are only a few conceptual approaches to address issues
in siting problems. Thus, because we are ultimately interested in the
applicability of the methodology and not in its theoretical structure, we
will explicitly consider the main components of siting and conceptual approaches for addressing each of them.

As mentioned in Section 2.1, the two basic components of siting studies
involve identifying candidate sites for the energy facility and selecting the
best site from them. Different procedures are used on these two components. Usually screening procedures of some sort are used to identify candidate sites within a large region of interest. Evaluation procedures are
then used to determine the desirability of each candidate site.

---

† Hobbs and Voelker [1978], Hobbs [1979], and Keeney *et al*. [1979a] characterize and appraise several approaches for power plant siting.

2.2.1   Screening Procedures to Identify
         Candidate Sites

Most siting studies begin with a relatively large region of interest within
which the client wishes to locate an energy facility. A region of interest
may be the service area of a utility company, the shoreline of the Pacific
states (e.g., for an LNG import facility), or the Great Plains for a central
wind or solar power plant. Usually the region of interest is orders of magnitude larger than the area required for the proposed facility. Thus, in
theory there are often thousands of possible sites for the facility. Many of
these are clearly impractical, and many others are obviously not as good
as other possibilities. Because it is infeasible and undesirable to evaluate
carefully all of the thousands of possibilities, we wish to eliminate quickly
and reasonably most of the obviously inferior sites to allow concentration
on potentially good ones. This motivates the screening procedures which
are intended to identify a manageable number of good candidate sites for
careful review and evaluation.

In practice, screening is often conducted in steps. The idea is that if a
substantial portion of the region of interest can be eliminated from further
consideration by screening with only a few criteria, a great deal of effort
can be saved since one need not collect additional data on this eliminated
area. To the extent possible, it is desirable to begin screening with criteria
that are more or less generally acceptable. The obvious choices stem from
legal considerations, such as a nuclear power plant that must be a specific
distance from an earthquake fault. Others include functional concerns,
such as availability of a water source for cooling or substantial sunshine
for solar power.

The formal approaches which have been used for screening can be
categorized into three basic procedures:

Exclusion Screening.   A site is excluded from the set of candidate sites
because it does not meet an exclusion criterion. The elimination of nuclear power sites within 5 miles of a town of over 3000 population is an example.

Inclusion Screening.   A site is included as a candidate site because it
does meet a particular criterion. The inclusion of any site within 1 mile of
a coal mine for a coal power plant is an example.

Comparison Screening.   Sites are rated on the degree of acceptability
based on a weighting scale combining more than one criterion. A cutoff
level is then determined to categorize sites as candidate sites or not.

Each of these procedures can be used sequentially with different criteria or concurrently with more than one criterion. Also, for any particular study, any combination of them can be used.

It is worth noting that in some sense the first two procedures are really the same. If one excludes areas within 5 miles of a town, it is by implication including areas more than 5 miles away. In another sense, comparison screening is also like the others. The distinction is that the criterion by which one screens (i.e., includes or excludes) is a combination of lower level criteria which are aggregated in some fashion.

There are some important weaknesses of the screening techniques used on most siting studies. One might categorize the weaknesses into two types: (1) use of unstated and implicit assumptions and value judgments in specifying criteria and (2) application of oversimplified criteria leading to the rejection of very good candidate sites. Perhaps the most serious, and easiest to miss, example of the former is the definition of the region of interest. By its nature, this definition eliminates all sites outside of this region but usually no attention or justification is given to this crucial screening decision. Indeed, it is often not thought of as screening. An example of the second weakness could result from eliminating sites not within 1 mile of a coal field. A site that is "almost perfect" in every other respect might be 2 miles away. It may have been much better than any site within 1 mile and yet it is eliminated. Sections 4.1 and 4.2 present more details on the use and misuse of screening in practice.

2.2.2 EVALUATION PROCEDURES FOR SITE SELECTION

As a result of the screening process, there are a number of candidate sites which have been identified. Typically the number will be between 5 and 15, but this need not be the case. The second stage of siting procedures involves evaluation of each of these sites to help identify a proposed (i.e., best) site. At this stage, because the candidate sites are specified, it is both necessary and feasible to obtain some site-specific data for the comparisons. This should be gathered from available literature and site visits by members of the siting team.

Information pertaining to each of the general concerns listed in Section 1.3 is relevant for the comparison. This means that for each candidate site, the siting team should amass information about the potential impact with respect to economics, the environment, socioeconomics, health and safety, and public attitudes. For each of these concerns, there could be data on a variety of aspects. Conceptually, one could arrange this information as indicated on Table 2.1, where, of course, each cell of the array would be filled in.

The evaluation of the candidate sites would then involve comparing the

TABLE 2.1

INFORMATION FOR EVALUATING CANDIDATE SITES

Sites	Potential siting impacts on general concerns				
	Economics	Environment	Socioeconomics	Health and safety	Public attitudes
Site S_1	Economic impact of selecting site S_1	Environmental impact of selecting site S_1	Socioeconomic impact of selecting site S_1	Health and safety impact of selecting site S_1	Public attitudes of selecting site S_1
Site S_2	Economic impact of selecting site S_2		$\cdots$		Public attitudes of selecting site S_2
.			.		.
.			.		.
.			.		.
Site S_J	Economic impact of selecting site S_J		$\cdots$		Public attitudes of selecting site S_J

impacts of each site, keeping in mind the siting features discussed in Section 1.4 which render this comparison particularly difficult. Essentially, each of the currently used evaluation approaches utilizes an array similar to Table 2.1. However, the approaches differ greatly in the degree to which the process is formalized. They can be categorized into six basic procedures. In increasing order of formality, the approaches are:

Favorability Selection. A site is considered good enough to be chosen as the proposed site on its merits alone, rather than in comparison with other sites. This corresponds to considering only one row of Table 2.1, the one describing the proposed site.

Qualitative Comparison. The full array is used in comparison, but the information in each cell consists mainly of a qualitative description of the possible impacts. The balancing of the impacts in the general concerns is also done qualitatively in order to select the proposed site.

Cost-Effectiveness Analysis. The economic concerns are carefully analyzed and other concerns are either qualitatively or quantitatively addressed. Usually "softer" aspects such as aesthetics and attitudes are omitted. An informed comparison is made to determine whether increased costs in changing from one site to another are worthwhile in terms of additional benefits.

Site Rating. From the information in the array, a rating is assigned to each site via an informal method. There is qualitative discussion to support the ratings, but the necessary value judgments are never made explicit.

Dominance. A site S_1 is ranked above S_2 if S_1 ranks higher than S_2 in each of the general concerns.

Cost–Benefit Analysis. Weights are assigned to each of the five general concerns and the several factors within each concern if this is appropriate. Each site is rated with respect to each concern, resulting in a rating in each cell. The overall site rating is assigned by multiplying the concern weight by the site rating in that concern and adding over concerns. To be more specific, if w_i is the weight of concern i and x_{ij} is the rating on concern i for site S_j, then the site ranking $r(S_j)$ for S_j is calculated from

$$r(S_j) = \sum_i w_i x_{ij}.$$

There are important weaknesses in most of the evaluations of sites which utilize these six approaches. A broad characterization of the weaknesses might be (1) lack of sufficient data, (2) use of unstated and implicit assumptions and value judgments, and (3) oversimplification of the value judgments.

In terms of Table 2.1, some of the columns may often be omitted with no justification for such action. Even if the columns are there, the data may be unreliable or coarse to the extent that it is useless in site comparison. There may be large acknowledged uncertainties and the cells may not include any information about this.

Each of the approaches must combine the information in each row of Table 2.1 to obtain the overall potential impact of each site, and then select the best site. This process necessarily requires value judgments. With favorability selection, qualitative comparison, cost-effectiveness analysis, and site rating, the manner in which this is done is seldom made explicit or understandable.

With dominance, any set of value judgments weighting the general concerns would lead to the same conclusion about eliminating the dominated sites from further consideration. However, if the screening process is conducted at all responsibly, it is quite unusual to eliminate all but one of the candidate sites using dominance arguments. This is particularly true because of the ever present uncertainties in siting problems. An approach other than dominance would need to be used to select a proposed site from those not dominated.

Cost–benefit analysis makes the necessary value judgments explicit, but they are unnecessarily simple. They imply that a constant linear substitution rate is appropriate between each pair of concerns (and factors) and do not account for any risk attitude. Actual siting studies described throughout Chapters 3–9 indicate that these assumptions are not generally justifiable. Furthermore, cost–benefit analysis usually omits consideration of uncertainties. As will be indicated in Section 2.3, such oversimplifications are not necessary.

2.2.3 Shortcomings of the Procedures

The screening and evaluating procedures discussed can be used in any combination to go from a region of interest to a proposed site. The question is, how do they measure up with regard to the siting requirements discussed in Section 2.1? In short, although there are differences among both the screening and the evaluation procedures, they simply do not provide what might be hoped for and what is needed with respect to the quality or the perception of the analysis. With regard to practicality, they are fine.

Taking the easy one first, practicality refers to expertise required to conduct the analysis and its cost. There is nothing conceptually or mathematically complex about the screening and evaluating procedures now in use and certainly the expertise is available. In the substantive fields providing information for these procedures, such as science and engineering, the professional sophistication required can be great; but there are many capable experts available and interested in such studies. Also, based on historical evidence of past siting studies, the cost, time, and effort required are not at all inordinate.

With respect to quality analysis, currently used approaches are needlessly inadequate. As mentioned above, we have (or can get) the substantive knowledge from the engineering and scientific fields. What is needed is a much better method for using this information. Paraphrasing from Section 2.1.1, quality analysis should be logically sound, defensible, and useful for decision making. Most existing siting studies are weak in these areas.

Analysis may not be logically sound for several reasons. First, the fundamental assumptions on which essentially all of the currently used approaches are based are never explicitly stated. Often, these assumptions, if made explicit, would be clearly recognized as inappropriate. Second, the judgments which are necessarily made in any specific study, and the reasons for them, are often not clearly articulated. And third, a great deal of clearly relevant information, such as uncertainties, is entirely missing in most studies.

For these reasons, it is very difficult to justify implications of an analysis. It is usually the case that the professional judgments and value judgments necessary to decide which sites are better are not brought into the open. But any enlightened observer realizes that such judgments have to be made and, hence, suspects either an incomplete or misleading analysis. Many of the reasons for screening criteria cutoffs are not sufficiently justified and may seem to be arbitrarily set to either include or exclude areas or sites for reasons other than those stated.

Many current analyses are not very helpful for decision making because they simply do not address many of the features of the problem. These features, covered in Section 1.4, must therefore be appraised informally by the clients, regulators, and others, since they cannot be neglected. Furthermore, because they are not formally included in the analysis, it is not possible to conduct sensitivity analyses of these features to gain potentially useful insights about what is and is not important in a particular study.

With regard to perception of the analysis, most enlightened people can understand what was done, but it is usually much harder to see why it was done. This relates to the abovementioned difficulty in justifying the analysis. Most siting studies do not explicitly address the concerns of the population affected by the potential site, with the result that many of these people suspect that the analysis was done to get the most for the clients, regardless of others. In many cases, this conception is simply not true. An analysis to back up the client's contention that others' attitudes were used and considered important in a decision would help to improve perception a great deal. Finally, because of the omissions and implicit considerations in most siting studies, they are often perceived as unfair and not rational or reasonable.

In summary, we strongly feel there exists a much better procedure for siting of energy facilities than those currently used. This procedure, decision analysis, measures up much better in terms of quality analysis and perception than the other procedures, because with decision analysis the complicating features of siting problems, as outlined in Section 1.4, are explicitly addressed and necessary assumptions are made explicit. The tradeoff which one must make in using decision analysis rather than other approaches is a modest decrease in practicality in order to improve the analysis and people's perception of it. Decision analysis requires a degree of substantive knowledge not needed by other siting approaches and therefore fewer individuals have expertise in the field. Also, because an effort is made to address more of the siting problem in terms of explicitly analyzing the problem features, the procedure is more costly. In our judgment, this tradeoff is definitely worthwhile. After reading through Chapter 9 about the theory, procedures, and siting case studies of decision analysis, you will be in a position to make this judgment for yourself.

2.3   Overview of the Decision Analysis Siting Procedure

Decision analysis is a systematic and logical procedure, based on a set of axioms for rationally analyzing complex decision problems. These fundamental axioms are formulated in a slightly different manner in von Neumann and Morgenstern [1947], Savage [1954], and Pratt, Raiffa, and Schlaifer [1964]. They are presented for our siting context in the Appendix following Chapter 10. Decision analysis is developed on the assumption that the attractiveness of alternatives (i.e., sites) to the client should depend on

1. the likelihoods of the possible consequences of each alternative site and

2. the client's preferences for those possible consequences.

What makes decision analysis unique is the way in which these factors are quantified and incorporated formally into the analysis of a problem. Existing information, collected data, models, and professional judgments are used to quantify the likelihoods of various consequences. Utility theory is used to quantify preferences.

The decision analysis approach attempts to consider systematically all the available relevant information and to use explicitly the preferences of the clients. This is done by breaking the problem into parts which are easier to analyze than the whole, and then putting the parts back together in a logical fashion. The crucial difference between decision analysis and other approaches that claim to help the decision maker is that decision analysis provides theoretically sound procedures for formalizing and integrating the judgments and preferences of experts and the client to evaluate alternative courses of action in complex decision problems. It is essential to exploit the experience, judgment, and knowledge of both professionals with training relevant to the problem and the individuals responsible for making decisions.

TABLE 2.2

STEPS IN THE DECISION ANALYSIS SITING PROCEDURE

Step 1. *Identifying Candidate Sites*
Selecting the region of interest
Choosing screening criteria
Determining candidate areas
Initial site visits

Step 2. *Specifying Objectives and Attributes of the Siting Study*
Specifying general concerns and relevant interest groups
Determining the objectives
Defining measures of effectiveness (attributes) for each objective

Step 3. *Describing Possible Site Impacts*
Quantifying impacts in terms of attributes
Quantifying uncertainty, using probability distributions
Assessing judgments of experts
Collecting data and updating estimates

Step 4. *Evaluating Site Impacts*
Determining the functional form of the multiattribute utility function
Assessing the single attribute utility functions
Assessing the value tradeoffs and the multiattribute utility functions

Step 5. *Analyzing and Comparing Candidate Sites*
Verifying the appropriateness of the decision analysis assumptions
Integrating the previous information to evaluate alternatives
Conducting a sensitivity analysis with respect to preferences and impact inputs
Reappraising assumptions made in the analysis

As listed in Table 2.2, the decision analysis methodology for siting can, for the purposes of discussion, be broken into five steps:

1. identifying candidate sites,
2. specifying objectives and attributes of the siting study,
3. describing possible site impacts,
4. evaluating site impacts, and
5. analyzing and comparing candidate sites.

The last step involves synthesizing the information obtained in steps 1–4 to evaluate and compare the alternative choices.

Figure 2.1 schematically illustrates the decision analysis siting process and indicates where each of the features of siting problems raised in Section 1.4 are addressed. As can be seen, each of the features is explicitly considered. Because of this, the quality of the analysis can be significantly increased over that of other approaches which do not address each of these features. This should help significantly in improving the perception of the analysis.

In the remainder of this section, we shall give a short overview of the

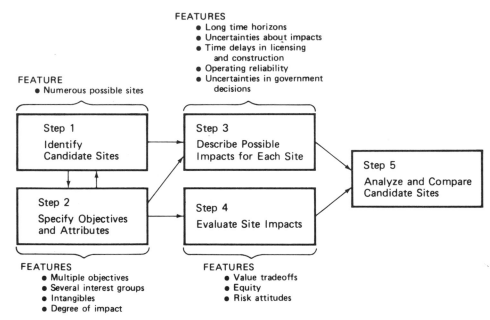

Fig. 2.1.   Schematic representation of the decision analysis siting procedure.
Note: Arrows indicate a major influence of the results of one step on another step.

decision analysis siting methodology. This will provide the reader with a feel for the entire process. The five steps are discussed in detail in Chapters 4–8.

2.3.1 STEP 1: IDENTIFYING CANDIDATE SITES

Once a decision has been made to build a particular energy facility, the region of interest is usually decided upon. This is an extremely important consideration because it eliminates any site outside the region from being a possible candidate for the facility. Hence, careful appraisal and justification of such a definition is necessary.

Next, a series of screening models is used to eliminate much of the region from further consideration. The purpose is to find a number of candidate areas in which the potential for identifying good candidate sites is deemed high. Some of the screening can be done by legal (e.g., regulatory) requirements, such as the distance from an active earthquake fault. In such a case, the value judgments about acceptability are implied by the law and need not be considered in the siting study. (Of course, such value judgments should always be subject to modification if new data or experience indicates such a change would be appropriate.)

Most of the screening criteria require value judgments to be made by the clients and/or their analysts. For instance, a nuclear power plant requires water for cooling. For one such case, discussed in Chapter 3, sites more than 10 miles from their source of cooling water were eliminated. This condition obviously uses the value judgment that 9.5 miles is acceptable and 10.5 unacceptable. It also implies that distance from the source is an appropriate measure for a screening criterion. Since the distance is really of interest only because of the cost of pumping water to a site, perhaps a model relating distance and other variables to cost and a screening criterion on cost would be more appropriate.

Another screening criterion in the same study is that the site be no more than 800 feet above the source of cooling water. Again, this consideration is for economic reasons. In some sense, the value judgment on acceptability of up to 800 feet should be consistent with the 10-mile distance discussed above. In fact, a model relating cost to both horizontal and vertical distance from the water source may provide a more useful screening criterion. Sections 4.2 and 4.3 have a detailed discussion of these topics.

The result of the formal screening is usually a number of relatively homogeneous areas, referred to as candidate areas, in which one suspects the best potential sites in the region of interest are located. Visits are made to each of these areas by members of the siting team, such as biolo-

gists, meteorologists, and demographers. Using this information plus
other easily available information (e.g., population statistics), the team
usually identifies one or more good candidate sites in each candidate area.
The process for this may involve the use of screening models, as discussed above, but on a more local level. Alternatively, because of the
homogeneity of a candidate area, it may be relatively easy for members of
the team to select the best sites within the area directly. For instance, two
adjacent sites 1 mile apart may be expected, in a specific case, to have the
same socioeconomic, environmental, and health and safety impacts, and
have the same effects on public attitudes. However, because of the geology of the two sites, it may be clear that one is much better on economic
grounds than the other due to construction costs. Such a professional
judgment from the siting team would be sufficient to eliminate the inferior
site.

A major distinction between decision analysis screening models and
other screening models is the degree of attention—both informal and
formal—paid to the value structure and the manner in which it is utilized
in the model. Not only should this be internally consistent, but it should
be consistent with the value structure used in the evaluation of candidate
sites once they are chosen. Once the candidate sites are evaluated, the decision analyst should reappraise each of the screening assumptions to
minimize the likelihood that excellent sites were eliminated in the
screening process. Chapter 4 contains more details on decision analysis
screening models.

2.3.2   STEP 2:   SPECIFYING OBJECTIVES
AND ATTRIBUTES OF THE SITING STUDY

Any siting procedure must specify objectives and attributes (i.e., measures of effectiveness) to measure the degree to which the objectives are
achieved by each of the candidate sites. The distinct aspect of the decision analysis approach is the degree of formality with which this specification is conducted.

The starting point for specifying clear objectives is the list of five general concerns discussed in Section 1.3. These are environmental impact,
economics, socioeconomics, health and safety, and public attitudes. The
question one wishes to answer in determining objectives and attributes is,
for example, what is the environmental impact of concern in a particular
problem and how do we measure it. The process of answering such questions is essentially a creative task. There are, however, several aids which
can be employed. Previous siting studies and legal and regulatory guidelines should be of significant help in articulating objectives. To the degree

that individuals outside of the siting team participate in the siting process (not the siting study) as intervenors, shareholders, or concerned citizens, they may contribute useful ideas.

From all of this information, an objectives hierarchy should emerge with broad objectives, one pertaining to each general concern, at the top and more detailed objectives further down. The lower-level objectives can be viewed as means to the higher-level ends. Holes in the hierarchy can be identified and filled in by following means–ends relationships.

For each of the lowest-level objectives in the hierarchy, we need to identify an attribute. Sometimes this is easy. For example, an obvious attribute for the objective "minimize construction costs" is millions of dollars. However, it is much harder to determine an attribute for an objective like "minimize visual degradation." This often requires construction of an attribute especially for the problem under consideration. Procedures to do this are discussed in detail in Chapter 5.

Now we must introduce a little formal notation for future convenience. Let $O_1, ..., O_i, ..., O_n$ be n lowest-level objectives with the associated attributes $X_1, ..., X_i, ..., X_n$. Furthermore, define x_i to be a specific level of X_i, so the possible impact of selecting a particular site can be characterized by the consequence $(x_1, x_2, ..., x_n)$. An example of an objective O_i is "maximize the economic benefit to the local town," and an associated attribute X_i may be "yearly tax paid to the town." A level x_i could then be 16 million dollars. The desired product of step 2 of the decision analysis siting procedure is the sets of objectives and attributes.

The manner in which the siting problem features indicated in Fig. 2.1 are addressed is as follows: The multiple objective feature is addressed in specifying O_1 to O_n. Some of these objectives can concern the public attitudes so that the several interest group feature is also included here. The intangibles are included by using objectives such as "minimize aesthetic disruption." The degree of impact on this and every other objective is measured by the attribute defined or constructed for the purpose.

2.3.3 STEP 3: DESCRIBING POSSIBLE SITE IMPACTS

It is impossible to describe the impacts of selecting a site for an energy facility in terms of one consequence. The reason is the inherent uncertainties about what the eventual consequence will be. Consequently, for each site, we want to describe the impacts by listing the various possible consequences and the probability that each might occur. This can be done formally by assessing a probability distribution function $p_j(x)$ for each site S_j. In some cases, the uncertainty about the consequences as measured on some attributes may be relatively small. Then it may be an appropriate simplification to omit the uncertainty on that attribute. Because one can treat $p_j(x)$ in general to include cases with no uncertainty [i.e., where $p_j(x)$ is simply x itself], we will use p_j throughout.

The assessment of the magnitude and the probabilities of the possible consequences should be accomplished, when possible, through the development and use of formalized models. Such models would define the possible levels of impact in terms of the attributes for each alternative site.

In the disciplines pertinent to the five general concerns, there are extensive bodies of literature on techniques for predicting impacts. In the conduct of any of these assessments, knowledgeable professionals in the various disciplines have to be involved and their familiarity with available impact-predicting techniques used. This is particularly important because the available techniques for predicting impacts for the different disciplines vary widely in analytic formality and scope of predictive capability.

There are several methods for quantifying probabilities in practice. One method is to use a standard probability distribution function and assess parameters for that function. For example, the parameters of a normal distribution can be the mean and standard deviation. Another technique, referred to as the fractile method, involves directly assessing points on the cumulative density function. Thus, in assessing $p_j(x_i)$, one is asked for a level x_i' such that the probability is p_j' that the actual level of X_i is less than x_i'. This questioning is repeated for several values of p_j' such as 0.05, 0.25, 0.5, 0.75, and 0.95. By fitting a common probability distribution to the assessed data, one obtains $p_j(x_i)$. A third procedure for direct assessment is appropriate when the possible impact is categorized into a number of distinct (discrete) levels. The expert or client is asked to specify the probability of each. In a short summary, all this sounds rather easy, but as indicated in Chapter 6, it is an involved process with many potential sources for large errors.

A factor which can greatly increase the complexity of the impact assessments is probabilistic dependencies among the attributes. This means that if two attributes are probabilistically dependent, knowledge about the impact measured on one of them will affect the estimated impact of the other. Clearly, there are such relationships across sites. For illustration, if there are just two objectives, one economic and one environmental, the best site economically is probably not the best environmentally, and *vice versa*. If it were, the choice of a site would be simple. Sometimes this may happen, but because one can usually get better environment at a price, the stated relationship of probabilistic dependence holds. Fortunately for siting studies, one is interested in probabilistic dependencies conditioned on a particular site. Since cost overruns at a site have no natural mechanism for influencing the site environmental impact, attributes for these concerns may be conditionally probabilistically independent. However, if there are conditional dependencies, analytical or simulation models which take them into account should be used.

In describing the possible impacts, the time at which the impacts might occur must also be indicated. Thus, the feature of long time horizons is addressed. The other four features—uncertainty, time delays in licensing and construction, operating reliability, and uncertainties in government decisions—are all handled by assessing the probabilities describing the possible effects each of these might have. Suggestions for doing each of these are found in Chapter 6.

2.3.4 Step 4: Evaluating Site Impacts

With the $p_j(x)$ for each candidate site S_j, we have a formal description of the possible impacts. In step 4 we wish to quantify the desirability of each of the possible consequences. Multiattribute utility theory provides the methods and procedures for doing this. A utility function u is assessed which assigns a number $u(x)$ to each possible consequence x. The utility function has two convenient properties:

1. $u(x_1', x_2', ..., x_n') > u(x_1'', x_2'', ..., x_n'')$ if and only if $(x_1', x_2', ..., x_n')$ is preferred to $(x_1'', x_2'', ..., x_n'')$ and

2. in situations involving uncertainty, the expected value of u is the appropriate index to evaluate alternatives.

These properties follow from assumptions first postulated by von Neumann and Morgenstern [1947] and repeated in the Appendix for our siting problem.

The preferences of interest in siting are those of the client. They are quantified by asking the client several questions about his or her value judgments. This process of determining the utility function can be broken into five steps:

1. introducing the terminology and ideas,
2. determining the general preference structure,
3. assessing the single-attribute utility functions,
4. evaluating the scaling constants, and
5. checking for consistency and reiterating.

For discussion purposes each of these is considered a specific step, though in reality there is considerable interaction among them, as indicated in Chapter 7.

The three features of siting problems addressed in this step of a decision analysis are value tradeoffs, equity, and risk attitudes. Each of these require value judgments which are made explicit in assessing u. The decision analysis approach focuses on eliciting and clarifying the necessary information about values and expressing it in a form useful for evaluating the alternatives. In addition, there are three important advantages in assessing the utility function used in decision analysis:

1. the models and procedures used are derived formally and on a sound theoretical basis,

2. the procedures systematically elicit the relevant information about value tradeoffs, equity, and risk attitudes with provision for consistency checks to insure accuracy, and

3. a sensitivity analysis of the client's value judgments can be conducted.

2.3.5 STEP 5: ANALYZING AND COMPARING CANDIDATE SITES

Once the problem is structured, the magnitude and associated likelihood of impacts determined, and the preference structure established, the information must be synthesized in a logical manner to evaluate the candidate sites. The basis for this evaluation is expected utility which, as pointed out earlier, follows from the axioms of decision analysis.

The calculation of expected utility is a mathematical computation involving the probability distribution p_j for each site S_j and the utility function u, all of which are available from the previous two steps. In general, this computation involves the summation (or integration) of the probability of each possible consequence multiplied by the utility of that consequence. This is the formal way of combining the likelihoods of and preferences for consequences. This results in an expected utility $E_j(u)$ for each site S_j which is

$$E_j(u) = \int_x p_j(x)u(x) \, dx.$$

The higher the $E(u)$ number, the more desirable is the alternative. Thus, the magnitude $E(u)$ can be used to establish a ranking that reflects the decision maker's preference for one set of consequences over other sets. Furthermore, one can transform the $E(u)$ numbers to obtain information about how much one candidate site is preferred to another. It should be remembered that the expected utility associated with choosing an alternative is directly related to the objectives originally chosen to guide the decision. The evaluation of the candidate sites derived from this method reflects the degree of achievement of the objectives.

In real-world situations it is extremely important that the client be able to examine the sensitivity of any decision to different views about the uncertainties associated with various levels of impact and to different value structures. With decision analysis, we quantify the impacts (even the subjective concerns), explicitly consider uncertainties, and develop a formal statement of the value structure. Without quantification, it would be difficult to conduct a sensitivity analysis. A useful way of presenting the results of a sensitivity analysis is to identify sets of conditions, in terms of uncertainties and preferences, under which various options would be chosen.

2.3.6   SUMMARY

All the significant features involved in a siting study are addressed within the decision analysis framework. Of special significance is the explicit treatment of uncertainties associated with the possible site impacts and the value judgments of the client. The client also plays an important role in determining what information is to be used in the analysis and how it should be gathered. Decision analysis provides a framework for using this information rationally and consistently. The inputs and the assumptions used in the evaluation of alternatives are clearly identified.

Experience in using decision analysis indicates that knowledgeable professionals, industry executives, and government officials are willing to address the difficult professional judgments and value questions that are necessary to focus meaningfully on the complex features of siting problems. We believe that the decision analysis approach proposed in this book will result in improved decision making and smoother regulatory processes for the siting of major energy facilities. In any specific study, the client has to judge the appropriateness of the approach in order to appraise and make use of the results.

2.4   A Perspective on Objective, Value-Free Analysis

A comment sometimes heard among groups of analysts, regulatory authorities, and managers of energy firms is that what is really needed to help the decision makers is objective, value-free analysis. This section, devoted to that topic, has a very simple message. There is no such thing as an objective or a value-free analysis. Furthermore, anyone who purports to conduct such an analysis is professionally very naive, stretching the truth, or using definitions of objective and value-free which are quite different from those commonly in use. If there ever was an objective, value-free analysis of a siting problem, the analysis and its implications would be of almost no use for helping to make any decision more responsibly. Let us explain.

The "Random House Dictionary of the English Language" defines "objective" as we are using it here as an adjective meaning "not affected by personal feelings or prejudice; based on facts; unbiased." For a siting problem to exist, the problem had to be identified and this required someone's personal feelings (i.e., judgments). The identification of alternative sites requires professional judgments also, and hence feelings and prejudices. The siting problem requires use of the experience of experts in order to focus on reasonably good sites. Specifying objectives (i.e., the noun meaning goals) and attributes entails judgments of perhaps many people, each of whom is naturally biased by his or her own personal and professional experiences. In describing possible site impacts, decisions must be made on how much data of what type to collect, how to use it, and what information to elicit from which experts. None of this can be done, nor should it be done, objectively. It is crucial to obtain, understand, and utilize the professional judgments of various experts.

The ultimate purpose of a siting study is to choose a site for a proposed facility. The analysis is conducted partially to help identify the best (or at least a good) candidate site. The concepts of best and good require value judgments. As indicated in Section 1.4, some of these involve value tradeoffs between the various objectives, equity considerations, and attitudes toward risk. Others are clearly necessary in the screening process and in selecting objectives and attributes. Aspects felt to be important—a value judgment—are included in the analysis. Another value judgment necessary for analysis is the acceptance of the assumptions on which the analysis is based.

In order to conduct any analysis of a siting problem, the candidate sites must be identified, the objectives and attributes specified, and the possible site impacts described and evaluated. The result is a formal mathematical model of the problem. At this stage, one might refer to the subsequent phase of the analysis as objective and value free. However, even deciding what sensitivity analyses to conduct and how to interpret the results of the study requires value judgments.

The point is that professional judgments and value judgments are absolutely necessary in essentially every step of an analysis in order to address the concerns and features of a siting problem. Objective, value-free analysis is undesirable because it simply avoids the problem. What is needed is a logical, systematic analysis that makes the necessary professional and value judgments explicit. The resulting analysis should be responsive to the client's needs and justifiable to the public and the regulatory authorities. Decision analysis is uniquely a methodology which provides for such analyses.

CHAPTER 3

THE WASHINGTON PUBLIC POWER SUPPLY SYSTEM NUCLEAR SITING STUDY

The purpose of this chapter† is to illustrate the decision analysis approach to siting of energy facilities. The specific problem involves a potential nuclear power facility in the state of Washington. This study, conducted by Woodward-Clyde Consultants in 1975, is one of the first which explicitly contained each of the five steps of decision analysis for the siting of an energy facility. This chapter provides a general overview of the study.

It will be clear to the reader that not all of the features of siting problems were explicitly addressed in this study. Several of the more important features were formally included, but some were necessarily dealt with informally or in a somewhat *ad hoc* manner. This was done for two reasons: some data were simply unavailable to the siting team and time pressures required concentration on features felt to be the most relevant. However, the analysis presented should provide a useful introduction to the approach.

Background for the Problem

The Washington Public Power Supply System (WPPSS) is a joint operating agency consisting of 21 publicly owned utilities in the state of

† This chapter is adapted liberally from Keeney and Nair [1977].

Washington. In 1974, WPPSS authorized a study to identify and recommend potential new sites in the Pacific Northwest suitable for 3000 MWe nominal capacity nuclear power generating stations that might be required after 1984. The study was conducted on the basis of existing information and field reconnaissance; no detailed site-specific studies were made. The purpose of the study was to recommend sites that would have a high likelihood for successful licensing and that would therefore be most suitable for the detailed site-specific studies necessary to select a nuclear power plant site.

The study team comprised several staff members from Woodward-Clyde Consultants: Gail Boyd, Steve James, Keshavan Nair, Gordon Robilliard, Charles Hedges, Ashok Patwardhan, Dennis McCrumb, Don West, Ram Kulkarni, and Wayne Smith. In 1975, I was a consultant to Woodward-Clyde Consultants, working on this study. In addition, the firm of R. W. Beck and Associates provided information on cost and transmission line considerations. Both the Public Power Council Siting Committee (Mr. William A. Hulbert, Jr., Chairman) and the WPPSS management (Mr. J. J. Stein, Managing Director) supported the use of decision analysis techniques in the siting process. Mr. David Tillson, siting specialist of WPPSS, monitored the contract and provided valuable information crucial for the study.

This chapter is organized to coincide with the five steps of decision analysis discussed in Section 2.3. Section 3.1 indicates how the candidate sites were identified using a screening process. Section 3.2 describes the objectives and the attributes used to evaluate the candidate sites. The probability assessments describing the possible impacts associated with each site are given in Section 3.3, and the assessment of the utility function used to evaluate these impacts is presented in Section 3.4. Section 3.5 presents the evaluation of sites using the information developed and the sensitivity analysis. The final Section 3.6 contains a brief critique of the analysis.

3.1   The Screening Process: Identifying Candidate Sites

The region of interest consisted of approximately 170,000 square miles, including the entire state of Washington, the basins of major rivers in Oregon and Idaho that flow into Washington rivers, and the major river basins of the Oregon coast. Since the purpose of the study was to find new sites, all areas within a 10-mile radius of the ERDA-Hanford reservation and other site areas for which electric generating facilities have been formally proposed or are under development were excluded. It was clearly impractical to evaluate every possible site in such a large area. Financial and time constraints required concentration on areas where the likelihood of finding candidate sites was high. The purpose of the screening process was to identify such candidate sites.

The first step in the screening process involved establishing the basis for identifying candidate sites. An extensive hierarchy of objectives pertaining to nuclear power plant siting was developed. These objectives pertained to health and safety, environmental, socioeconomic, and economic concerns. Criteria defining a required level of achievement on each objective were established to identify areas for further consideration. Examples of the specific screening criteria are given in Table 3.1.

Note that some of the criteria for inclusion result from the rules of regulatory agencies, e.g., distance from a capable earthquake fault or location with respect to a protected ecological reserve. Other considerations are functional in nature, e.g., the accessibility of an adequate supply of cooling water. There are also considerations related to cost for which the project team, in consultation with representatives of WPPSS, established minimum levels of achievement, e.g., distance from railroads, waterways, and rugged terrain. In addition, considerations relating to environmental aesthetics were included. Examples of such considerations are exclusions of areas of scenic beauty or unusual ecological character that have not been designated as legally protected areas.

Once screening criteria were specified, those parts of the region of interest where a criterion was satisfied were identified and plotted on an appropriate map. Overlay techniques were used to produce composite maps which specified candidate areas meeting all the criteria. A field reconnaissance team composed of experienced engineers, geologists, and environmental scientists visited those areas meeting all the screening criteria. On the basis of their observations and published information, these experts identified nine candidate sites within the candidate areas for further consideration.

3.2 Specifying Objectives and Attributes

To help in identifying those characteristics that would delineate the suitability of locating a nuclear power facility at one site rather than another, detailed descriptions of the sites were developed. The information gathered included data on the area, location, present use, and ownership of the site; the quality and quantity of water available and its location relative to the site; details of natural factors, including geology, topography,

TABLE 3.1

EXAMPLES OF SCREENING CRITERIA USED TO IDENTIFY CANDIDATE SITES

Concern	Consideration	Measure	Criteria for inclusion
Health and safety	Radiation exposure	Distance from populated areas	Areas more than 3 miles from places populated by more than 2500 people Areas more than 1 mile from places populated by less than 2500 people
	Flooding	Height above nearest water source	Area must be above primary floodplain
	Surface faulting	Distance from fault	Areas more than 5 miles from capable or unclassified faults more than 12 miles in length
Environmental impact	Thermal pollution	Average low flow	Rivers or reservoirs yielding 7-day-average, 10-year-frequency low flow greater than 50 ft^3/s
	Sensitive or protected environments	Location with respect to ecological areas	Areas outside of designated protected ecological areas
Socioeconomic impact	Tourism and recreation	Location with respect to designated scenic and recreational areas	Areas outside of designated scenic and recreational areas
Economics	Routine and emergency water supply and source characteristics	Cost of cooling water acquisition	Rivers or reservoirs yielding 7-day-average, 10-year-frequency low flow greater than 50 ft^3/s
		Cost of pumping water	Areas less than 10 miles from water supply Areas less than 800 ft above water supply
	Delivery of major plant components	Cost of providing access for major plant components	Areas within 25 miles of navigable waterways

flooding potential, and volcanic considerations; population in the vicinity of the site; vegetation and wildlife in the area and fish in the streams; access to various transportation modes for construction and operation of the facility; the existence of a local work force; and potential socioeconomic effects of the construction phase. As a result of this, plus information gathered during the screening process, approximately 30 potential objectives with associated attributes for evaluating sites were identified.

It was unlikely that each of these objectives would be significant in the evaluation process. Hence, each one was qualitatively examined (and, in some cases, preliminarily quantitatively examined) to determine the reasonableness of keeping it in the evaluation process. Three general concepts were used for this:

1. The site-dependent variation of the impact in terms of an attribute. For instance, even though yearly manpower costs for plant operation may be significant, it might be omitted from consideration if these costs are nearly identical for all sites.

2. The significance of the impact in terms of an attribute as it relates to impacts as measured by other attributes. For example, the annualized capital cost of a nuclear power plant is in the range of 200 to 300 million dollars for these sites, and the annual revenue loss from adverse effects of plant operation on fish is in the range of 0 to 500 thousand dollars. Under these conditions, the contribution of the latter to the relative preferences of the sites could be neglected.

3. The likelihood of occurrence of significant impacts as measured by an attribute. If one combines the magnitude of impact with the likelihood of its occurrence, the resulting "weighted" impact can be relatively insignificant. Consider, for example, that adverse effects on crops could cost as much as 9 million dollars per year. However, considering the near-zero probabilities of such extreme losses, the "weighted" impact would be in thousands of dollars rather than in millions. Such an impact is considered insignificant.

The examination of possible objectives was evolutionary in nature. Preliminary estimates were made of possible impacts and their probabilities. Using these, some objectives were disregarded. Estimates of the remaining impacts were updated on the basis of field visits, and a few more objectives were discarded. In addition to the examples above, some possible impacts which were formally examined and then discarded included the impact on nearby residents of fog from the cooling towers, the effect of such fog on boat transportation on the Columbia River, and the possibility that the water vapor would freeze on highways, making driving par-

TABLE 3.2

Attributes and Ranges Used in Evaluating the Candidate Sites

General concern		Attribute	Range	
			Worst	Best
Health and safety	X_1:	Site population factor	0.20	0
The environment	X_2:	Loss of salmonids		
		Y: Number of salmonids in stream (1000's)	100	0
		Z: Percentage lost	100	0
	X_3:	Biological impacts at site	(Constructed scale described in Table 3.3)	0
	X_4:	Length of transmission intertie (to 500 kV system) through environmentally sensitive areas	50 miles	0
Socioeconomics	X_5:	Socioeconomic impact	(Constructed scale described in Table 3.4)	
Economics	X_6:	Annual differential site cost (1985 dollars, 30-year plant life)	$40,000,000	0

ticularly dangerous. By this process the list of attributes in Table 3.2 for evaluating candidate sites was generated.

For each of the attributes, a measurement scale was established and ranges of possible impacts were determined. The attributes can be grouped into two classes: those which have a natural scale and those which have a constructed scale. These two types of scales are discussed in detail in Section 5.4. A natural scale is one for which the basic measure is quantified. Each point on such a scale is clearly specified. For example, attribute X_6 has a natural scale since it is quantitatively defined as differential costs in terms of dollars. The attributes measured with natural scales were X_1, site population factor; X_2, impact on salmonids; X_4, environmental impact of the transmission intertie; and X_6, annual differential site cost. The levels of X_3, biological impact, and X_5, socioeconomic impact, were represented on constructed scales, defined in Tables 3.3 and 3.4, for which a number of specific points were qualitatively defined. A level of impact could occur in the interval between points on the constructed scale; however, only the specific points were clearly defined. The definition of points of the scales was made by describing levels of the various components of the attribute. This process will become clearer with what follows.

3.2.1   CLARIFYING THE ATTRIBUTES

Attribute X_1, the site population factor, is an index developed by the former U.S. Atomic Energy Commission (see Kohler *et al.* [1974]) to indicate the relative human radiation hazard associated with a nuclear facility. The site population factor for site S, denoted SPF(S), is defined by

$$\text{SPF}(S) = \sum_{r=1}^{50} P(r)r^{-1.5} \Big/ \sum_{r=1}^{50} Q(r)r^{-1.5}, \qquad (3.1)$$

where r is distance in miles from site S, $P(r)$ is the population living between $r - 1$ and r miles of S, and $Q(r)$ is the population that would live between $r - 1$ and r miles of S if there were a uniform density of 1000 people per square mile. The $r^{-1.5}$ is meant to account for the decrease in radia-

TABLE 3.3

CONSTRUCTED SCALE FOR BIOLOGICAL IMPACTS AT THE SITE

Scale value	Level of impact
0	Complete loss of 1.0 sq mile of land which is entirely in agricultural use or is entirely urbanized; no loss of any "native" biological communities.
1	Complete loss of 1.0 sq mile of primarily (75%) agricultural habitat with loss of 25% of second growth; no measurable loss of wetland or endangered species habitat.
2	Complete loss of 1.0 sq mile of land which is 50% farmed and 50% disturbed in some other way (e.g., logged or new second growth); no measurable loss of wetland or endangered species habitat.
3	Complete loss of 1.0 sq mile of recently disturbed (e.g., logged, plowed) habitat plus disturbance to surrounding previously disturbed habitat within 1.0 mile of site border; or 15% loss of wetlands and/or endangered species.
4	Complete loss of 1.0 sq mile of land which is 50% farmed (or otherwise disturbed) and 50% mature second growth or other community; 15% loss of wetlands and/or endangered species.
5	Complete loss of 1.0 sq mile of land which is primarily (75%) undisturbed mature "desert" community; or 15% loss of wetlands and/or endangered species habitat.
6	Complete loss of 1.0 sq mile of mature second growth (but not virgin) forest community; or 50% loss of big game and upland game birds; or 50% loss of wetlands and endangered species habitat.
7	Complete loss of 1.0 sq mile of mature community or 90% loss of productive wetlands and endangered species habitat.
8	Complete loss of 1.0 sq mile of mature, virgin forest and/or wetlands and/or endangered species habitat.

[a] Three main impacts captured by this scale are those on native timber or sagebrush communities, habitats of rare or endangered species, and productive wetlands.

TABLE 3.4

CONSTRUCTED SCALE FOR SOCIOECONOMIC IMPACTS

Scale value	Level of impact
0	Metropolitan region, population 100,000. No significant impact.
1	Semiremote town, population 250. Self-contained company town is built at the site. As many as half of the plant construction force continue to commute from other areas. Some permanent operating personnel continue to commute. Cultural institutions are overloaded, very little change in the social order. Public debt outstrips revenues by less than six months over previous levels.
2	Remote town, population 250. Self-contained company town is built at the site. Most of the work force moves into company town. Most permanent operating personnel begin to assimilate into the community. Cultural institutions are impacted, significant changes take place in the social order. Growth of the tax base due to permanent operating personnel is orderly, but public debt outstrips revenues by more than six months but less than a year over previous levels.
3	Semiremote city, population 25,000. About half of the plant construction force in-migrates and seeks housing in the city. Most of new growth is in mobile homes. All city systems (law enforcement, sewer, water, schools, code enforcement) are taxed to the limit. Outside financial assistance is required. Cultural institutions are impacted, social order is slightly altered. Permanent operating personnel are easily assimilated into community; tax base grows significantly, but lags in assessment; planning, and capital improvements in construction produce a boom-town atmosphere. Public debt outstrips revenue growth by one to two years.
4	Remote city, population 25,000. Most workers locate in the city. All city systems are impacted. Land-use patterns are permanently disrupted. Growth outstrips planning activities and regulatory systems. Assessment falls behind. Revenue–debt lag is greater than two years.
5	Semiremote town, population 1500. Many workers commute from outside areas. Permanent operating personnel and some workers seek housing in the city. New growth is predominantly mobile homes, with much permanent construction as well. New construction in service establishments and expansion of commercial facilities. Town has basic planning and land-use regulatory functions established, but these are overwhelmed by magnitude of growth. Assessment and enforcement lag two years or more; community facilities are impacted. Land-use patterns are permanently disrupted. Cultural institutions are severely impacted; social order is permanently altered. Much growth occurs in unincorporated areas, untaxable by town.
6	Remote town, population 1500. Most workers try to locate in or near the town. Most growth in unincorporated areas. City systems are impacted; lack of regulation in unincorporated areas impacts rural development patterns, which in turn severely impacts the cultural institutions and social order of the small town. Tax base cannot expand to meet demand for capital improvements.
7	Remote city, population 10,000. Severe impact due to attractiveness to large numbers of plant workers. Basic services and established planning, assessment, and enforcement procedures are sufficient to provide the framework for rapid growth but insufficient to handle the magnitude of such growth. Massive imbalances in long-term city finances occur, leading to several-year lags in revenues to debts. City size and bonding experience probably do not permit revenue financing, so the ''bust'' portion of the cycle is virtually inescapable.

tion exposure hazard as a function of distance. The purpose of the denominator in (3.1) is to allow one to interpret an SPF = 0.1, for example, as equivalent to a uniform distribution of 100 (i.e., 0.1 × 1000) people per square mile within 50 miles of the site.

Two separate indices were required to measure adequately the impact on salmonids. These are the number of fish in a stream, defined by attribute Y, and the percentage of fish lost, defined by attribute Z. The reason for using these indices, rather than simply the number of fish lost, is that the genetic composition of the population of salmonids in each stream is distinct. Therefore the loss of 2000 fish in a stream of 2000 is a more important loss than 2000 fish in a stream of 50,000. For the Columbia River (which has an annual escapement greater than 300,000), only the number of salmonids lost is important, since it is virtually impossible that a large percentage of these fish would be lost due to a specific nuclear power plant and because the fish in the Columbia are indigenous to several different streams that flow into the Columbia. See Section 6.2.3 for more details concerning the measurement of salmonid impact.

Because attributes X_3 and X_5 were meant to describe many detailed possible impacts, it was necessary to develop constructed indices for each of them. The constructed index for biological impacts shown in Table 3.3 was developed by two experienced ecologists on the siting team. It is a qualitative scale of potential short- and long-term impacts which could result from the construction and operation of a power plant on a site. The impacts range from 0 for no impact to 8 for maximum impact. Site visits and general reconnaissance showed that the biologically important characteristics (aside from aquatic resources) of the regions were:

1. virgin or large, mature second-growth stands of timber or "undisturbed" sagebrush communities,
2. known or potential habitats of endangered species,
3. wetland areas, though most comprise small swamps.

The constructed index for socioeconomic impact, attribute X_5, was created by a sociologist/planner associated with the siting team. The scale, presented in Table 3.4, considered the effects of rapid population growth, overloading of municipal service systems, impacts on cultural institutions, alteration of the social order, increase demand for capital improvements, changes in the tax base, impacts on municipal administrative services, alteration of land use patterns, and revenue lags in public financing of capital projects. These considerations are the primary components of what is commonly termed a "boom–bust" cycle.

The length in miles of the transmission intertie line running through environmentally sensitive areas is measured by attribute X_4. Attribute X_6

is the annual differential cost between sites in terms of 1985 dollars assuming a 30-year plant life. The discount rate used was 8.4%. Costs such as the major plant components are not included in attribute X_6 since these would be the same for all sites. The differential is calculated relative to the lowest-cost site for which the "differential cost" is set at zero.

3.3   Describing Possible Site Impacts

The consequences associated with site development at each site can be characterized by the levels which the six attributes of Table 3.2 would assume should a power plant be constructed on that site. To account for the uncertainty associated with estimating the levels of the attributes, probabilistic estimates were made.

3.3.1   FORM OF PROBABILITY ASSESSMENTS

The estimation of the possible impacts at each site was accomplished in three ways. Attribute X_1, site population factor, and attribute X_4, length of power transmission intertie passing through environmentally sensitive areas, were assumed to be deterministic, since each was known with a high degree of certainty. For attributes X_3 and X_5, measured by constructed indices, the probabilities that the impact would fall within ranges specified by two adjacent impact levels were assessed. The probabilistic estimates for attributes X_2 and X_6 were quantified by assessing the parameters—the mean and standard deviation—for a normal probability distribution.

Assessing the probabilities individually over each attribute implicitly assumes that probabilistic independence existed between the attributes. After initial assessments, the project team concluded that it was reasonable to assume that *conditional* on any alternative, the probabilities associated with the level of any attribute were independent of the level of any other attribute. Thus, for example, the probability of various levels of biological impact was independent of the level of impact on salmonids *given* a particular site.

3.3.2   THE ASSESSMENTS FOR EACH ATTRIBUTE

The probabilistic assessments for each site were based on existing information, site visits, and data developed during the study. Each attribute for each site was assessed by specialists in each of the relevant disciplines. Thus, the assessments represent the professional judgment of individuals based on their expertise and on all information currently available concerning the candidate sites. The resulting data are illustrated in Table 3.5. Let us briefly mention how this was done for each attribute.

The Site Population Factor. To calculate the SPF, a number of people residing in concentric rings with centers at the candidate sites was needed. Since people residing close to the candidate sites receive more weight in the SPF calculations, it was considered necessary to obtain more accurate counts in this region. Therefore, using detailed maps, houses within 5 miles of the candidate sites were counted and an average of three people per house was assumed.

For distances greater than 5 miles from the sites, maps were used to identify cities. The population of each was obtained from census data. However, the populations of towns and cities are generally given for incorporated areas only. The unincorporated population in each county was assumed to reside near the incorporated areas rather than, for instance, uniformly over the county. Therefore the town and city populations were proportionally scaled up to equal the total population for each county. These scaled-up estimates for each city were used when calculating the SPF.

Special consideration was also necessary when a city or town fell on a ring boundary. If the population was less than 100,000, it was assumed that all the population resided in the ring closest to the site. This assumption will yield a higher SPF than actually exists. For cities with a population greater than 100,000, it was assumed that the population was evenly distributed within the city. In these cases, the proportion of the area within each ring was used to estimate the population within that ring.

Impact on Salmonids. Estimates of the annual escapement of adult salmon were made using historical records (Wright [1974]). The possible reduction in the annual spawning escapement of salmonids was quantified with a normal probability distribution. The average losses and standard deviations were assessed using the professional judgments of biologists based on their understanding of losses associated with construction of the cooling water intake structure, intake and discharge of cooling water, and storage impoundments for cooling water. The impact on salmonids is dependent on the proportion of the river flow used for cooling water. Since the cooling water requirements remain approximately constant for all candidate sites, the impact is determined by the size and characteristics of the river supplying cooling water. The salmonids which could be entrained are those passing the intake along the edge of the river. To be conservative, it was assumed that the concentration of salmonids along the edges was higher than in the middle. The estimates of losses due to en-

TABLE 3.5

Attribute	Site			
	Lewis 1	Lewis 2	Lewis 3	Grays Harbor
X_1 (SPF)	0.057	0.040	0.025	0.048
X_2 (loss of salmonids)				
$\quad Y$ (1000s)	75	75	75	5.5
$\quad Z$ (percent)	8	8	8	15
$\quad \sigma_Z$	4	4	4	7.5
X_4 (length of intertie through sensitive environments, miles)	1	1	7	6
$X_6 \quad \bar{x}_6$	2.035	0 (4.929)	1.535 (15.343)	1.933 (7.613)
$\quad \sigma_{x_6}{}^b$ (differential cost $\$10^6$/year, 1975 dollars)	0.51	0 (1.23)	0.38 (3.84)	0.48 (1.90)

	Range	Prob.	Range	Prob.	Range	Prob.	Range	Prob.
X_3 (biological impact at site)	1–2	0.9	1–2	0.9	1–2	0.8	2–3	0.2
	2–3	0.1	2–3	0.1	2–3	0.2	3–4	0.8
X_5 (socio-economic impact)	1–2	0.2	1–2	0.25	1–2	0.3	2–3	0.2
	2–3	0.65	2–3	0.55	2–3	0.45	3–4	0.5
	3–4	0.15	3–4	0.1	3–4	0.15	4–5	0.3
			4–5	0.1	4–5	0.1		

[a] X_1 and X_4 were treated deterministically, ranges were used for X_3 and X_5, and normal distributions were used for X_2 and X_6.

[b] Figures in parentheses include additional costs associated with elimination of possible liquefaction potential.

trainment were made assuming the use of newly developed intake structures designed to minimize or virtually eliminate entrainment (i.e., Raney Well). The effect of construction of the intake structure and storage impounds would primarily result in loss of spawning and juvenile rearing areas. More details on the salmonid assessments are found in Section 6.2.3.

Biological Impact at Each Site.   The scale for assessing the biological impact at each candidate site was presented in Table 3.3. The ecologists were asked to assess the probabilities that the impact would fall between adjacent intervals on this scale. To help in thinking about their responses,

Probability Assessments for Impacts at Candidate Sites[a]

	Site			
Wahkiakum	Clatsop	Linn	Benton	Umatilla
0.044	0.023	0.052	0.011	0.013
17	5	3	430	365
15	15	15	1	1
7.5	7.5	7.5	0.5	0.5
12	1	0	0	0
12.347	17.713	4.834	10.936	11.423
3.09	4.43	1.21	2.73	2.86

Range	Prob.	Range	Prob.	Range	Prob.	Range	Prob.	Range	Prob.
3–4	0.2	3–4	0.2	1–2	0.3	0–1	0.1	0–1	0.7
4–5	0.5	4–5	0.5	2–3	0.6	1–2	0.5	1–2	0.3
5–6	0.3	5–6	0.3	3–4	0.1	2–3	0.4		
1–2	0.2	2–3	0.1	2–3	0.2	2–3	0.1	1–2	0.05
2–3	0.45	3–4	0.55	3–4	0.5	3–4	0.4	2–3	0.6
3–4	0.2	4–5	0.3	4–5	0.2	4–5	0.4	3–4	0.2
4–5	0.15	5–6	0.05	5–6	0.1	5–6	0.1	4–5	0.15

qualitative descriptions for each site developed from site visits were used. Section 6.2.3 presents details on these assessments.

Environmental Impact of Transmission Intertie. The length of power transmission intertie passing through environmentally sensitive areas (i.e., land that was not clear-cut, cultivated, or urbanized) was used as a proxy variable to measure adverse environmental impacts. This length was assessed from field visits to each of the sites. Since the values for this attribute were known with a high degree of certainty, this attribute was treated as a deterministic variable.

Socioeconomic Impact. A series of considerations was required to make subjective evaluations of the likely socioeconomic effects of a nuclear plant on communities near each site. First, for each candidate site, the percentage of the plant construction labor force likely to immigrate was estimated. This was superimposed over the existing characteristics of

communities near each of the nine candidate sites. Existing characteristics of communities included: population size, travel time from site to
labor supply, age of community, type of public financing for which the
community is likely to be eligible (based primarily on size and age), size of
the corporate area, role of the community in the region, and generalized
land use patterns (used also to subjectively evaluate the tax base). The
major plant-related condition superimposed over the existing community
characteristics was the presence or absence of a company town built at
the site. No candidate sites were located within incorporated areas, and
the assumption was made that payments in lieu of taxes would not be
made to any municipal corporation.

Annual Differential Site Costs. The economic comparison does not
include a detailed estimate of the total cost of a plant at each of the candidate sites, but is considered to be a representative evaluation of the differential costs of construction and plant operation associated with each
site. Differential costs are measured relative to the least expensive site,
Lewis 2. The comparison was based on current (1975) bid prices which
were escalated to a proposed bid date of 1980 (on-line date in 1985) using
an 8.4% average annual rate of escalation. Allowances for contingencies,
interest during construction, and bonding cost were included in the differential costs. The differential capital costs were converted to an annual
cost expressed in 1985 dollars using an appropriate factor for cost of
bonds and an estimated plant life. This nonescalatable annual cost plus
the annual differential costs of operation formed the basis for the economic comparison of the sites. The cost estimates were developed using
"standard power plant arrangements" at each of the candidate sites.

Site visits indicated that a potential for liquefaction of existing foundation materials under earthquake loading existed at the Lewis 2, Lewis 3,
and Grays Harbor sites. Because the likelihood of liquefaction at these
sites cannot be ascertained without site-specific studies, two cost estimates were made for each: one if the elimination of liquefaction potential
is not necessary, and one if it is found to be necessary. The method for
eliminating the potential for liquefaction that was used to arrive at cost estimates was to remove the liquefiable foundation materials and replace
them with suitable compacted fill. These additional costs were incorporated in the capital costs associated with site grading and are reflected in
the annual differential site costs.

The primary cost estimates were average values. The uncertainty in
these estimates was represented by a normal probability distribution, and
it was assumed that the standard deviation was equal to one-fourth the
average values. Few data were available to justify this assumption so we
were particularly careful to check the cost estimates in the sensitivity
analysis of Section 3.4.

3.4 Evaluating Site Impacts

The next step in the analysis was to evaluate the relative desirability of the consequences $x = (x_1, x_2, ..., x_6)$ where x_i is defined as a specific amount of attribute X_i, $i = 1, 2, ..., 6$. Thus, for instance, x_6 may be 8 million dollars, a specific amount of the annual differential cost attribute X_6. We wanted to determine the utility function $u(x_1, x_2, ..., x_6)$ over the six attributes of Table 3.2.

The value structure discussed below was initially provided by various professionals at Woodward-Clyde Consultants based on their perception of the WPPSS point of view. For each attribute the utility function was assessed using the most knowledgeable members of the siting team (i.e., the "experts"). The structural assumptions for the multiattribute utility function and the value tradeoffs were jointly assessed by key members of the project team. The resulting value structure was discussed and verified with WPPSS and seemed to represent their views. Other points of view were considered by conducting sensitivity analyses.

As mentioned in Section 2.3.4, the process of determining the utility function can be broken into five steps:

1. introducing the terminology and ideas,
2. determining the general preference structure,
3. assessing the single-attribute utility functions,
4. evaluating the scaling constants, and
5. checking for consistency and reiterating.

At this stage in the analysis, the siting team was familiar with the terminology and our purpose. In practice, as described below, consistency checks and reiteration of steps occur through the utility assessment process.†

3.4.1 DETERMINING THE GENERAL PREFERENCE STRUCTURE

The first important step in selecting the form of the utility function involves investigating the reasonableness of preferential independence and utility independence conditions. Provided certain conditions are appropriate, the six-attribute utility function is expressible in a simple functional form of the six single-attribute utility functions. Let us illustrate with ex-

---

† The remainder of this section contains details about the assessment of the utility function. The reader uninterested in these details may proceed to Section 3.5.

amples of how one checks for these conditions. More details on these independence conditions are found in Section 7.2.

Two attributes (X_i, X_j) are *preferentially independent* of the other attributes if the preference order for (x_i, x_j) combinations does not depend on fixed levels of the other attributes. Consider differential cost X_6 and impact on salmonids X_2. The siting team was asked what level of X_6 would make $(x_6, 100\%$ of a stream with 100,000 salmonids lost) indifferent to (40 million, 0%), given that the other four attributes were at their best levels. The answer obtained was 20 million. It then examined the same question with the other attributes at their worst levels. It was still felt that an appropriate response for x_6 was 20 million. By considering other pairs of indifferent points, we established that the value tradeoffs between (X_6, X_2) would be independent of the level of the other attributes. Since the project team had been exposed to concepts of preferential and utility independence, they were in a position to state, after an initial series of questions of the above type about the attributes, that in general, the value tradeoffs between any two attributes did not depend on the levels of the other attributes. Thus each pair of attributes was considered preferentially independent of the others.

Attribute X_i is defined to be *utility independent* of the other attributes if the preference order for lotteries on X_i does not depend on fixed levels of the other attributes.† This implies that the conditional utility functions over X_i are the same regardless of the levels of the other attributes. To establish whether X_3 (biological impact) was utility independent of the other attributes, we assessed the conditional utility function for X_3, assuming the other attributes were at fixed levels. The assessment was conducted using the techniques described in the subsequent section. It was decided that the relative preference for lotteries involving uncertainty only in the consequences for X_3 did not depend on the other attributes. Thus, attribute X_3 was utility independent of the other attributes.

The above independence conditions, which were verified as appropriate, allowed us to conclude that the structure of the utility function had to be either

$$u(x) = \sum_{i=1}^{6} k_i u_i(x_i) \tag{3.2}$$

or

$$1 + ku(x) = \prod_{i=1}^{6} [1 + kk_i u_i(x_i)], \tag{3.3}$$

---

† A lottery is defined by specifying a set of consequences which may occur and the probability of each.

where u and the u_i are utility functions scaled from zero to one, the k_is are scaling constants with $0 < k_i < 1$, and $k > -1$ is a scaling constant. Equation (3.2) is the additive utility function and (3.3) is the multiplicative utility function. More details about these, including suggestions for assessment, are found in Section 7.2.

The result states that the multiattribute utility function can be completely defined by the individual attribute utility functions u_i and the value of the scaling constants k_i. For reference, the multiplicative utility function turned out to be the appropriate one for this study, as we will show later. Although only one utility independence assumption is necessary to conclude that either (3.2) or (3.3) holds, this condition was verified for all the other attributes as a consistency check.

3.4.2 ASSESSING THE SINGLE-ATTRIBUTE UTILITY FUNCTIONS

The assessment of the utility functions with natural indices—that is, u_1, u_2, u_4, and u_6—was done, using the standard 50–50 lottery technique discussed in Section 7.5. For instance, by considering preferences between a series of specified levels of X_6 and a 50–50 lottery yielding either a 0 or a 40 million dollar differential cost, each with probability 0.5, it was decided that WPPSS would be indifferent to a specified level of 22 million dollars. Thus, since utility is a measure of preference, the lottery and 22 million dollar level must have equal expected utilities. Consistent with (3.3), we set the origin and scale of u_6 by letting the utility of the worst point, 40, equal 0 (see Table 3.2) and the utility of the best point, 0, equal 1. Equating expected utilities leads us to $u_6(22) = 0.5$, which gives us another point on the utility curve. From this, the exponential utility function in Fig. 3.1 was specified. By examining the implications of this utility function for additional choice situations, it was decided that it was appropriate for evaluating the various sites.

For the constructed scales, a modified assessment technique was required. In order to achieve meaningful utility assessments for these attributes, only the defined points on the scales were used. For instance, with biological impact, the biologist member of the team was asked, "For what probability p is a biological impact of magnitude 4 (see Table 3.3) equivalent to a lottery yielding a p chance at level 0 and a $(1 - p)$ chance at level 8." By trying several values of p, we found that $p = 0.6$ was the indifference value. We then set $u_3(0) = 1$ and $u_3(8) = 0$ from which it followed that $u_3(4) = 0.6$. Questioning continued in this manner until the utility of each of the defined points on the subjective scale was fixed. A number of

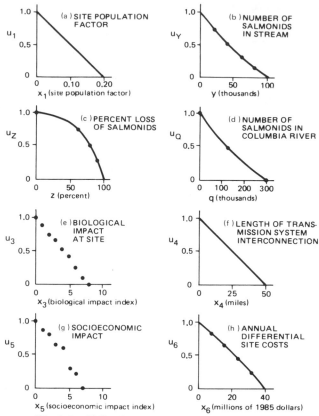

Fig. 3.1.   Utility functions for the WPPSS attributes.
Note: (a) $u_1(x_1) = 1 - 5x_1$ (b) $u_Y(y) = 0.773(e^{0.0083(100-y)} - 1)$ (c) $u_Z(z) = 1.039(1 - e^{0.0327(z-100)})$ (d) $u_Q(q) = 0.7843(e^{0.00274(300-q)} - 1)$ (f) $u_4(x_4) = 1 - x_4/50$ (h) $u_6(x_6) = 1 + 2.3(1 - e^{0.009x_6})$

consistency checks were used which resulted in some changes to the original assessments.

The adjusted utility functions assessed for each individual attribute are shown in Fig. 3.1. Details of the assessment of the utility functions u_2 and u_3 are given by Keeney and Robilliard [1977]. The assessment of u_2 was particularly interesting because two separate measures—the numbers and the percentage lost—were required to describe adequately the possible impact on salmonids. Let us define Y as the number of salmonids in a stream (in thousands) and Z as the percentage lost. Because attribute X_2 is a composite of Y and Z, we will define $x_2 \equiv (y, z)$. If a stream has less than 100,000 salmonids, a utility function u_2 was found to be

$$u_2(x_2) \equiv u_2(y, z) = u_Z(z) + u_Y(y) - u_Z(z)u_Y(y), \qquad y \leqslant 100,$$

where u_Y and u_Z are illustrated in Fig. 3.1b and c. For streams with more than 300,000 salmonids, an appropriate utility function was

$$u_2(x_2) \equiv u_2(y, z) = 0.568 + 0.432u_Q(q), \qquad y \geq 300,$$

where Q is defined as the number of salmonid lost and $q = 0.01yz$. The utility function u_Q is shown in Fig. 3.1d. There are no streams with between 100,000 and 300,000 salmonids in the areas involved in our study so the discontinuity in u_2 between y equal to 100 and 300 is not a difficulty.

3.4.3 EVALUATING THE SCALING CONSTANTS

The scaling constants were assessed in two steps by five members of the project team. The first step was ranking the ranges of attributes in order of importance and the second involved quantifying the magnitude of each k_i.

To establish the ranking of the k_is, the first question asked was: "Given that all six attributes are at their worst level as defined in Table 3.2, which attribute would you most like to have at its best level, assuming that the other five attributes remain at their worst levels?" The answer to this question identifies the attribute whose k_i value should be the largest. A similar question was asked considering only the remaining five attributes. This process was repeated until the complete ranking of the k_is was determined.

It was the consensus that if all attributes were at their worst levels and only one attribute could be moved to its best level, the single attribute which should be moved was attribute X_6, annual differential site cost. This represents changing annual differential site costs from 40 million dollars per year for 30 years to 0 dollars per year. It should be noted that if the worst value of the differential site cost were smaller than 40 million dollars, some other attribute might have been moved first. Of the remaining five attributes, the site population factor X_1 was most desired at its best rather than worst level.

The order in which the remaining attributes were moved from their worst to their best levels was X_2, X_5, X_4, and X_3. This ordering implies

$$k_6 > k_1 > k_2 > k_5 > k_4 > k_3. \qquad (3.4)$$

The next step was to establish the actual values of scaling constants. This was accomplished by assessing specific value tradeoffs between attributes. These value tradeoffs measure how much one is willing to give up on one attribute to gain a specific amount on another attribute. For example, the tradeoff between attribute X_6 and X_1 was established from the following considerations:

1. In establishing the relative ranking that $k_6 > k_1$, it was found that site A, with an annual differential site cost of 0 dollars and a site population factor of 0.20, was preferred to site B, with an annual differential site cost of 40 million dollars and a site population factor of 0, given that all other attributes are fixed at the same levels for both sites A and B.

2. Therefore, there must be a site C, with a site population factor of 0.20 and an unspecified annual differential site cost between 0 and 40 million dollars, which is indifferent to site B above, given again that all other attributes are fixed at identical levels for both sites B and C.

The project team's response was that if site C had an annual differential site cost of 5 million dollars, it would be indifferent to site B. This implies that the project team was willing to incur an increase in annual differential site cost from 5 to 40 million dollars in order to move a site from a sparsely populated area (SPF = 0.20) to an uninhabited area (SPF = 0). This assessed value tradeoff is represented pictorially in the top graph of Fig. 3.2.

The remaining value tradeoffs assessed for other pairs of attributes are also shown in Fig. 3.2. The implications of these value tradeoffs are that one is just willing to incur:

1. an increase in annual differential site cost of from 20 to 40 million dollars in order to save all the salmonids in a river of 100,000 salmonids,

2. an increase in annual differential site cost of from 35 to 40 million dollars in order to avoid laying the new transmission intertie lines through 50 miles of environmentally sensitive areas,

3. an increase in annual differential site cost of from 31 to 40 million dollars in order to eliminate completely the severe socioeconomic impact of a full boom–bust cycle (i.e., change level 7 to level 0 on the constructed scale in Table 3.4),

4. an increase in annual differential site cost of from 39 to 40 million dollars in order to eliminate completely an extreme biological impact over one square mile (i.e., change level 8 to level 0 on the constructed scale in Table 3.3).

In order to check the consistency of the value tradeoffs, several other tradeoffs not involving cost were empirically established. These are shown in the inserts of Fig. 3.2. They proved to be consistent with the original assessments. The implications of these tradeoffs are, for example, that one is just willing to accept a loss of all salmonids in a river of

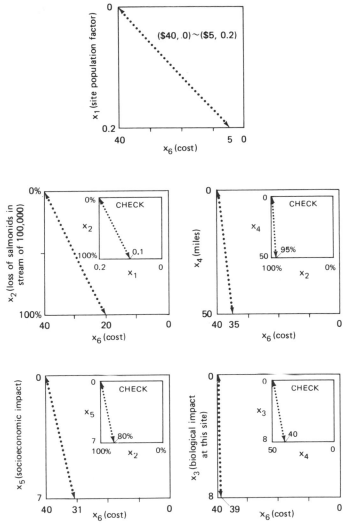

Fig. 3.2.   Value tradeoffs made in the assessment of the WPPSS scaling constants.
(The assessed pairs of indifference points are connected by arrows. Value tradeoffs in the
inserts are consistency checks.)

100,000 in order to move the site from a sparsely populated area (SPF = 0.2) to a less populated area (SPF = 0.1).

The final step in the assessment of scaling constants involved deter-mining a probability p such that option I, a consequence with 0 differential cost and all other attributes at the worst levels of Table 3.2, and option II, a lottery yielding either all attributes at their best levels with probability p or all at their worst levels with probability $1 - p$, are indifferent. After considering several levels of p, the siting team's response converged to $p = 0.4$. With this and all the above information about values, calculations are all that are needed to determine the scaling constants.

By definition, when all attributes are at their best levels, $u = 1.0$, and when all attributes are at their worst levels, $u = 0.0$. Therefore, the expected utility of the lottery in option II above is

$$p(1.0) + (1 - p)(0.0) = p = 0.4.$$

From (3.3), the utility of the sure consequence in option I is k_6. Since the two options are indifferent, their expected utilities must be equal, so

$$k_6 = p = 0.40. \tag{3.5}$$

The assessed tradeoffs between cost and each of the other attributes are used to express all other scaling constants in terms of k_6. Since k_6 is known, the other k_i values can be determined.

Consider the calculation of scaling constant k_1, associated with attribute X_1, the site population factor. By definition, the indifference points of the tradeoff assessments must have equal expected utilities. Thus, from the indifference point of the assessed value tradeoff in Fig. 3.2a, we know that

$$u(x_6 = \$40, \quad x_1 = 0) = u(x_6 = \$5, \quad x_1 = 0.2), \tag{3.6}$$

where we have not bothered to specify levels of the other attributes. However, because of the preferential independence conditions previously verified, we know that (3.6) is valid for all values of the attributes X_2, X_3, X_4, and X_5. In particular, assume that the other attributes are at their worst levels, such that $u_2(x_2) = u_3(x_3) = u_4(x_4) = u_5(x_5) = 0$. Then using (3.3), the utilities in (3.6) are equated by

$$1 + kk_1 = 1 + kk_6(0.895)$$

which simplifies to

$$k_1 = 0.895k_6.$$

Since we know from (3.5) that $k_6 = 0.40$,

$$k_1 = 0.895(0.40) = 0.358.$$

The remaining scaling constants can be calculated in an analogous manner, yielding the set

$$k_6 = 0.400, \qquad k_1 = 0.358, \qquad k_2 = 0.218,$$
$$k_5 = 0.104, \qquad k_4 = 0.059, \qquad k_3 = 0.013. \tag{3.7}$$

The constant k is calculated from (3.3) given the k_i values. If (3.3) is evaluated with all attributes at their best values (i.e., all utilities are 1.0), then k is the solution to

$$1 + k = \prod_{i=1}^{6} (1 + kk_i), \qquad -1 < k \neq 0. \tag{3.8}$$

Using (3.7) and (3.8), the unknown k is calculated to be

$$k = -0.325. \tag{3.9}$$

The multiattribute utility function (3.3) is completely specified by the k_is in (3.7), the k in (3.9), and the single-attribute utility functions in Fig. 3.1.

3.5   Analyzing and Comparing the Candidate Sites

Since the costs of eliminating liquefaction potential are significant, and since site-specific information could eliminate this uncertainty, it was considered appropriate to analyze the problem once including potential liquefaction costs and then excluding them. The results would provide guidance on whether it would be worth obtaining definitive information on liquefaction potential. For example, if the sites that are ranked high without considering liquefaction potential are ranked very low when considering liquefaction potential, then it may be appropriate to obtain site-specific information.

A small computer program was developed for evaluating the sites and conducting sensitivity analyses. Because of the utility independence assumptions verified before selecting the utility function (3.3) and the assumption of probabilistic independence conditional on each alternative, it was appropriate to calculate certainty equivalents, attribute by attribute, for each of the alternatives. This gave us a six-attribute vector representing the "equivalent certainty impact" of each site. These were examined for dominance. No strict dominance existed, but there were several cases of "almost" dominance (e.g., one alternative preferred to another on all but one attribute). Thus, without introducing the full power of multiattribute utility, we were in a position to specify a reasonable ranking of the sites. In particular, the less preferred sites were easily identifiable. We proceeded to the utility analysis.

3.5.1 RANKING RESULTS BASED ON BEST ESTIMATES

The expected utility of each site was first calculated using the best estimates of all inputs for both the liquefaction and no-liquefaction cases. This resulted in two preferential rankings given in Table 3.6. Both the rankings and expected utilities indicate which sites are better than others when considering all six attributes. The differences in expected utilities for each site result from changes in all six attributes for the sites. However, it is easier to consider the significance of the difference in expected utility in terms of only one attribute. For ease in interpreting this significance, the differential cost of an "equivalent" site with attributes X_1 through X_5 at their best levels is shown for each site. This equivalent site is one with the same expected utility as the real site with which it is associated. Note, for instance, that the differences between the sites ranked one and five for both the liquefaction and the no-liquefaction cases are equivalent to approximately 9 million 1985 dollars per year—a rather substantial amount.

3.5.2 SENSITIVITY ANALYSIS

The purpose of the sensitivity analysis is to investigate how the ranking of the alternatives changes if the inputs to the decision analysis differ from the best-estimate values. Sensitivity analyses were conducted both with and without costs associated with liquefaction potential. For each of these conditions, the sensitivity of the scaling constants in the multiattribute utility function and of certain changes in the possible consequences were examined.

Changes in the Scaling Constants. The best-estimate values of the scaling constants, k_i, $i = 1,2, \ldots, 6$, are given by (3.7). In the sensitivity analysis, the value of each k_i was increased and then decreased as much as possible without changing the order of these k_is. For example, k_1 was the second largest k_i value based on the best-estimate values. The adjacent values were $k_6 = 0.400$ and $k_2 = 0.218$. Therefore, two sensitivity runs were performed to investigate the influence of k_1 values of 0.399 and 0.219, which represent the range that maintains the same order of the k_is. The range for k_6 was varied from 0.358 (i.e., the value of k_1) to 0.500.

The analysis indicated the rankings of the sites remained essentially unchanged for all the changes in the k_i factors. Specifically, in the case

TABLE 3.6

Best–Estimate Ranking of Nine Candidate Power Plant Sites

	Without liquefaction potential				With liquefaction potential		
Rank order	Site	Expected utility	Differential cost of "equivalent" site[a]	Rank order	Site	Expected utility	Differential cost of "equivalent" site[a]
1	Lewis 3	0.921	10.85	1	Lewis 1	0.894	14.60
2	Lewis 2	0.920	10.98	2	Lewis 2[b]	0.887	15.53
3	Lewis 1	0.894	14.60	3	Linn	0.854	19.89
4	Grays Harbor	0.868	18.06	4	Wahkiakum	0.843	21.30
5	Linn	0.854	19.89	5	Grays Harbor[b]	0.827	23.35
6	Wahkiakum	0.843	21.30	6	Lewis 3[b]	0.822	23.98
7	Umatilla	0.812	25.22	7	Umatilla	0.812	25.22
8	Benton	0.811	25.34	8	Benton	0.811	25.34
9	Clatsop	0.808	25.71	9	Clatsop	0.808	25.71

[a] An equivalent site is one of equal utility with all attributes at their best levels except for costs (in millions of 1985 dollars per year).
[b] With these sites, additional site grading costs associated with correction of possible liquefaction potential is included in the analysis.

where no liquefaction potential was assumed, there were no changes in the ordering of the best six sites. When liquefaction was assumed, there were a few changes between the sites ranked five and six depending on the specific changes in the k_is. However, the sites ranked one through four were invariant in this case.

Changes of Selected Consequences.   The sensitivity of the rankings in Table 3.6 to the estimates of the differential costs and salmonid impacts were investigated. Specifically, we investigated the implications due to the following factors: increases in differential site costs of 20 and 50%, a change from 0.25 to 0.50 in the coefficient of variation† of the normally distributed site costs, and the elimination of a scheme to prevent entrainment of salmonids at the cooling water inlets.

For the case including liquefaction potential, there were no changes in the ranking of the six best sites for any of the variations mentioned. Assuming no liquefaction potential, Lewis 2 replaced Lewis 3 as the best site for 20 and 50% increases in the costs. These were the only changes in the ranking of the six best sites of Table 3.6. In both cases, there were some changes in the rankings of the three worst sites.

3.5.3   CONCLUSIONS AND RECOMMENDATIONS

The results of the ranking process indicated that six of the nine candidate sites were superior to the other three under all reasonable variations of the preference structure and assessed consequences. The six sites include the Lewis sites, Grays Harbor, Linn, and Wahkiakum. Considering both rankings (i.e., with and without liquefaction), the three sites recommended for detailed site-specific evaluation were Lewis 1, Lewis 2, and Linn. If liquefaction potential was studied first and found not to exist at Grays Harbor and Lewis 3, then the three sites recommended for site-specific studies were Lewis 2, Lewis 3, and Grays Harbor. In interpreting these recommendations, it should be noted that the Lewis sites were located close to each other.

Site-specific studies should concentrate on obtaining information to satisfy regulatory agency requirements. The most important of these are the geological, seismological, and geotechnical studies necessary to identify and classify lineaments and landslides. Additional studies to identify potential major environmental, socioeconomic, or cost impacts, and to refine some of the cost data utilized in the ranking process should be conducted. Because of the site visits that have already been made, a lower order of effort is required for these studies.

---

† An alternative way to state this assumption is that the standard deviation of site costs increases from 25 to 50% of the average estimated costs.

The sites were identified and ranked on the basis of criteria described in this chapter. There are several factors which were not considered in this study but could have a significant bearing on the selection of a specific site. These include political and legal considerations, the necessity for geographic distribution of plants, the future requirements of multiple plants at a site, and the reliability of the transmission grid.

The evaluation process was based on the judgments and preferences of the project team, the results of which were discussed with WPPSS. It was recommended that further assessments be conducted to include explicitly the preferences and judgments of members of WPPSS. It may also be desirable to include explicitly the preferences and judgments of the general public.

The preferential ranking of the nine candidate sites is presented in Table 3.6. However, if the most preferred site is selected for construction, the best site for the next plant is not necessarily the second best site in the original ranking. This is so because of the influence of the selected site on the desirability of the remaining sites. Procedures could be developed to rank the next best site after selecting one site from the nine considered.

3.6 Appraisal of the Analysis

Any analysis of any siting problem requires the professional judgment of the siting team as indicated in Section 2.4. On each particular problem, there are also a variety of exogenous factors such as time available for the study and accessibility of data which influence the character of the study. The project team must always balance the factors which are addressed formally in the model and those that are addressed informally outside the model when determining what might be desirable and what is feasible for the case. In this section, to give a flavor of the types of professional judgments required, we comment on those made in the WPPSS study. The application of the five steps of the decision analysis siting methodology (see Section 2.3) to the WPPSS nuclear siting study was presented sequentially in Sections 3.1–3.5, so we will arrange our appraisal of the analysis accordingly.

3.6.1 IDENTIFYING THE CANDIDATE SITES

The screening process for identifying candidate sites was conducted attribute by attribute. A big assumption is implicitly made in this process when areas are excluded because they fall just over a cutoff level for one criterion. Thus, utilizing this approach may disqualify potential areas that are adequate with respect to several criteria but are just barely inadequate on one or two. However, such an approach provides a mechanism for rapidly focusing attention on candidate areas which have higher probabilities of containing acceptable potential sites. We considered the advantages (particularly in terms of time) in this study of applying each screening criterion individually to override the disadvantage of possibly disregarding some candidate areas. Furthermore, after evaluating the candidate sites, we reconsidered the appropriateness of the screening criteria to see if there were any obvious inconsistencies. That is, had any of the screening criteria been slackened (i.e., more areas included), it seems unlikely that additional candidate sites would have been identified which would rank among the top five in Table 3.6. Section 4.3 discusses hierarchical and sequential screening models which are more dependable in differentiating between "good" and "bad" candidate areas.

3.6.2 SPECIFYING OBJECTIVES AND ATTRIBUTES

This step of the analysis addresses four of the siting features: multiple objectives, several interest groups, intangibles, and degree of impact. There were clearly multiple objectives in the problem as listed in Table 3.2. A reasonable degree of care was taken in selecting them and omitting others. The attributes X_1, site population factor, and X_5, socioeconomic impacts, were meant to address the major concerns and possible impacts on the local communities. However, none of the attributes attempts to measure the attitudes of public groups, such as local citizens, which would probably be greatly interested in the siting problem. Hence, such groups may feel that since they had no direct input to the analysis, their views were not properly considered.

Intangibles were considered. The attribute X_4, length of the intertie, was intended to describe the environmental and aesthetic impact of the line. And, as seen from Table 3.3, some of the aspects included in X_5, the socioeconomic impact attribute, pertain to intangible concerns. It would be desirable, however, to clarify the meaning of words such as "impacted" and "disrupted" used in defining levels of the socioeconomic scale.

Each of the scales for an attribute explicitly addresses the degree of impact feature. The scales for X_1 and X_4 are proxy scales (see Section 5.3), meaning that they do not measure health and safety and transmission line environmental impact directly, but rather relate to them in some nonquantified manner. With X_1 for instance, it would be possible to model the meteorological conditions at a site; the receptiveness of humans to radiational doses, and dose–response relationships to describe the health and safety impact more directly. The use of attribute X_6, annual differential cost, is not as good as one would like for two reasons. The first is that the base cost from which the differential is measured matters, but it was not available in this study. Second, it assumes discounting at a constant rate is appropriate, a point we shall return to in Section 3.6.4.

3.6.3 Describing Possible Site Impacts

As stated in Section 3.3, it was reasonable to assume that probabilities associated with the different attribute levels were conditionally independent, given any candidate site. The deterministic assessments of X_1 and X_4 seemed appropriate for the current time. However, no attempt was made to forecast the population growth (or decline) in the future around the potential sites. Our judgment was that, aside from plant-induced differences, the growth rates at all sites would be similar enough so that the distinction would probably not influence the implications of the study. The internal assessments made for the two constructed indices were justifiable.

The probabilistic assessments for salmonid impact and cost were not done as thoroughly as we would have liked. A normal distribution was used for salmonid losses with the average percentage lost and the standard deviation assigned using the judgments of an experienced biologist based on stream size and annual escapement. The annual escapement was taken from historical records. A probabilistic model including variables of annual escapement, water flow in the stream, and likelihood of thermal releases, and so on might have given us better estimates of salmonid impact.

Of greater concern was the basis for the cost estimates. A normal distribution was assumed here with the mean representing the results of an independent economic study of the facility. We assumed a standard deviation of 25% of the mean in our base estimates. Because of the numerous factors which influence costs of a nuclear power plant and because of the well-known uncertainties in these costs, a probabilistic model of the economics would have been very desirable. This is particularly true in light of the fact that the economic attribute is given the greatest weight in the utility function used for evaluating sites. Such a model could have been used to address explicitly features such as time delays in licensing and construction and operating reliability, which were not explicitly included in the analysis.

3.6.4 Evaluating Site Impacts

The utility function assessed for this study carefully addressed the value tradeoff and risk attitude features of siting studies. The equity feature was not directly considered because attitudes of interest groups were not included as attributes. However, equity is indirectly addressed for local groups in the assessment of value tradeoffs involving health and safety and socioeconomic impacts, for environmentalist groups with the value tradeoffs involving environmental impacts, and for the utility company stockholders with value tradeoffs involving costs.

It may not be appropriate to obtain a net present value by discounting at a constant rate of 8.4% over the lifetime of a project. Mainly because of the large uncertainties in costs and competing uses of capital, WPPSS may strongly prefer a cost stream (i.e., costs over time) with a higher net present value if the stream represents a comparatively stable economic situation. Stability is entirely neglected by discounting. Procedures discussed in Section 7.3 can account for concerns such as economic stability.

Perhaps a more serious concern is that the value structure of WPPSS was not assessed. The utility function used in the study was determined by the siting study team taking the viewpoint of WPPSS. In this regard, it does not seem inappropriate to have experts assess the individual utility functions for health and safety, environmental impact, and socioeconomics, since an expert's knowledge is needed to interpret the relative implications of the different attribute levels. However, for the cost attribute X_6 and the value tradeoffs among attributes, it would be much better to have direct assessments from WPPSS, the client. In the absence of such direct assessments, the cost utility function and the value tradeoffs of the study team were discussed with WPPSS, and they appeared to be reasonable representations of WPPSS preferences. The sensitivity analysis indicated that possible deviations would almost certainly be of little significance.

3.6.5 ANALYZING AND COMPARING THE CANDIDATE SITES

There is little appraisal one can offer of the actual analysis, except to note that the interpretation of the results is subject to the weaknesses as well as the strengths mentioned in this section. This analysis did explicitly address many of the major features of siting studies in an unambiguous manner. Thus, it is easy to determine the influence these features should have on the results.

With respect to the sensitivity analysis, two additional aspects might have been informative. One would involve a more detailed sensitivity analysis of the uncertainties in the costs. This analysis was not made since it required a probabilistic model of costs, and to build one would have been a major undertaking. Second, the siting team could have taken the viewpoint of other interested groups (e.g., environmentalists) and assessed value tradeoffs as these groups might have done. A site ranking from these various viewpoints would be interesting.

The study reported here was to help establish an inventory of sites for possible future use. It was presented to WPPSS in late 1975. They reviewed it but took no action because of a nuclear referendum on the Washington State ballot in November 1976 which would have potentially halted construction of new nuclear power plants. The referendum did not pass. In 1977 and 1978, WPPSS updated the socioeconomic and economic data. These results were substituted into the model to update the ranking. Furthermore, WPPSS went to local communities near candidate sites to inform the public of the study and to elicit their input. As of 1979, because of continuing construction delays, cost escalation, and inflation uncertainties at five other WPPSS nuclear power plants (Business Week [1979b]), the timetable for possible use of the sites evaluated in this study has been pushed further into the future.

CHAPTER 4

IDENTIFYING CANDIDATE SITES

The first step in an energy facility siting study is to identify a reasonable number of candidate sites. The set should contain the best feasible sites for the proposed facility. However, because of the complexity of the problem, it is usually not possible to pick the best among these sites without a formal analysis. This difficulty exists in spite of the fact that analysis often indicates that the relative desirability of the candidate sites varies greatly.

The purpose of identifying a manageable set of candidate sites for careful evaluation is to increase the likelihood of finding a better proposed site as a result of the study. The candidate sites are carefully evaluated and this process requires significant effort and expense. For this reason, it would be infeasible to evaluate every possible site with such care. If each site was individually examined, the data and evaluation would be of such poor quality that it would be difficult to identify the best site (or even the better ones). At the other extreme is the intuitive selection of one candidate site only and careful analysis to "justify" it. Here too, because of the high probability of missing better sites in the intuitive selection process, the likelihood of finding the best site is less than need be the case.

What is needed is a balanced approach which responsibly and rather quickly surveys all the possible sites and provides a reasonable mechanism for selecting the set of candidate sites. The correct balance of effort for any particular problem should be chosen by the siting team in conjunction with the client.

Section 4.1 discusses the use of screening to identify candidate sites.

Section 4.2 illustrates the weaknesses of most screening techniques. These weaknesses are overcome by the decision analysis screening models introduced in Section 4.3. The task of implementing the decision analysis screening models is discussed in detail in Section 4.4.

4.1   The Use of Screening

The mechanism by which one can rapidly identify candidate sites is screening. The overall objective of this screening is the same as that of the entire analysis, to identify the best possible site for the proposed facility and to provide documentation and justification for the approach used. There are three main levels of screening used in most siting studies. The first is informal and implicit, the selection of a region of interest; the second is formal and explicit, the use of screening criteria to identify candidate areas; and the third is informal and explicit, the use of professional judgment to select candidate sites within candidate areas.

4.1.1   Selection of the Region of Interest

The region of interest is defined as the total area in which one is willing to site the proposed facility. By definition every site outside the region is excluded. The decision on the region of interest is usually made by the client at the beginning of a study or before it begins, in an informal and implicit manner. Hence, it is often not subject to the minimum degree of scrutiny that other screening decisions receive. Unfortunately, the region of interest is frequently considered as a given condition for a problem, and yet it can be very crucial.

The region of interest is commonly chosen to coincide with either a political district, such as a state; a service area, such as that of a utility company; or a geographical area, such as the watershed region for a particular river. This situation was the case with the WPPSS study discussed in Chapter 3. The region of interest in the WPPSS study was the state of Washington, the major river basins in Idaho and Oregon which are tributaries to the Columbia River, and the major river basins of the Oregon coast.

In some cases, the region of interest is defined by a proximity criterion to the anticipated market for a product or to the existence of needed raw resources. An example of the former is a region of interest which is within 100 miles of the load center for the electricity from a power plant. An example of the latter situation could be a region of interest for a proposed oil refinery which would be within 2 miles of a major Middle Atlantic States port.

4.1.2 IDENTIFICATION OF CANDIDATE AREAS USING SCREENING CRITERIA

Two examples of screening criteria are the following: a site must be more than 5 miles from an active fault, and a site must be outside designated scenic and recreational areas. Each screening criterion involves two parts, an attribute and a cutoff. The attribute is necessary for measuring or determining a characteristic of any potential site and the cutoff level for defining what is acceptable for a candidate site and what is not. Often the attribute can be measured by a scale, such as distance from an active fault, and other times it is a more qualitative consideration, as is the case with a scenic area. Screening criteria are at the same time both inclusive and exclusive. That is, each criterion divides any region into two parts, one which is acceptable for a site and the other which is not.

Usually, many screening criteria are formally and explicitly applied to any specific situation. A site must satisfy all of these to be further considered. In other words, a site which fails any of the criteria is eliminated from consideration. Let us consider the development of screening criteria and their application.

Developing Screening Criteria. The bases for screening criteria come directly from the general concerns listed in Section 1.3. It is clear that the best potential sites are as satisfactory as possible in terms of health and safety, environmental impact, socioeconomic impact, economics, and attitudes of the public and others. Each screening criterion is meant to relate to at least one of these concerns. For example, earthquake faults are of concern for their implications on health and safety and economic costs. Being at least 5 miles away from such a fault may be considered safe enough and may also imply a manageable cost for design.

The specific cutoff level is usually determined from one of two sources. The first is a governmental or regulatory guideline specifying a cutoff. The California Energy Resources Conservation and Development Commission [1977] discusses 61 screening criteria which they feel help to identify areas not suitable for siting energy facilities. The other source is a cutoff set by the client or siting team. An example in the WPPSS siting study is that the power plant should be within 10 miles of its cooling water supply. In setting such cutoffs, value judgments must be made about acceptability. These judgments should later be reappraised in light of the implications of the study.

The screening criteria appropriate for one study may well be inappropriate for another similar study. The criteria depend on social, political, technological, and financial conditions. Since these conditions are different in different regions and since they change with time, screening criteria will also change. Furthermore, screening criteria also depend on the time and effort available for the total study, since this influences the appropriate screening effort. The point is that although the general concerns remain the same in different siting studies, the screening criteria, even for similar facilities, may differ greatly.

Application of Screening Criteria. In order to use screening criteria efficiently, it must be possible to determine easily whether each potential site in the area being considered does or does not satisfy the particular criterion.This has several implications.

Criteria should be used for which there are readily available data since, in general, site-specific visits are not feasible or worthwhile at this stage. It is easy to screen using criteria with mapped data (e.g., floodplains, distances), populations, and physically measurable quantities (e.g., weather information). It is much more difficult to determine derived effects, such as pollution concentrations, as opposed to pollution emissions, and socioeconomic impact, as opposed to jobs created. Criteria for which there is significant uncertainty, such as those involving future populations, are also not desirable for screening.

As a result, many screening criteria involve what we will refer to in Chapter 5 as proxy attributes. That is, the scale for the criterion is an attribute which relates in some way to the general concerns, but it does not directly measure any aspect of them. The distance from a proposed refinery to a harbor is a proxy attribute, since it relates to cost concerns but does not measure costs directly in any way.

The screening criteria are usually applied sequentially as follows. A first criterion is used for the entire region of interest, resulting in part of that region being eliminated from further consideration. Next, a second criterion is applied to the remaining acceptable region. Hence, data pertaining to the second criterion need not be obtained for the area rejected by the first criterion. Next, other criteria are sequentially applied to the remaining acceptable area. Overlay maps are used to illustrate which areas remain acceptable after the application of each criteria. Because of the sequential nature of the screening, a judicious ordering of the applications of the criteria can be very time saving. Roughly speaking, criteria which reject the largest area and which have easily obtainable data should be used first.

The process above can be carried out until a reasonable set of candidate sites remains. Alternatively, the process often results in a set of candidate areas. These areas are each quite homogeneous. Candidate sites within these areas are then identified by professional judgments.

4.1.3 THE USE OF PROFESSIONAL JUDGMENT
TO SELECT CANDIDATE SITES

A siting team comprises experts in various disciplines concerning the environment, economics, health and safety, socioeconomics, and public attitudes. However, there are no overall experts who might be in a position to survey an entire region of interest and pick a set of candidate sites or even perhaps the best site. With two good sites separated by a large distance, and consequently different in many respects, it is extremely difficult for anyone (or the entire siting team) to select the better one using informal methods. This proves to be the case even when the subsequent analysis indicates that the sites are quite distinguishable in terms of overall desirability. However, for two sites near each other in the same candidate area, it is very likely that the differences are quite small. Then it is an easy task for the appropriate expert to identify the better site. Such a choice can effectively be made in an informal, but explicit, manner.

To illustrate, consider a candidate area of 100 square miles which is very homogeneous in nature. And let us suppose that one wishes to locate a nuclear power plant which would require approximately one square mile of land. The study team may review the data on the candidate area, fly over it, and then spend a day or so on the ground. As a result, it may be clear that for any candidate sites in the area, the impacts on health and safety, socioeconomics, and public attitudes would be essentially identical. Consequently, these concerns would have no impact on the relative desirability of sites within the candidate area.

Much of the area may be covered by a thick mature forest with several animal and plant species and the remaining part may be farmed land. Consequently, from an environmental viewpoint, a site on farm land is preferred. It may also be the case that the only economic aspect that seems relevant for differentiating sites within the area is distance from the major river designated as the cooling source. Assume that the plowed fields are closer to the water. Thus, from an economic viewpoint, a field site is also preferred to a forest site. By viewing the field, it may be clear that all sites within it are more or less equivalent, so a site at the far end is chosen which might require slightly less foundation preparation. Hence, this site is chosen as a candidate site. Had the environmental and economic concerns been in conflict within the candidate area, two candidate sites, one representing the favored site in each case, could have been chosen.

4.2 Weaknesses of Standard Screening Techniques

There are several major weaknesses of screening techniques as they are now used in most siting studies. Some of these have to do with the way the techniques are sometimes used, and others have to do with the concepts of the techniques themselves. Included in the former category is the lack of sufficient data to screen accurately or poor judgment in selecting a cutoff level. Although clearly important, these misuses of screening are not inherent in the technique itself. The weaknesses inherent in a technique will be the focus of this section.

Essentially all of the weaknesses of commonly used screening techniques have to do with the value judgments necessary to use the techniques. Often in practice these value judgments are unstated and implicit. Worse, though, is the fact that these value judgments may not be recognized or understood by the clients or members of siting team. In either of these cases, as elaborated in Sections 4.2.1–4.2.5, the values expressed are inappropriate for the screening problem. The first weakness pertains to all the screening procedures discussed in Section 4.1, the other four to the use of screening criteria.

4.2.1 UNSTATED AND UNCLEAR ASSUMPTIONS

The purpose of screening is to identify good sites rapidly. Sites are considered good if they measure up well in terms of the general concerns for siting, but screening is usually done using criteria which do not directly measure the goodness of some general concern. Consequently, two types of assumptions are needed for screening. One type pertains to the values for the different levels of achievement of the general concerns, the other to the relationship between the levels of the screening criteria and the general concerns.

When the region of interest is chosen at the beginning of a study, it is often presented with little justification. The assumptions about how the particular choice of a region of interest relates to the general concerns is unclear. As stated in Section 4.1.1, the region of interest is often selected without the use of screening criteria. It would be easy enough to use screening criteria for this purpose, alleviating the reliance on unstated and unclear assumptions and making it possible to conduct a justifiable sensitivity analysis.

The same types of issues arise to a lesser degree in the use of screening criteria to identify candidate areas and in the use of professional judgment to select candidate sites from candidate areas. In the former case, the relationship between the screening criteria and the general concerns should not be too difficult to identify since the screening criteria are usually one-dimensional. Thus, with an explicit value structure for achieving general concerns, one could readily determine an appropriate cutoff level for acceptability. The practice of using professional judgment directly in selecting candidate sites from within candidate areas appears reasonable and justifiable. However, in this case, especially because there is no quantified data to back up the choices of candidate sites and because the reasons for the choice could be different in every candidate area, it is particularly useful to provide a thorough qualitative reason for each choice.

4.2.2 SCALE SELECTION FOR SCREENING CRITERIA

In addition to the previous assumptions, the selection of a scale for screening criteria has an additional one. This assumption is that all circumstances leading to a similar scale level are equally preferred. Because most screening scales concern proxy indices, this is seldom precisely true. Furthermore, this assumption is usually not made explicit and, hence, is somewhat difficult to appraise. In addition, with only minor adjustments or alterations of the scale, much better screening criteria can be developed.

To illustrate this, consider the criterion in Table 3.1 stating that for the WPPSS study, any nuclear power plant should be more than 3 miles from any place populated by more than 2500 inhabitants. This has several implicit value judgments. There is no reference to prevailing wind direction, and yet the screening was for possible radiation exposure. For both normal operations and accidents, the wind carries the radiation to individuals. A release 6 miles upwind from a town could be much worse than one 2 miles downwind. However, the latter would eliminate a potential site, whereas the former would not.

With this same criterion, there is the reference level of 2500. This implicitly assumes that both 2500 and 200,000 people located 6 miles from a facility are equivalent. There is no mention of the size of the population other than 2500. On the other hand, there is another screening criterion stating that any nuclear power plant should be more than 1 mile from any populated place (of less than 2500). The two criteria together begin to address the factor of population density.

However, it does seem in this case that the population density, wind consideration, and distance from the proposed facilities could easily be integrated into an index of "public radiation exposure." This would clearly be a better indicator of the radiation hazard than the screening scales used. There is a price that one must pay for this, in terms of the time and effort needed to determine the level of public radiation exposure for each possible site in the potential areas in consideration at the stage in the screening when the criterion is applied. With intelligent application, the task should not be too time consuming.

With some screening criteria, there must be additional interpretation of the meanings of words used. Again in Table 3.1, one screening criterion states any site should be outside of areas designated as scenic and recreational areas. The words "designated," "scenic," and "recreational" all need to be defined. For instance, whose designations are to be used.

4.2.3 CUTOFF SELECTION FOR SCREENING CRITERIA

Each screening criterion consists of an attribute and cutoff level which divide areas into those which are acceptable and those which are not. The big assumption here is that, up to some point, the site is acceptable according to a criterion, and then suddenly it is not. In general, we expect more of a gradation with indices. Using the WPPSS criterion of more than 3 miles from a populated place of over 2500 people, it seems unreasonable that 3.1 miles is definitely acceptable and 2.9 miles unacceptable. This is especially true in light of the implicit assumptions of the scale referred to previously.

The effect of a cutoff is shown in Fig. 4.1. The acceptability indicator has an implied desirability index which is a step function. We also plot two situations which might represent the actual desirability curve as a function of levels on the screening attribute. The degree to which the implied curve approximates the actual curve indicates the degree to which the cutoff assumption is appropriate in the strict sense.

In applying cutoff levels, there is clearly the advantage of simplicity. And there is no way to avoid making the value judgment about acceptable/unacceptable at some stage. As suggested in the appraisal of the

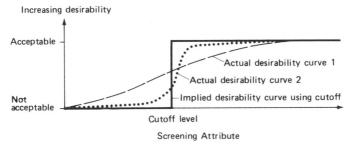

Fig. 4.1. Implicit assumptions of the cutoff level.

WPPSS screening in Section 3.6.1, combining attributes before the screening often results in more justifiable assumptions. This is discussed in more detail in Section 4.3.2. However, regardless of the level at which screening criteria are used, the judicious choice of a cutoff can help considerably.

Following from the basic premise that it is a more serious error to screen out a very good site than to keep in a poor site, the cutoff level should be relaxed (i.e., to exclude less area) when the actual desirability curve is more like curve 1 in Fig. 4.1. This is all the more reasonable since there are several other screening criteria which may more appropriately eliminate the same area and consequently focus the attention on a few of the best sites. Even though the effect is similar, the analysis is more justifiable since the screening was more conservative. And, perhaps more importantly, what may turn out to be the best site may have been kept in the list of possibilities for further evaluation.

4.2.4 Consistency among Screening Criteria

As screening criteria are currently used, there seems to be little checking on their internal consistency. By this we mean that the cutoff levels for different attributes are not set such that just missing any cutoff level is equally undesirable. This can be more precisely illustrated by means of the following notation.

Let us suppose that we have two screening attributes Z_1 and Z_2 and that specific levels of these attributes are defined by z_1 and z_2. We shall designate the cutoffs respectively as z_1^c and z_2^c, where we shall in this case accept sites with $z_1 \leq z_1^c$ and $z_2 \leq z_2^c$. The screening criteria are defined to be consistent if and only if the desirability of a level z_1^c on scale Z_1 is equal to the desirability of z_2^c on Z_2. If this is not the case, we are imposing a higher standard for site acceptability using one screening criteria than another. The implication of this inconsistency is that the screening will accept some sites which are worse than others which are rejected, given of course that screening is done independently and sequentially, attribute by attribute.

Figure 4.2 illustrates this. The screening attributes are drawn such that increasing levels are less desirable, and so that levels above one another on the two attributes are equally desirable. Thus, a level z_1^c on Z_1 is as desirable as z_2' and Z_2, for instance. To be more precise, this means a site characterized by the two levels z_1^c and z_2^* is equally as desirable as one characterized by z_1^* and z_2', where z_1^* and z_2^* represent the best levels of the respective screening attributes. Finally, the cutoff levels z_1^c and z_2^c are as indicated, so a higher standard is implied by z_2^c. Consider two potential

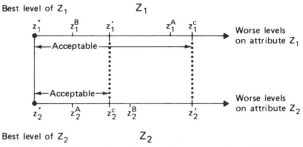

Fig. 4.2. Implication of inconsistent screening criteria.

sites A and B, which are respectively described by the pairs of attribute levels z_1^A, z_2^A and z_1^B, z_2^B indicated on the figure. Site A is clearly acceptable, whereas site B is rejected because of its level on screening attribute Z_2. However, note that if one compared the two areas on both attributes at once, site B is better than site A. They are equally desirable comparing z_1^B to z_2^A, and z_2^B is preferred to z_1^A. This circumstance could not occur if the cutoff levels for Z_1 and Z_2 were consistent.

The example in Section 4.2.5 illustrates the idea of consistency in a realistic situation.

4.2.5 NO VALUE TRADEOFFS

Because screening is done attribute by attribute, there is no formal way of addressing the value tradeoffs which are a crucial part of siting studies. This can lead to accepting a site which is just barely acceptable on all of the several screening attributes and rejecting a site just over the cutoff level on one of them but excellent on all the others. There is no mechanism to account for the compensating character of one attribute for another.

The basic idea is illustrated in Fig. 4.3 using two screening criteria and the notation introduced in Section 4.2.4. Site D is acceptable although it is near the cutoffs for attributes Z_1 and Z_2. Site E is unacceptable even though it is barely rejected by criterion Z_1 and measures up excellently in terms of Z_2. One might expect that site E is better than site D, given just the information with regard to the attributes Z_1 and Z_2.

The situation depicted in Fig. 4.3 can occur whether or not the criteria cutoffs are consistently set. The criteria are consistently set if sites described by points F and G in the figure are equally desirable.

As a realistic example of this situation, consider two screening criteria from the WPPSS study, included to keep economic costs reasonable.

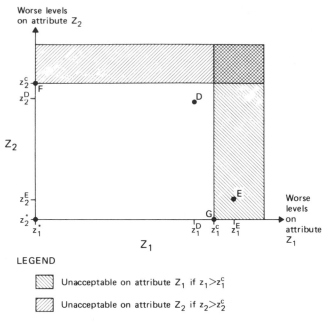

Fig. 4.3. Screening without addressing value tradeoffs.

These are that the site should be (1) within 10 miles of the water supply and (2) less than 800 ft above the water supply. For consistency, the additional cost of pumping water up 800 ft should be equal to the additional cost of transporting water 10 miles horizontally. Clearly a site 9.9 miles away and 750 ft above the water supply costs more than one 10.1 miles away and 100 ft above the water supply. However, the present screening techniques would accept the former and reject the latter. In Section 4.3 we will introduce decision analysis screening models which do account for the value compensation among different screening criteria.

4.3   Decision Analysis Screening Models

There are three major distinctions between decision analysis screening models and standard screening models. With decision analysis models:

1. the value judgments are clearly stated, explicit, and quantified,
2. the value judgments are appropriate for the problem in the sense that the weaknesses discussed in Section 4.2 are either completely avoided or at least alleviated, and

3. the screening is carried out on attributes more closely related to the fundamental objectives of the siting.

The decision analysis screening models can be categorized into two types, although there are certainly many similarities between them. The compensatory screening models take into account the value tradeoffs between screening attributes and screen higher in the attribute hierarchy than standard screening models, but still use cutoff levels to screen. The comparison models also include value tradeoffs, but screen out areas only when the comparison between two areas indicates that one of these is not a contender for having the best site. On any particular siting problem, a combination of these two approaches can be used.

4.3.1 EXPLICIT VALUE JUDGMENTS

A prime goal of any decision analysis screening model is to insure that all the value judgments necessary for the screening are explicit and clearly articulated. Of particular concern here are those judgments requiring qualitative justification, such as the selection of a region of interest or a screening scale.

With regard to the region of interest, it should be unnecessary to define this in a restrictive manner at the beginning of a study by using some political or geographical area. The only reason for doing this is the intuitive feeling that areas outside the region of interest would not contain viable candidate sites. However, if in fact this is the case, the screening models discussed in Sections 4.3.2 and 4.3.3 would clearly indicate it. This would eliminate the need to begin with a region of interest, create a stronger justification for the candidate sites eventually identified, and provide the information for a sensitivity analysis of the areas eliminated.

As an example, in the WPPSS study, suppose the entire area of the U.S. and Canada were considered as the region of interest. Since the purpose of the proposed facility is to generate electricity for the public utilities in Washington, a screening criterion on cost based on its relationship to transmission distance could probably screen out most of the areas excluded from the region of interest defined in the current study. The result may be essentially the same, but the process by which it is reached differs greatly.

Whenever scales or cutoff levels are chosen, they should be clarified and justified. This comment is meant to apply to the criteria required by governmental regulations, such as distance from a fault, as well as those chosen for functional or economic reasons. Such justification essentially means that the relationship between the scale and its cutoff level and the more fundamental objectives of the proposed facility should be clarified. Thus, reviewers and others interested are in a better position to appraise the screening.

Decision analysis screening models, as well as the standard screening models, rely a great deal on the professional judgments of experts in the various disciplines relevant to siting. In particular, for the reasons mentioned in Section 4.2.1, it seems appropriate to use professional judgments directly in selecting candidate sites from within the candidate areas. It is, however, particularly important to justify these choices, which includes making the required professional and value judgments explicit. This task is easier if the candidate area is small enough to include only a few potential sites. If some of the initial candidate areas are large, the application of a formal screening model to these areas may determine much smaller candidate areas. The range of screening criteria for such a model should be greatly restricted relative to the overall screening model because the initial candidate area should be much more homogeneous than the overall region.

4.3.2 Compensatory Screening Models

Let us illustrate the idea of a compensatory screening model within the context of the WPPSS study. In Table 3.1, because of the economics of pumping water, sites more than 10 miles away from and more than 800 ft above the water supply were screened from further consideration. This is illustrated in Fig. 4.4a.

Suppose the following were calculated. The annual cost of pumping the cooling water needed for the proposed WPPSS facility 10 miles was 5 million dollars and the annual cost of pumping that water 800 ft was 4 million dollars. Furthermore, let us assume it is reasonable to make costs proportional to the distance or height in the respective cases, so it would cost 2 million dollars to go 4 miles, and so forth. Thus a simple expression relating total pumping cost c to the distance d and the height h of the facility from the water supply is

$$c = 0.5d + 0.005h, \tag{4.1}$$

where c is in millions of dollars annually, d is in miles, and h is in feet. If we chose to screen out areas with more than 7 million dollars in annual pumping cost, we would get the result in Fig. 4.4b.

Comparing the two parts of Fig. 4.4 indicates several interesting facts. In each case, areas are included which are excluded in the other case. With the former screening, a potential site (point A in Fig. 4.4a) with 9 million dollars annual pumping cost is included and one (point B) with just

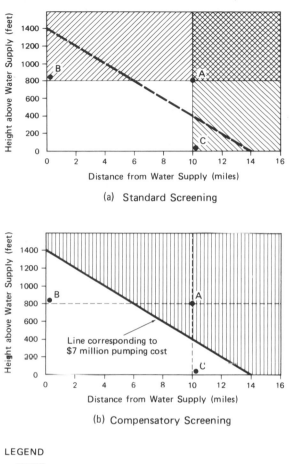

(a) Standard Screening

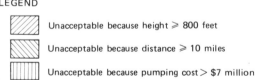

(b) Compensatory Screening

LEGEND

Unacceptable because height ≥ 800 feet

Unacceptable because distance ≥ 10 miles

Unacceptable because pumping cost > $7 million

Fig. 4.4. Comparison of compensatory screening and standard screening.

over 4 million per year is excluded. This situation is reversed as illustrated with the compensatory screening model. The latter clearly makes more sense since the reason for screening is costs.

In the terminology of Section 4.2.4, the difficulty is that the standard model in Fig. 4.4a is not consistent. With regard to distance of pumping, 10 miles, which corresponds to a cost of 5 million dollars, must be incurred before a site is unacceptable. For height above the water supply, only 800 ft, which corresponds to just over 4 million dollars, is grounds for exclusion.

There are two interesting features of this example worth stressing. First, the compensatory model screens higher in the screening attribute hierarchy since it combines two screening attributes. The higher level attribute, pumping cost, is more closely related to the economics concern than the lower level screening attributes of height above and distance from the water supply. Second, the manner in which the two screening attributes were combined required professional judgments rather than value judgments. The required judgments were the costs associated with pumping the water various distances both horizontally and vertically. These were then incorporated in the simple model of (4.1). A value judgment was required for selecting the cutoff level of 7 million dollars. This turned out to be within the 4–9 million dollar cutoff range implied by the standard screening.

The generalization of the example to include several screening attributes is conceptually straightforward. Let us assume we have screening attributes $Z_1, Z_2, ..., Z_n$, all relating to the cost concern. What we want to do is construct a simple model of the relationship and then screen on costs. That is, we wish to specify a function $c(z_1, z_2, ..., z_n)$ which assigns a cost to any set of levels $z_1, z_2, ..., z_n$ of the respective screening criteria. This model requires professional judgments and data for its construction. The more screening attributes that are combined, the less likely it is that any of the weaknesses of the standard screening models discussed in Section 4.2 will have a significant influence on the study.

There is another class of compensatory screening models which requires value judgments, rather than professional judgments and data, in order to combine lower level screening criteria. To construct an example and illustrate the simultaneous use of the different kinds of compensatory screening models, we will again use the WPPSS information in Table 3.1. Furthermore, suppose we have constructed the water supply cost model (4.1). For current purposes, let us assume these were the only costs appropriate for screening (i.e., they varied greatly over potential sites in the region of interest). Thus, we could roughly specify the differential costs for any site. In addition to costs, suppose we use the screening attribute of miles from the site to a place populated by more than 2500 people. What we want to do is construct a compensatory screening model on the two screening attributes: water supply cost and distance from the plant to more than 2500 population.

There are no professional judgments or data which can identify indifferent combinations of those two attributes. Value judgments are necessary.

Clearly, less expensive sites further from the population are better. Using standard screening, the cutoff levels of 3 miles and 7 million dollars may have been established as illustrated in Fig. 4.5a. As indicated, a site A with almost 7 million dollars in water supply costs and only 3.1 miles from 2500 people is acceptable, whereas a site B costing 7.1 million dollars and 9 miles from 2500 people is excluded from further consideration.

Using value judgments, the indifference curve indicated in Fig. 4.5b

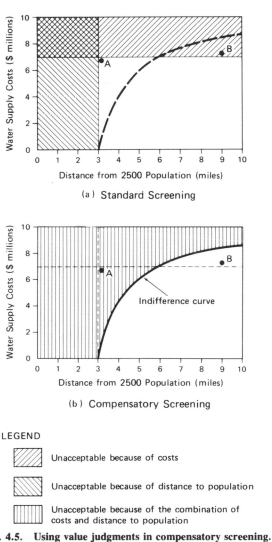

(a) Standard Screening

(b) Compensatory Screening

LEGEND

Unacceptable because of costs

Unacceptable because of distance to population

Unacceptable because of the combination of costs and distance to population

Fig. 4.5. Using value judgments in compensatory screening.

might be assessed. This curve might then be established as the cutoff level for acceptability. Use of the indifference curve is consistent in the sense of Section 4.2.4, whereas in this example, the cutoffs for the standard model are not consistent. If they had been, the indifference curve shown would need to be asymptotic to the 7 million dollar cost line. Only then would the undesirability of 7 million in costs equal that of being 3 miles from 2500 population.

Notice in the figure that (as was the case in Fig. 4.4) some areas included with each model are excluded by the other. In particular, site A is unacceptable with the compensatory screening and site B is acceptable. This is because consistency and value tradeoffs are included with compensatory screening.

There is the additional value judgment of where to draw the cutoff. In this problem context, because of a regulation, all sites must be 3 miles or more from 2500 population. Thus, the indifference curve can not cut the distance axis at less than 3 miles. However, an indifference curve could have been selected which cuts the axis at more than 3 miles.

The general case for compensatory screening using indifference curves is the following. Given the screening attributes $Z_1, Z_2, ..., Z_n$ with respective levels $z_1, z_2, ..., z_n$, one determines an acceptable cutoff level by specifying one combination of the criteria, such as $(z_1', z_2', ..., z_n')$, which is the boundary between acceptable and unacceptable. Then, using additional value judgments, the indifference curve through the point $(z_1', z_2', ..., z_n')$ in the screening attribute space is constructed. Sites corresponding to descriptions as desirable as or more desirable than $(z_1', z_2', ..., z_n')$ are acceptable and others are excluded.

The compensatory screening models do not suffer from problems of inconsistency or of not addressing the value tradeoffs, both of which are often present in standard screening models. They do still require the selection of a cutoff level. This level is, however, applied at a higher level in the screening objectives hierarchy so that the degree to which it screens out desirable sites is significantly decreased.

4.3.3   COMPARISON SCREENING MODELS

The idea of a comparison screening model can best be introduced with a simple hypothetical example involving three screening attributes Z_1, Z_2, and Z_3. Let us use z_1, z_2, and z_3 as specific levels of the respective attributes and suppose the best possible levels of these attributes are z_1^*, z_2^*, z_3^* and the worst possible levels are z_1^0, z_2^0, z_3^0. Now as a first step in developing the screening model, a value function over the three attributes must

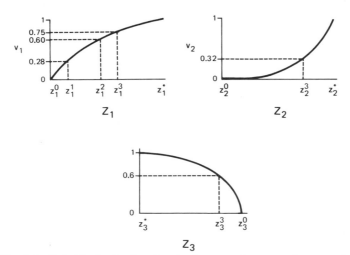

Fig. 4.6. **Individual value functions in the comparison screening example.**

be assessed using the value judgments of the siting team.† The value function v assigns a number to each combination of z_1, z_2, and z_3, such that higher numbers represent preferred circumstances. The origin and scale of v are arbitrary so we can set

$$v(z_1^*, z_2^*, z_3^*) = 1 \quad \text{and} \quad v(z_1^0, z_2^0, z_3^0) = 0, \tag{4.2}$$

which will insure that the value of any combination of screening levels (i.e., any site in the region of interest) is bound between 0 and 1.

In many situations it will be reasonable to use an additive value function. For instance, to continue our example, let us suppose that

$$v(z_1, z_2, z_3) = 0.6v_1(z_1) + 0.25v_2(z_2) + 0.15v_3(z_3), \tag{4.3}$$

where v_1, v_2, and v_3 are each value functions scaled from 0 to 1 by

$$v_i(z_i^*) = 1 \quad \text{and} \quad v_i(z_i^0) = 0, \qquad i = 1, 2, 3, \tag{4.4}$$

as illustrated in Fig. 4.6. The value judgments of the siting team are used to select the form of (4.3), to assess the v_i, and to specify the scaling constants 0.6, 0.25, and 0.15.

The value function (4.3) is our screening model. We begin by collecting data on one attribute, say Z_1, for all locations in the region of interest. Let us suppose some locations have a z_1' level of Z_1 such that $v_1(z_1')$ is 0.95. For

---

† Chapter 7 provides detailed discussions about the types of value functions and their assessment.

such a location the minimum possible level for the overall v is found from (4.3) to be

$$v(z_1', z_2^0, z_3^0) = 0.6(0.95) + 0.25(0) + 0.15(0) = 0.57. \qquad (4.5)$$

At the same time a location with the z_1 level of Z_1 such that $v_1(z_1) \leq 0.28$ can have an overall value no higher than 0.57 since even if attributes Z_2 and Z_3 are at their best levels, from (4.3),

$$v(z_1, z_2^*, z_3^*) \leq 0.6(0.28) + 0.25(1) + 0.15(1) = 0.57. \qquad (4.6)$$

Thus we can consider the level z_1^1 in Fig. 4.6 such that $v_1(z_1^1) = 0.28$ as a derived cutoff that will be updated later. For now a location with $z_1 \leq z_1^1$ can be screened from further consideration.

In the screening, we want to retain several possible sites for consideration since the model is somewhat rough. Thus, one might argue that the screening at this first stage should be done below 0.57, say at 0.5, to allow some flexibility. However, since it is unlikely that many other areas would be perfect on the attributes yet to be used in the screening, a margin of error is *de facto* included even though we screened using the 0.57.

The next phase of the screening might be with attribute Z_2, for example. Only the remaining areas not screened using Z_1 are appraised in terms of Z_2. Suppose the best level of Z_2 is z_2'' such that

$$v_2(z_2'') = 0.8. \qquad (4.7)$$

In general, the site which is best on Z_2 will not be the best one on Z_1. So let us assume that for the site with z_1', the level of Z_2, call it z_2', is such that $v_2(z_2')$ is 0.4, and the level of Z_1, labeled z_1'', corresponding to the site with $z_2'' = 0.8$ is 0.7. Then clearly, the minimum possible overall values for these two sites are

$$v(z_1', z_2', z_3^0) = 0.6(0.95) + 0.25(0.4) + 0.15(0) = 0.67 \qquad (4.8)$$

and

$$v(z_1'', z_2'', z_3^0) = 0.6(0.7) + 0.25(0.8) + 0.15(0) = 0.62. \qquad (4.9)$$

However, suppose a review of the screening data indicated that a location characterized by z_1^+ and z_2^+ had $v_1(z_1^+) = 0.9$ and $v_2(z_2^+) = 0.68$, which corresponded to a minimum possible overall value of

$$v(z_1^+, z_2^+, z_3^0) = 0.6(0.9) + 0.25(0.68) + 0.15(0) = 0.71. \qquad (4.10)$$

The resulting value in (4.10) allows us to screen additional areas. First, just using Z_1 again, the cutoff z_1^1 can be increased. Now any location with a level z_1 of Z_1 such that $v_1(z_1) \leq 0.6$ can have an overall value no greater than 0.71 since even with attributes Z_2 and Z_3 at their best possible levels (note this is now z_2' rather than z_2^* for Z_2), from (4.3),

$$v(z_1, z_2', z_3^*) \leqslant 0.6(0.6) + 0.25(0.8) + 0.15(1) = 0.71. \quad (4.11)$$

Thus, we can consider the level z_1^2 such that $v_1(z_1^2) = 0.6$ to be a new derived cutoff level. This is indicated in Fig. 4.6.

For this same reason, we can also check to see if any locations can be eliminated because of their level of Z_2 alone. In this example, even if Z_2 is at its worst level z_2^0, it is still possible to have an overall value of 0.72 if Z_1 and Z_3 are at their respective best possible levels z_1' and z_3^*. Since this is larger than the 0.71 in (4.10), it is not yet possible to screen using Z_2 alone.

There are, however, many possible combinations of z_1 and z_2 which could eliminate locations from further consideration. Any pair z_1 and z_2 such that

$$0.6v_1(z_1) + 0.25v_2(z_2) < 0.56 \quad (4.12)$$

could be eliminated since the corresponding v would necessarily be less than the 0.71 from (4.10). Using the value functions from Fig. 4.6, the derived exclusion area is illustrated in Fig. 4.7 in both the v_1, v_2 space and the space of the attributes Z_1 and Z_2.

The last phase of our screening example is to collect data for attribute Z_3 only for the areas still in consideration. Let us assume that the best level of Z_3 in the remaining area is z_3^*. Furthermore, suppose a review of the remaining area indicates that the highest actual overall v is 0.84. On the last screening attribute, we do not want to eliminate all locations with v less than 0.84 since this would leave us perhaps only one site. Here it is more appropriate to relax the screening criterion to say 0.8, a level which must be determined using the judgments of the siting team. Using this value, it follows that the cutoff on Z_1 can again be increased. Any location with a level z_1 of Z_1 such that $v_1(z_1) \leqslant 0.75$ can have a value no greater than 0.8. Even with Z_2 and Z_3 at their best levels z_2' and z_3^*, we find from (4.3) that

$$v(z_1, z_2', z_3^*) \leqslant 0.6(0.75) + 0.25(0.8) + 0.15(1) \leqslant 0.8. \quad (4.13)$$

Thus, the level z_1^3 such that $v_1(z_1^3) = 0.75$ is an updated derived cutoff level on Z_1. This is illustrated in Fig. 4.6.

Now it is possible to screen on Z_2 individually. If the level of Z_2 is z_2 such that $v_2(z_2) \leqslant 0.32$, then the overall v is less than or equal to 0.8 even with Z_1 and Z_3 at their best levels z_1' and z_3^*, respectively, because from (4.3)

$$v(z_1', z_2, z_3^*) \leqslant 0.6(0.95) + 0.25(0.32) + 0.15(1) = 0.8. \quad (4.14)$$

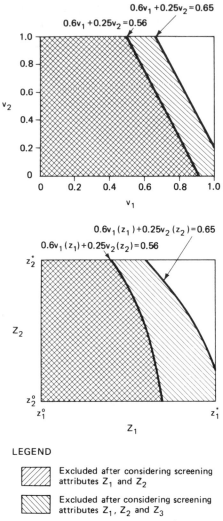

Fig. 4.7.   **Derived exclusion regions in the comparison screening example.**

As above, we can define z_2^3, where $v(z_2^3) = 0.32$, to be the derived cutoff level on Z_2 illustrated in Fig. 4.6.

It is also possible to screen using Z_3. If z_3 is such that $v_3(z_3) \leqslant 0.6$, then the location can be eliminated. From (4.10), the highest value on the first two attributes has $v_1(z_1^+) = 0.9$ and $v_2(z_2^+) = 0.68$. Thus, for any z_3, from (4.3)

$$v(z_1^+, z_2^+, z_3) \leqslant 0.6(0.9) + 0.25(0.68) + 0.15(0.6) = 0.8. \quad (4.15)$$

Hence, the derived cutoff level z_3^3, where $v_3(z_3^3) = 0.6$, is shown in Fig.

4.6. Note that because of the decreasing function v_3, this cutoff means that locations with $z_3 \geqslant z_3^3$ are excluded.

It is of course possible to screen on each pair of attributes. Let us illustrate again with Z_1 and Z_2. Now, any combination of z_1 and z_2 such that

$$0.6v_1(z_1) + 0.25v_2(z_2) \leqslant 0.65 \qquad (4.16)$$

can be excluded from further consideration because the corresponding v would be less than the 0.8 cutoff. The corresponding region excluded from consideration is illustrated in Fig. 4.7.

For similar reasons, locations characterized by a z_2 and z_3 such that

$$0.25v_2(z_2) + 0.15v_3(z_3) \leqslant 0.23 \qquad (4.17)$$

can be excluded since, in combination with the best level of Z_1, i.e., $v_1(z_1') = 0.95$, the overall v would be less than 0.8. Also, combinations of z_1 and z_3 with

$$0.6v_1(z_1) + 0.15v_3(z_3) \leqslant 0.6 \qquad (4.18)$$

can be excluded.

Graphical Illustration of the Comparison Screening Model. Although the description is perhaps a little tedious, the concept of the comparison screening model is simple. Imagine a map of the entire region of interest laid flat on the floor of a room. Because of our scaling convention, we initially assume the minimum value for any point on the map is zero, which is the level corresponding to the floor.

After collecting the data on screening attribute Z_1, we can imagine each location rising from the floor to an elevation corresponding to the minimum value that location could eventually have after including data on the remaining screening attributes. We have a contour map of hills and valleys where hills represent higher valued locations. If the value of any location is low enough, it can be cut off from further consideration. If we flooded the room with ink up to the cutoff level, then any location covered with ink is excluded.

Next, data on the second screening attribute is incorporated and the remaining locations all increase in elevation. The result is a new contour higher in the room. More ink is allowed in the room up to a second cutoff level. And the procedure keeps repeating until the last screening attribute. At this point, the areas remaining represent the candidate areas to be considered in the rest of the siting study. Additional local formal screenings or professional judgments may be used to identify candidate sites from these candidate areas.

An illustration of the comparison screening process for our hypothetical example is shown in Fig. 4.8. The illustration, of course, assumes that the region of interest is one-dimensional, but the idea should be clear. The

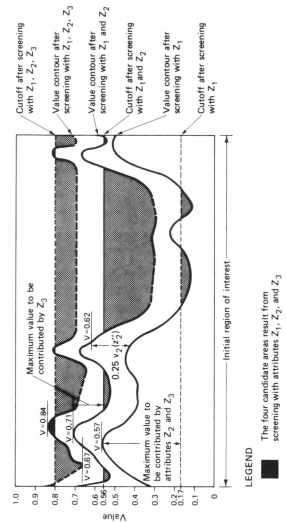

Fig. 4.8. Illustration of the comparison screening example.

Cutoff after screening with Z_1, Z_2, Z_3

Value contour after screening with Z_1, Z_2, Z_3

Value contour after screening with Z_1 and Z_2

Cutoff after screening with Z_1 and Z_2

Value contour after screening with Z_1

Cutoff after screening with Z_1

Maximum value to be contributed by Z_3

$V = 0.62$

$V = 0.84$

$V = 0.71$

$V = 0.57$

$V = 0.67$

$0.25\ v_2(z_2'')$

Maximum value to be contributed by attributes Z_2 and Z_3

Initial region of interest

Value

1.0
0.9
0.8
0.7
0.6
0.56
0.5
0.4
0.3
0.2
0.17
0.1
0

LEGEND

The four candidate areas result from screening with attributes Z_1, Z_2, and Z_3

highest point on the first contour is 0.57, and the maximum which Z_2 and Z_3 could contribute is 0.4. Thus, anything below a value of 0.17 on the first contour is eliminated. Two sections of the initial region were eliminated in this case because these could never overtake the best location even if attributes Z_2 and Z_3 were the best possible in those regions and the worst possible elsewhere.

Next the data was collected on the second attribute Z_2, and the second contour for remaining areas was drawn in. Its high point corresponded to a value of 0.71, which had $v_1 = 0.9$ and $v_2 = 0.68$ as provided in (4.10). Note that the value above the 0.57 height on the first contour is 0.67 from (4.8), and that the location marked $v = 0.62$ had the best level of Z_2 as indicated in (4.9). Given the high value of 0.71, and since the maximum contribution of Z_3 is 0.15, any point below $v = 0.56$ was screened from further consideration.

Applying screening criterion Z_3 resulted in the third value contour as illustrated for the areas remaining after screening on Z_1 and Z_2. After the cutoff level of $v = 0.8$ was set, there were four separate candidate areas to keep in the siting problem for further analysis.

Generalization of the Model. The use of the comparison screening model is the same for n screening attributes as for three. Given Z_1, Z_2, ..., Z_n, one constructs a value function scaled from 0 to 1 over all the possible levels of the attributes. There is no requirement that this value function be additive as was the case in our example. Data is then collected sequentially for the attributes, and any area which can should be excluded at the first possibility. This may not occur after only one attribute. In particular, no elimination can occur until enough screening attributes have been considered to give a value for some location above 0.5, half-way on the value scale.

It is important to point out that the comparison screening model does not suffer from any of the weaknesses of standard screening models referred to in Section 4.2. It explicitly considers the value tradeoffs in a consistent manner, and the cutoffs are derived to insure that, subject to the other data and value judgments, no sites are excluded which could have been real contenders.

4.4 Implementation of Decision Analysis Screening Models

There is a good bit of art in applying the decision analysis screening models. The basic idea is to apply them using screening attributes in a sequential manner. After each iteration, some areas are eliminated, so the data to be collected for the next attribute are for a smaller area. The ordering of the attributes becomes important because it can save time and effort. The order should depend mainly on two factors: how easy it is to collect the data for each screening attribute and how much area might be screened out as a result of that data. This will be considered in more detail separately for the compensatory and comparison models.

It is important to recognize that various combinations of these models can be employed on the same problem in a variety of ways. First, the order in which the screening attributes are applied cannot affect the identity of the areas left for consideration. Thus, the model can be applied differently in different parts of the overall region of interest. One order may be best near the coast and another in mountainous areas.

Different screening models may be used in different areas. For instance, suppose the first screening had identified several candidate areas. It may be useful to utilize a separate model in each of these areas. It is not even necessary that these different models be consistent with each other in the sense that the cutoff levels in each be "equally good." Rather, we wish to identify the best candidate sites in each candidate area to be included in the candidate site evaluation.

4.4.1 SELECTING THE SCREENING ATTRIBUTES

The basis for developing the screening attributes lies in the general concerns as discussed in Section 4.1.2. The screening attributes should be defined to avoid any implied value judgments which do not seem reasonable. In particular, one should be aware of all the possible problems indicated in Section 4.2.2. In addition, the screening attributes should be selected so that

1. the data needed to assign a level of the attribute to each location can be easily collected,
2. there is little uncertainty in the level of an attribute appropriate for each location, and
3. the relationship between the screening attribute levels and the fundamental objectives is clear.

It is desirable to construct screening models to have screening attributes which correspond as much as possible to the attributes used in the final evaluation model.† This will greatly facilitate the later checks neces-

---

† There should be significant interaction between the screening process for choosing candidate sites and the specification of objectives and attributes for evaluating candidate sites. The siting team should be pursuing both steps 1 and 2 of a siting decision analysis simultaneously. Preliminary results from one step will aid in conducting the other.

sary to decrease the likelihood that the screening model eliminated any real contenders.

With screening criteria one need not be complete. Omitting entire domains of concern from screening could not result in the exclusion of good sites. However, the more screening criteria which can be appropriately used, the greater the area which can be eliminated. This will facilitate the search for a "best" site.

4.4.2 MAKING THE JUDGMENTS NEEDED IN THE SCREENING MODELS

Once the screening attributes are selected, there are various possible ways in which they can be used in the overall screening process. These possibilities are appraised in an informal manner and depend greatly on the difficulty involved in making the judgments necessary for the screening models. Three basic types of judgments might be necessary:

1. specifying indifference curves,
2. specifying a value function, and
3. selecting a cutoff level.

Let us discuss these in sequence.

Specifying Indifference Curves. Only one indifference curve is needed to implement a compensatory model. That is the curve which includes the cutoff level as one of its points. Thus, the cutoff establishes a basis from which to construct the indifference curve. The circumstances are different depending on whether a model relating the screening attributes to more fundamental attributes is built or value judgments are used.

In the former case, the entire set of indifference curves is specified by the model itself. An example illustrating this is the pumping costs of water characterized by (4.1). In such a case, the indifference curves over the screening attributes of distance from the water supply and height above the water supply simply are the curves of equal cost. In the example, these curves are linear, although this need not be the case. One of these curves is illustrated in Fig. 4.4.

Figure 4.5 illustrates a case where value judgments are needed to construct the indifference curve. The basic idea for constructing such a curve is to begin with a point such as A in Fig. 4.9, which is taken from the example being discussed. Then we wish to find a few combinations of distance Z_1 and cost Z_2 which have the same preference as A: a distance of 3 miles and a zero water supply cost. We will denote a point on Fig. 4.9 as (z_1, z_2) so A is (3, 0). Suppose we fix the distance at 4 miles and search for a water supply cost z_2' such that $(4, z_2')$ is indifferent to (3, 0). Members of

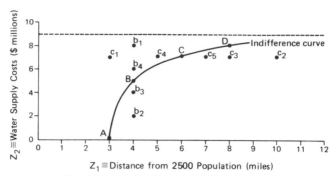

Fig. 4.9.   Constructing an indifference curve.

the siting team may consider 8 million dollars and quickly conclude (4, 8) is worse than (3, 0). This would imply z_2' is less than 8. Considering (4, 2), it may be obvious that this is preferred to (3, 0). Again (3, 0) is less preferred than (4, 4), but (3, 0) is preferred to (4, 6). Finally it is decided that (3, 0) and (4, 5) are indifferent. This sequence of tried points is indicated by b_1, b_2, b_3, b_4, and B in Fig. 4.9. This indicates the way in which one converges to the indifferent point B. One begins by asking easy questions and then converges by asking progressively harder ones. The indifferent pair means that the siting team would be willing to allow water supply costs to increase from 0 to 4 million if they could move from 3 to 4 miles away from a population of 2500.

The same type of convergence process might be followed to search for a level z_1' such that $(z_1', 7)$ is indifferent to both A and B. This may begin with (3, 7), denoted by c_1 in Fig. 4.9 and eventually lead through c_2, ..., c_5 to C which is (6, 7). The point C is then indifferent to A and B. Note that c_1 had to be less preferred than A and B since it is closer to the population and more costly than B. Possibilities like this for continually checking for consistent assessments of the value judgments are good practice. When inconsistencies are found—and good analysis will certainly identify them—parts of the procedure need to be redone after discussing the inconsistency. That is one of the purposes of formal analysis.

Finally, for this example, it may be decided that it would not be worth quite 9 million dollars to have a site 20 miles away from a population of 2500 if one existed 3 miles away with no additional water cost. This means that the indifference curve through A, B, and C must be asymptotic to costs at 9 million dollars. This allows us to draw in the curve shown in Fig. 4.9 and used earlier in Fig. 4.5. By simple questioning, it is easy to check if the other points on the curve do, in fact, seem indifferent to A, B, and C. For instance, D which represents (8, 8) is one such point which could be considered. In MacCrimmon and Toda [1969] and MacCrimmon and Wehrung [1977], there are several techniques discussed which may be helpful for assessing indifference curves.

Specifying a Value Function. In using the comparison screening model, we need to obtain a value function $v(z_1, z_2, ..., z_n)$ which assigns a value v to each possible combination of the screening attributes $Z_1, Z_2, ..., Z_n$. The greater values correspond to preferred combinations of the levels of the screening attributes. Chapter 7 contains details on the theory and procedures for assessing value functions. Here we will simply outline the idea.

One begins by trying to determine some qualitative preference attitudes which are appropriate for evaluating $(z_1, z_2, ..., z_n)$ combinations. This, it is hoped, breaks down the form of v into something simple like the additive value function

$$v(z_1, z_2, ..., z_n) = \lambda_1 v_1(z_1) + \lambda_2 v_2(z_2) + \cdots + \lambda_n v_n(z_n), \quad (4.19)$$

where v_i, $i = 1, ..., n$ are single-attribute value functions and the λ_i, $i = 1, ..., n$ are scaling factors. These v_i and λ_i are then assessed, a relatively easy task. The value function assigns values such that higher values indicate preferred locations.

It is important to recognize that if one has a value function, all the indifference curves over $(z_1, z_2, ..., z_n)$ can be generated. Points of equal value, as indicated by the value function, must be on the same indifference curve. If one has the set of indifference curves, it is easy to determine the direction in which preferences increase. From this, one can construct value functions. This is important when compensatory models are used within general concerns for screening and a comparison screening model is used across general concerns.

Selecting a Cutoff Level. In compensatory models, the selection of a cutoff level is important. When there is a value judgment needed for the cutoff level (that is, when no legal restriction prevails), the level should be set as strictly as possible so as to screen out the most area, and yet not so strict as to eliminate viable candidate sites. Hence, these cutoffs must be established by considering the range of possibilities on this and other screening attributes, as well as the impact this attribute has on the fundamental objectives. Let us illustrate this with the pumped water cost from the example in Fig. 4.4.

Soon after beginning to collect data on the distance from and height above the water supply, it should be evident what ranges in cost would be expected, if they are not already known. Assume for this case that several potential sites are very near the water and thus have a pumping cost very close to zero. It may then be reasoned that a difference of, say, 7 million dollars in pumped water costs would be very difficult to overcome with the other attributes. From this value judgment, the cutoff of 7 million dollars as illustrated in Fig. 4.4 might be established. In Section 4.4.5, we discuss the reappraisal of the appropriateness of the cutoff level after the evaluation of the candidate sites resulting from the screening process.

When the cutoff level for a screening attribute is required by an existing law or regulation, the process is much simpler. However, even in this case, it is worthwhile to include an appraisal of the screening criterion. As in Fig. 4.5, even though the legal requirement may set the cutoff, it is obviously permissible for the client or siting team to make it stricter. This is done by the indifference curve in Fig. 4.5. The area rejected using the compensatory screening, but accepted with standard screening, results from strengthening the screening on miles from the population as the water supply costs increase. This screening is much more in line with the eventual evaluation model.

The setting of cutoffs with comparison screening models is discussed in detail in Section 4.3.3. The cutoffs are set as a function of the difference in preferences between good existing sites, as partially evaluated on screening attributes for which data has already been utilized, and all other sites. Any site which can not make up the difference on the remaining criteria is eliminated from further consideration. Then data from another screening attribute is incorporated into the model, and the cutoffs are strengthened in light of this new information. It is a characteristic of the model that, once established, the cutoffs are strengthened at each iteration. The process for making these judgments is inherent in the model.

As data on the last screening attribute is added to the comparison model, a value judgment is needed about what level of value to set as the final cutoff. This should be done to preserve a reasonable set of candidate areas (or candidate sites themselves, if screening for sites) in which to search for candidate sites.

4.4.3   IMPLEMENTING THE COMPENSATORY
SCREENING MODEL

There are a few natural choices in most problems for sets of screening attributes which should be combined into a compensatory screening model. The most obvious of these are sets of attributes which influence costs. Here it should not be too difficult to use professional judgments and data to provide an actual model of costs. This model should include as many of the attributes influencing costs as is feasible. However, even if several screening attributes are included, judicious ordering of the data collection can save much time.

To illustrate this, let us reconsider the simple example in Fig. 4.4 concerning pumping cost. Let us suppose we have built the model (4.1) and decided to screen out options with a greater than 7 million dollar pumping cost as indicated in Fig. 4.4b. It is not necessary to calculate the costs as a function of both distance and height for all the area under consideration. From the figure, it is clear that any location more than 14 miles from the water supply is excluded, and this can be done irrespective of the height above the water supply. Similarly, the height can be used individually to eliminate any remaining region more than 1400 ft above the water supply.

In the remaining area, we may wish to use the combined distance and height for screening. However, we may consider other single screening attributes and come back to this later. It is also possible to screen on one attribute conditioned on the other. This provides a simple procedure for eliminating more of the region from further consideration. For instance, it may be that much of an area which remained in consideration after the single-attribute screening is more than 800 ft over the water supply. From Fig. 4.4b, it is clear that areas above 800 ft can be excluded if they are more than 6 miles from the site. After the single-attribute screening, the conditional screening would have left the acceptable areas in terms of distance and height as shown in Fig. 4.10b. Compared with Fig. 4.10a, this clearly shows what was additionally excluded as a result of the conditional screening. Figure 4.10b also indicates the combinations which have not yet been screened by a full application of the pumped cost screening criterion.

Other natural choices for compensatory screening models would be groupings within the same general concern as outlined in Chapter 1. While with costs there are functional relationships to build models based on data and professional judgment, these other models may have to rely more on value judgments of the siting team in addition to data and professional judgments. For instance, a socioeconomic impact model may be constructed which indicates the relative desirability of the distance of the proposed facility from communities of various sizes. We could then screen on combinations of distance and community size. In this instance, we would have to be careful because having more than one community within a reasonable distance from the proposed site could have the effect of mitigating the boom–bust cycle at each.

It may be that instead of defining a cutoff for exclusion with the compensatory model, we could use it for constructing a value function to be

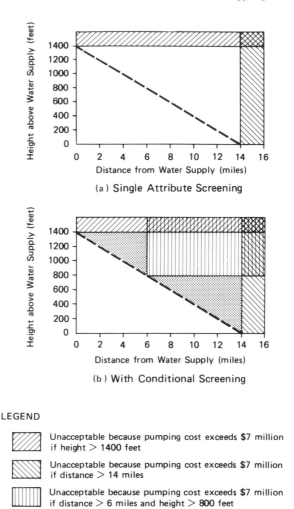

Fig. 4.10.   **Illustration of conditional screening.**

used in a comparison model. After all, the cutoff is just an indifference curve. If we can construct the complete set of indifference curves, which is actually done by the pumping cost relationship in (4.1), we have a value function for pumping cost. Higher costs have lower values and by an appropriate positive monotone transformation, this will give us a value function for pumped cost scaled from 0 to 1.

4.4.4 IMPLEMENTING THE COMPARISON SCREENING MODEL

Much of the flavor of implementing the comparison model is given in Section 4.3.3. The screening attributes should be applied in an order which takes advantage of the ease of obtaining the data and the perceived amount of the region which may be excluded in earlier iterations. After each iteration, one should update the screening on each of the attributes individually, then screen in pairs, and so on. However, if it is easier to collect data on the next attribute than to screen with certain pairs of attributes, this should be done. It is always appropriate to come back later and do the screening on combinations of attributes over a more restricted area. This could result in saving time and effort.

4.4.5 VERIFICATION OF THE SCREENING CRITERIA

The final product of the screening is a set of candidate sites. These sites are then evaluated as described in the WPPSS siting study in Chapter 3. The final result is the expected utility of each site as indicated in Table 3.6. Knowing the utilities of the best sites, we can reconsider all the excluded areas within the original region of interest to verify whether or not the screening was justified. This concept can best be illustrated by using the WPPSS evaluation.

From the results in Table 3.6, it may be obvious to the client that no site with a utility less than a specified level would really be in final contention for the proposed site. In the WPPSS case, let us assume that no location with a utility less than 0.85 (on the 0 to 1 scale) or violating regulations has to be considered. To verify the screening, we wish to show that any site in the region excluded by screening would necessarily have a utility less than 0.85 or be violating a regulation.

Using the utility function (3.3), which is a special type of value function used in the evaluation, it is easy to calculate levels of the evaluation attributes which would force the utility of a site to be less than 0.85. This results in, for example, a differential cost of more than 20.4 million or a site population factor greater than 0.083. These levels alone, even if all the other attributes were perfect, would force a site to have a utility less than 0.85. Using a differential cost model or a site population factor model, many areas could likely be excluded in the state.

Returning to the utility function, we would find that the combination of differential cost greater than 11 million dollars and a site population factor greater than 0.04 would also result in a utility less than 0.85. This could be used to eliminate more areas. This process continues with other attributes and other considerations. Then the legal requirements are included, for example, a site must be more than 3 miles from 2500 population and 5 miles from a fault.

If the final result is that this verification process eliminated all the areas previously screened, the screening is justified. If there are areas previously screened, but not eliminated by the verification, the areas should be carefully considered as candidate sites to be added to those already evaluated. Then the verification process should be repeated until all the screening is justified.

Clearly, more time and effort are required for decision analysis screening than for standard screening. The benefit is that it is more likely that an excellent site will be identified, since it is less likely that such sites will be inappropriately excluded in screening. The verification of screening criteria helps to insure this. Consequently, and because thorough documentation of the screening is available, the screening process is easier to justify. As a final consequence, the chances that the chosen site will be licensed by appropriate authorities can be significantly increased.

CHAPTER 5

SPECIFYING OBJECTIVES AND ATTRIBUTES OF SITING STUDIES

At this point in the siting study, we shall assume that a set of candidate sites has been identified. The next step, described in this chapter, is to specify the desired objectives of siting the facility and to establish attributes useful for measuring the degree to which the objectives are achieved. However, in practice, because of the many interrelationships between identifying the candidate sites and selecting objectives and attributes to evaluate them, this second step of siting decision analysis is well underway by the time the set of candidate sites is determined.

Before proceeding, let us give a formal definition of the problem. There is a set of candidate sites S_1, ..., S_j, ..., S_J to be evaluated. We need to specify objectives O_1, ..., O_i, ..., O_n of the siting problem. For each objective O_i, it is necessary to identify an attribute X_i with a scale to indicate the degree to which objective O_i is achieved. A specific level of achievement measured on X_i will be indicated by x_i.

To illustrate this notation, one objective, say O_1, of selecting a nuclear power plant site is to minimize the power plant cost. An attribute for this objective could be cost measured in millions of dollars. Strictly speaking, the attribute is cost and its scale is millions of dollars. However, since we shall always refer to the attribute and its scale together, no confusion can exist if we refer to both notationally as X_1. A specific level of X_1 might be $x_1 = 1500$ meaning 1500 million dollars.

As a result of structuring the problem as indicated, it is possible to describe the consequences of importance at a particular site by the vector

$x \equiv (x_1, \ldots, x_n)$. Then analysis would involve the following. We would model the consequences of selecting each candidate site S_j as $p_j(x)$, where p_j could be either a probability distribution over the possible consequences of S_j or a deterministic relationship between an alternative S_j and its consequences. Then a measure of the desirability of the various xs would be needed. This measure of preference should be a value function $v(x)$ indicating the order of preferences of the xs if p_j is deterministic.† Then the decision maker should choose the alternative resulting in the highest value. If p_j is probabilistic, the preference indicator should be a utility function $u(x)$, which allows one to compare alternatives on the basis of expected utility, the candidate site associated with the highest expected utility being preferred.

Because the deterministic situation is a special case of the more complex (and more typical) probabilistic situation, most of that which follows in this book focuses on the probabilistic case. In this situation, our basic problem can be schematically illustrated as in Fig. 5.1. For each candidate site S_j, the expected utility $Eu(S_j)$ is calculated from $p_j(x)$ and $u(x)$. Given the attributes $X_1, \ldots, X_n$, the decision problem becomes one of determining $u(x)$ and $p_j(x)$ for each alternative.

5.1 The Objectives Hierarchy for Evaluating Sites

An objective, as we use the term, has two features. It identifies a general concern (or a part thereof) and an orientation for preferences. Examples of objectives are "minimize pollution," "maximize the distance from a scenic highway," or "optimize the socioeconomic impact." When the orientation is somewhat in doubt, the general term optimize serves well to specify an objective.

We do not include the following as objectives: keep differential costs under 20 million dollars annually, locate the facility within 10 miles of a port, and eliminate social disruption. Whenever a cutoff is defined, as in all these cases, the sites are necessarily divided into two groups, those which do meet these cutoffs and those which do not. The statement "eliminate social disruption" essentially has a cutoff level at zero disruption. When such statements are used in the screening models, we refer to them as screening criteria. Since they do not distinguish between, for example, sites which are 1 and 9 miles from a port, such guidelines are not too useful in the evaluation phase of the decision analysis. Objectives do not have this shortcoming.

† A complete discussion of value functions and utility functions is found in Chapter 7.

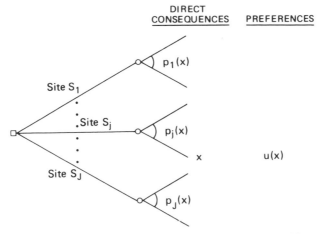

Fig. 5.1. Schematic representation of the basic siting problem.

The following are also not objectives: costs, local taxes paid, and jobs created. These may each be indicative of a concern in a siting study, but no orientation for preferences is given. They may be appropriate attributes for the objectives: minimize costs, maximize local taxes paid, and create as many jobs as possible.

5.1.1 STRUCTURING THE OBJECTIVES

The most reasonable way of structuring objectives is an objectives hierarchy. Such a hierarchy starts with the overall objectives at the top and lists more specific objectives at each lower level. A sample hierarchy for a proposed oil refinery in a harbor is illustrated in Fig. 5.2, where, for brevity, one or two words are used to indicate the objective. An objectives hierarchy has several advantages:

1. The top levels are fairly standard for most siting studies since they come directly from the general concerns discussed in Chapter 1. This indicates the scope of concern.
2. It helps to insure that no large holes (missing objectives) will occur at lower levels, since such holes should be fairly obvious.
3. The major objectives provide a basis for defining lower level objectives, since the latter are means for achieving the higher level objectives.
4. The lower level objectives are more specific and, as such, it is easier to identify reasonable attributes for them.
5. Situations where redundancy or double-counting might easily occur can be identified.

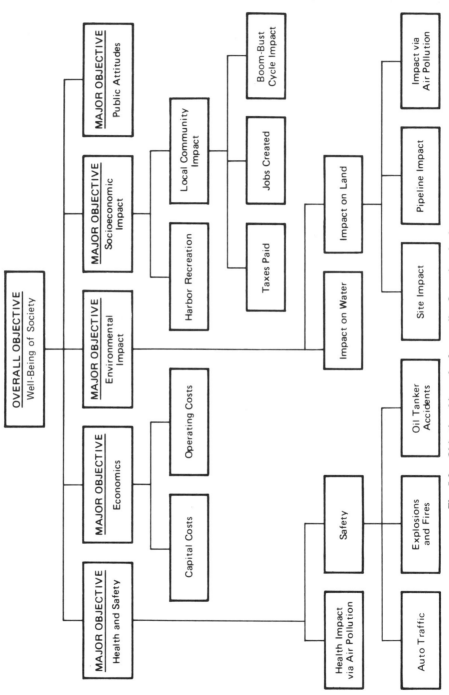

Fig. 5.2. Objectives hierarchy for an oil refinery in a harbor area.

6. Attributes must be defined only for the objectives at the bottom of any branch of the hierarchy.

7. The hierarchy provides a basis from which to develop and appraise screening criteria.

Let us briefly consider these advantages in terms of the objectives hierarchy in Fig. 5.2. The top objective is a parenthood (where did that term come from?) objective. The description "well-being of society" is short for "maximize the well-being of all those in society affected by the proposed project." It is essentially impossible to measure directly societal well-being. Consequently, well-being is broken into five major objectives which correspond to the general concerns discussed in Section 1.3.

With regard to advantage 2, if one began to fill out the objectives hierarchy and the only structure under, say, environmental impact concerned impact on land, the omission of water impacts should be very obvious. It may be, however, that an omission is appropriate, as will be discussed in Section 5.2. The point is that since such an omission would be blatant, it would immediately be identified and its appropriateness decided upon.

The means–ends relationships in the objectives hierarchy should be clear. For instance, the means causing impacts on the local community include the taxes paid, the jobs created, and the implications of the boom–bust cycle. Likewise, the impact created by a pipeline would be the means causing impacts from the proposed project on flora and fauna.

For an objective such as "optimize the socioeconomic impacts" of the project, there is no obvious attribute. It is much easier to determine attributes for "maximize the local taxes paid," "maximize the jobs created," and so on. However, sometimes it is still very difficult to define a measure for the lowest-level objectives such as "minimize the impact of the boom–bust cycle." Some suggestions of what to do in these cases are discussed in Section 5.4.

The issue with respect to advantage 5 of an objectives hierarchy is somewhat subtle. In Fig. 5.2, note the health impacts via air pollution and the land environmental impacts via air pollution. If these are measured by an air pollution emission level, it would be possible to combine them. Alternatively they can be kept separate as long as the value assigned to different air pollution levels with respect to each one does not inadvertently affect the assessment of the value for the other. If this circumstance did occur, the air pollution impact would be "double-counted."

Note that there are 15 lowest-level objectives in the hierarchy of Fig. 5.2. It is perfectly alright to have a major objective such as public attitudes be one of those lowest-level objectives. The 15 are those for which we would need to define attributes, if indeed all 15 meet the test of relevancy discussed in Section 5.2. Objectives which are not lowest-level are measured by the collection of attributes corresponding to all the objectives below them in the hierarchy. For example, the safety objective would be measured by the attributes pertaining to auto traffic, explosions and fires, and oil tanker accidents.

As mentioned in Section 2.3, the five steps in a siting decision analysis are not carried out in a sequential manner. Work is often simultaneously proceeding on more than one step and several iterations of each step are necessary in any thorough siting study. The objectives hierarchy specifies in detail what is meant by the general concerns for a specific problem. As a result, it provides a more thorough basis for selecting screening criteria as discussed in Chapter 4 than do the general concerns alone. On the other hand, as a result of the screening, certain objectives may not be necessary for the candidate site evaluations because the candidate sites are all about equally desirable in terms of those objectives. There is significant interplay between screening and the objectives hierarchy.

5.1.2 THE OBJECTIVES HIERARCHY IS NOT UNIQUE

For any particular siting problem, the objectives hierarchy is not unique. By eliminating some of the lowest-level objectives in Fig. 5.2, we shall still have a hierarchy appropriate for the problem. The appropriate level to which the hierarchy should be extended depends on many factors such as the ease of identifying attributes, the availability of data to describe sites in terms of these, who is the client for the study, and what use will be made of the study. Such issues are discussed in detail in Keeney and Raiffa [1976].

However, even if there are the same number of lowest-level objectives, two objective hierarchies for the same problem can be very different and yet reasonable as illustrated in Fig. 5.3. For instance, a different categorization under the major objective "health and safety" in Fig. 5.2 would be into "impacts of normal operations" and "impacts of accidents." As indicated in Fig. 5.3, the former may be broken down into sickness and fatalities and the latter into injuries and fatalities. Another categorization might have specified "sickness," broken further into temporary and chronic, along with injuries and fatalities. These are the reasons for interest in air pollution, auto traffic, explosions and fires, and oil tanker accidents. The latter are proxy for the former, as discussed in Section 5.3.

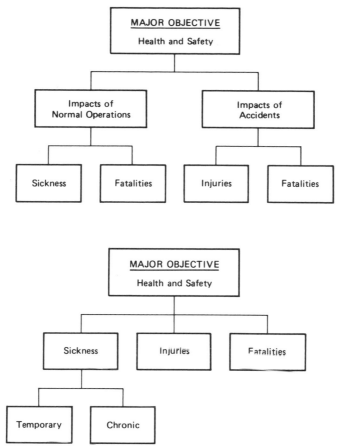

Fig. 5.3. Two alternative characterizations of the health and safety objectives for the oil refinery problem.

5.1.3 IDENTIFICATION OF OBJECTIVES

The identification of the objectives for a siting problem is essentially a creative process. There are, however, several devices which can be of considerable help in priming the process. These might be categorized into three areas: searching through relevant literature, exploring relationships in the objectives hierarchy itself and interrelationships between objectives and attributes, and modeling of the value structure for the problem. The starting point for each of these is the list of major objectives of siting indicated in Fig. 5.2.

There are two general types of literature which are particularly relevant to specifying objectives. These are other siting studies and laws and regulations. With respect to previous siting studies, one should consider the history of siting studies as a whole to be partly an evolutionary process which should result in a better model for decision making as time goes on. With more experience, we should learn from mistakes as well as develop methods for overcoming shortcomings which were obvious in earlier studies even at the time they were conducted. The review and appraisal for existing studies provides the mode for this learning.

Both federal and state laws and regulations are important in helping select objectives for the siting of energy facilities. The National Environmental Policy Act of 1969 is the milestone law which provides a basis for much of that which follows. The term "environment" as it is defined in this act is very broad. It includes many aspects of the environmental, socioeconomic, and health and safety general concerns as defined for use in this book in Section 1.3. Details of aspects to be addressed in siting studies as a result of such laws are found in various regulatory publications. Each federal or state agency requiring approval for specific facilities will have its own documents on information needed for licensing consideration. See, for example, Papetti, Hammer, and Dole [1973], U.S. Nuclear Regulatory Commission [1976, 1977, 1979b] and California Energy Resources Conservation and Development Commission [1977].

As a result of a complete literature search, we should have a good list of possible objectives and ideas of attributes for several of these. The objectives will range greatly in specificity and breadth. Some will address fundamental concerns, such as potential facilities, and some will concern the processes causing these, such as air pollution. Some will address location of impacts, such as impact on water, and others will indicate the result of the impact, such as fish killed. The problem is to put this list into a consistent and useful objectives hierarchy. Part of this task involves following the means–ends relationships through the hierarchy. In doing this, redundancies can be eliminated and "holes" filled in. Sometimes it will be difficult to determine an attribute for some objective. This might suggest a change to another objective which still captures the same basic concern. In Section 5.5, we suggest several desirable properties of the final set of objectives and attributes for a problem. In working toward these, the actual objectives can evolve quite a bit.

Once we are at the point of having a "first-cut" objectives hierarchy and associated attributes, a utility function can be assessed to quantify the value structure. One major reason for doing this is to provide a value model for evaluating the candidate sites that addresses the issues raised in Chapter 1. The second is to promote strong thinking about the problem structure, in particular about the appropriateness of the objectives and attributes. Often at this stage, even though we have been careful, significant potential improvements are identified. Sometimes we find that some objectives simply are not so important and it may be useful to drop them from the study. In other cases, it is very difficult to assess the preferences over some attribute because the attribute or situation is not well defined or the questions are inherently ambiguous or extremely difficult. Such was initially the case concerning salmonid impact in the WPPSS study (see Chapter 3). We had tried to quantify preferences over the attributes number of salmonid lost and percent of salmonid lost. This proved very difficult because it implied that the number of salmon in the river kept changing when preferences were assessed with one attribute fixed. Then, the attributes were changed to the number of salmonid in the river and the percentage lost for all streams, except for the larger Columbia River where only the number of salmonid lost was used.

At this stage in the evolutionary process of siting we should be relatively assured that important concerns in the comparisons of sites are not omitted because no one thought them important. Rather, the omission of such concerns is more likely to be the result of not using the available information and techniques. Careful attention to the generation of objectives for a specific problem should minimize the likelihood of omission. The devices discussed here promote such careful attention.

5.2 Selecting Objectives to Include in the Analysis

The main purpose of this section is to discuss reasons for including or excluding possible objectives from a siting study. In this sense, it is the same problem we considered in screening the region of interest to identify candidate sites. Here we are screening all possible objectives to get a set of objectives.

As indicated in the WPPSS study of Chapter 3, the siting team must make a number of decisions about which objectives to include. Basically the guideline for inclusion is whether or not the inclusion is likely to affect the outcome of the evaluation of the candidate sites. Section 5.2.1 elaborates on reasons for the inclusion of objectives.

5.2.1 Reasons for Including Objectives

There are three circumstances, all of which must prevail in order for an objective to be included in an analysis. To ascertain if these circumstances hold, one must be thinking, at least informally, of some attribute or measure for the potential objective. The following three circumstances can be appraised in a sequential manner for each objective.

1. There must be a difference in the degree to which the objective might be achieved by at least two candidate sites.
2. This difference must be significant relative to the other differences between the candidate sites.
3. The likelihood of this difference must be large enough to justify inclusion.

These three circumstances were used to appraise possible objectives for the WPPSS study in Section 3.2. Let us suggest some other possible examples.

In some sections of the United States, there is always the chance that archaeological ruins may be found upon excavating an area. One such area is the Southwest, where Indian ruins are not uncommon. In the pumped storage siting study discussed in Chapter 9, archaeological features were considered relevant to the objectives. However, it was felt that the likelihood of actually finding such features at each of the candidate sites was equivalent. Therefore, no objective for such considerations was included.

Many energy facilities require about 1 sq mile of land. In a specific case, some of the candidate sites may replace agricultural land and other "unproductive" land. Even though there may be a difference in the amount of agricultural land taken from production, the maximum difference of 1 sq mile may not be significant enough to include in the study.

If an objective passes the first two hurdles, it still may not be worth including formally in the evaluation of the alternatives. Suppose one is evaluating different transmission line corridors for a 365 kV electricity line. It may be that the chance of sabotage on those lines is not judged to be exactly equivalent. And if sabotage meant a blackout for a city, the differences in potential consequences for the different corridors could be great. However, if the probability of sabotage is 0.001 on the safest line over the lifetime of the project and 0.002 for the least safe, it may be appropriate to formally disregard sabotage in the study. Even if the implications of a blackout are equivalent to a cost of 10 million dollars, multiplying the range in the probabilities for the sites, which is 0.001, by the cost, gives 10 thousand dollars. Accounting for risk attitudes, this could not be worth more than about 20 thousand dollars, which would be insignificant relative to other siting factors.

This example should make it clear that the screening of alternatives can have a major effect on what the appropriate objectives are. For example, if all inland river sites for the WPPSS nuclear power plant were eliminated by screening, then there would be no need to consider salmon in the streams in the subsequent evaluation. In another situation, if all the candidate sites for a prototype solar power plant are screened to avoid the necessity of moving people, then such a consideration obviously has no role in a site comparison.

5.2.2 REASONS FOR EXCLUDING OBJECTIVES

There are basically three general reasons for excluding an objective from a siting study. These are:

1. The objective fails to meet the criteria for inclusion discussed in Section 5.2.1.
2. The client wishes to drop the objective from the study.
3. It is impractical to gather the necessary data for evaluating the candidate sites in terms of the attribute.

Because of their nature, if any of these circumstances prevails, the objective should be dropped. By definition, the first reason occurs if and only if the criteria for inclusion of an objective are not met. The second and third reasons have priority over the reasons for inclusion. In this way, both of them restrict the domain of a study. Let us suggest some examples.

In a coal power plant siting study, the client may wish to exclude the difficulties of actually purchasing the land needed for a site from the evaluation phase. Of course, areas where site purchase might be extremely difficult or impossible, such as national parks, might be excluded by the screening. This would create some homogeneity among the candidate sites with respect to purchase difficulties. Yet large differences may remain. The client's reasons for excluding land purchase may range from (1) a desire to avoid all designation of specific property until a site is actually purchased in order to avoid artificially inflating the land prices, to (2) land purchase is handled by another part of the client's organization, to (3) the client does not want outside groups to be involved in land purchase aspects.

There are two types of data needed once objectives are determined, possible impacts in terms of probabilities or point estimates and preferences in terms of utilities. If either of these is impractical to obtain for a particular objective, then by reason 3 the objective must be excluded. A major cause of this impracticality involves difficulties in identifying an attribute about which to collect the data and assess preferences. This particular difficulty is addressed in the following sections of this chapter.

For the same coal-fired power plant siting, there may be the feeling that an important objective is "minimize the amount of visual degradation due to pollutant emissions." It may be very difficult to identify a good attribute for this objective, since it would depend on the original visual quality of the environment, the amount of emissions, and the number of individuals affected.† Even if it is decided that a reasonable attribute is the number of people who feel that the emissions degrade the visual environment, there may be no practical means of responsibly estimating this number. Finally, given responsible estimates of the number of people affected and a single-attribute utility function over the number of people is assessed, the objective still may need to be disregarded. In order for an attribute to be useful, one must be able to make value tradeoffs between it and other attributes. On the other hand, even if this is extremely difficult, a broad sensitivity analysis over these tradeoffs may prove helpful.

As indicated earlier, any objective omitted from a siting evaluation because of either reason 2 or 3 restricts the domain of the study. It is important to recognize that the domain of the siting problem does not concurrently shrink. Thus the siting study will be addressing only a part of the problem. This *per se* is not necessarily bad. However it is of paramount importance to make certain that such omissions can be clearly understood by individuals reading and appraising the study. The results of the study must then be informally integrated in the mind of the client with the implications of the various sites in terms of omitted objectives in order to arrive at appropriate conclusions on site desirability.

5.3   Selecting Attributes for Measuring Objectives

Decision analysis requires the establishing of attributes to indicate the degree to which objectives are achieved. Attributes are either selected from existing measures, such as cost or number of people, or constructed for a specific situation, such as indices of environmental impact or aesthetic degradation. For siting problems, the process of defining these attributes should be logically consistent and systematic. At the same time, it is inherently subjective; it must encompass professional judgment, knowledge, and experience. Value judgments are also necessarily made in selecting attributes.

To refresh our memory with the notation used at the beginning of this

---

† Duke *et al.* [1977] has an interesting discussion about visual quality attributes.

chapter, we will assume that objectives O_1, ..., O_n have now been defined. The concern now is to develop a set of attributes X_1, ..., X_n on which to measure the achievement of these objectives. Attributes can be categorized according to two criteria: (1) how they measure achievement and (2) the type of scale used. An attribute can directly or indirectly indicate the achievement of an objective. For instance, if an objective is to "decrease sickness from air pollution," a direct measure might be the number of people sick due to air pollution, whereas an indirect measure might be the amount of air pollutant emissions. The indirect measure has some relationship to the objective, but that relationship may never be formalized and may not be precisely known. The two types of scales are natural and constructed. Slightly oversimplified, the natural scales are those with widely understood numerical indices and the constructed scales those without. This should become clearer in Section 5.4.

5.3.1   DIRECT ATTRIBUTES†

A direct attribute measures the degree to which an objective is achieved. The main advantage of a direct attribute is that it is easier to interpret than an indirect one. Knowing the attribute level clearly indicates the degree of achievement.

If X is a direct attribute for objective O, then the siting team needs only to assess probabilities $p_j(x)$ over X for each site S_j and a utility function $u(x)$ over X levels. Then the component of expected utility $Eu(S_j)$ due to objective O can be directly calculated as

$$Eu(S_j) = \int p_j(x)u(x), \qquad (5.1)$$

where $\int$ is meant to represent summation or integration, whichever is appropriate.

Unless there is a good reason, direct attributes are preferable to indirect ones. Two reasons for opting to forego a direct attribute are: no obvious direct attribute exists, or it is too difficult to determine probabilities of impacts with regard to that attribute. What obvious measures are there for the socioeconomic impact or the environmental impact at a proposed nuclear power plant site? The objectives are to minimize the detrimental socioeconomic and environmental impacts, but how can they be measured?

---

† This section assumes conditional probabilistic independence between attributes for each site and utility independence to illustrate the concepts in as clear a fashion as possible. These assumptions are discussed in Chapters 6 and 7, respectively.

5.3.2 Proxy Attributes

To illustrate what can be done in these cases, let us focus on the objective O for which no obvious direct attribute X exists. In such a case, one often resorts to using a scale which relates to the achievement of objective O, but which does not directly measure it. Such a scale is referred to as a proxy attribute (rather than an indirect attribute). This proxy attribute, which we shall label as Y with specific levels y, can be used in place of the direct attribute in evaluating alternatives. In general, of course, there may be a set of direct attributes replaced by a set of proxy attributes, but the basic ideas can best be illustrated in the situation where only X is replaced by Y. We shall return to the more complicated case later.

Using Y requires special consideration in the evaluation phase of analysis. To indicate this, assume that both a direct attribute X and a proxy attribute Y were identified. To assess a utility function $u'(y)$ over the attribute Y, one should think about two considerations: the likelihood of various levels of x being associated with a particular level of y, and the preferences of various levels of x. The first consideration could be quantified by the probability distribution $p'(x|y)$, which indicates the probabilities of various x levels, given that the level of Y is y. The second consideration could be quantified by the utility function $u(x)$. The utility function $u'(y)$ could then be derived from

$$u'(y) = \int u(x)p'(x|y)\,dx, \qquad (5.2)$$

where $\int$ again represents either summation or integration, whichever is appropriate.

In actuality, because no obvious X exists (which necessitates the use of Y in the first place), the above process is done informally by directly assessing $u'(y)$.

Now to evaluate alternative site S_j, it is necessary to specify the possible y levels which may result from choosing S_j. This could be represented by the probability distribution $p'_j(y)$, so the component of expected utility due to objective O is

$$Eu'(S_j) = \int p'_j(y)u'(y). \qquad (5.3)$$

To summarize what we have done, refer to Fig. 5.4. The part of the problem represented by (5.2) was conducted informally when the utility function $u'(y)$ for the proxy attribute Y was assessed. This was formally

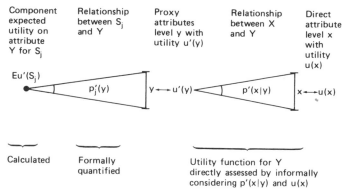

Fig. 5.4. Relationship between direct and proxy attributes.

combined with $p_j'(y)$ to use in calculating the expected utilities for evaluating the alternatives.

As an illustration of the previous process, assume objective O is to minimize the dangerous health impacts of an accidental release of radiation from a proposed nuclear power plant. Possible measures for X may be the number of deaths due to radiation release or the number of deaths plus injuries. However, it may not be appropriate or possible to use these measures in an analysis. Alternative proxy attributes Y may be the number of people living within 5 miles of the proposed plant, the total person-rem dose given an accident, or the site population factor. Each could serve as an indicator of the health impacts of an accidental release of radiation.

5.3.3 DIRECT PREFERENCE MEASUREMENTS

The proxy attribute Y may be closely or loosely related to a direct measure of O. However, one special case is worth noting. When the Y scale directly indicates whether site S_1 is better than S_2 in terms of achieving objective O, proxy attribute Y is referred to as a direct preference scale. Levels on this scale are simply component preferences. It may be that this direct preference scale is such that Eu' is itself equivalent to y. In any case, there must be a positive monotonic relationship between Eu' and y so that p_j' in Fig. 5.4 is a deterministic one-to-one relationship between Eu' and y. Said another way, if Y is direct preference scale, it must be a value function for achievement on the associated objective.

5.4 Natural and Constructed Scales for Attributes

To measure either the direct attribute X or the proxy attribute Y, there are basically only two types of scales, natural and constructed (subjective). In practice, the distinction between scale types is not always completely clear, but by discussing two extreme cases, the possibilities for cases in between should be obvious.

Natural scales are those which have been established and enjoy common usage and interpretation. For instance, costs in millions of dollars and numbers of deaths are examples of natural scales. Natural scales tend to have relevance to several problem contexts.

Constructed scales, on the other hand, are often developed specifically for a problem being addressed. For instance, one of the objectives of power plant siting may be to minimize the detrimental aesthetic impact. Because no natural scale of aesthetic impact exists, it is necessary to construct one. This may entail describing or pictorially illustrating several different degrees of aesthetic impact. These can be designated as integer levels, say from 0 to 10. One can construct the scale so that degradation between adjacent levels is uniform if it is necessary or useful to have a continuous scale. A constructed scale must have its key points defined to convey its meaning to individuals other than those who constructed it.

As society determines more and more ways to measure characteristics, the number and diversity of natural scales grow. For instance, before the decibel scale, one can imagine that a constructed scale would have been required to measure loudness. After significant use, what was once a constructed scale takes on many of the features of a natural scale. The GNP (gross national product) scale, for example, was constructed to aggregate several factors to indicate the economic health of the country. After years of usage, the meanings of specific GNP levels stand essentially on their own.

One special constructed scale is simply the dichotomy of whether or not an objective is met. For instance, one objective of a siting study may be to minimize the inconvenience to a citizen group (e.g., Indians) whose land might be used for a nuclear power plant. A measure might be whether private land would be needed for the particular plant site.

In many studies (e.g., Kalelkar *et al.* [1974] and Keeney and Nair [1977]), the term subjective scale or subjective index was used for what I have chosen to refer to here as a constructed scale. My reasoning is that the implication of such usage is that natural scales require no professional (subjective) judgment. This is somewhat misleading. Although no professional judgment is needed in constructing natural scales, professional judgment is used to decide if, in fact, a proposed natural scale is appropriate.

In this regard, a major "subjective" judgment in choosing attributes often involves whether to construct a direct scale or to use a proxy natural scale. This is especially true when the stated objectives are truly fundamental rather than means objectives to achieving the fundamental ends. For example, the fundamental objective of a pollution control program for coal power plants might be to reduce detrimental health impacts, whereas a means objective would be to reduce SO_2 pollution. To measure the former directly would probably require a constructed scale, whereas the natural scale of tons of SO_2 emitted per year might be a good proxy attribute for health. Which scale is more appropriate would depend on the specific problem.

In the remainder of this section, the main considerations for selection and construction of scales will be discussed. To illustrate the ideas, simple scales will be used. By this we mean scales where either attribute X is used to measure objective O or a proxy attribute Y is used in place of X. Other possibilities, including the use of several proxy attributes in place of one direct attribute, are also discussed.

5.4.1 SELECTING NATURAL SCALES

With natural scales, the main issues are: what subjective value judgments are implied by the scale, and does a knowledge of the level of the scale provide relevant information with respect to objective O for the decision. To illustrate, suppose objective O is to minimize the possible fatalities of a nuclear power plant accident. Two possible natural scales would be (1) the number of deaths and (2) the reduction in the average lifetime of exposed citizens. These scales have very different implicit value judgments. Consider the implications of the death of a 70 year old and a 5 year old. Using the number of deaths, each counts the same, namely one. However, suppose the death of the 70 year old eliminates 10 years of expected life, whereas the death of the 5 year old eliminates 70 years of expected life. Hence, using the measure of reduction in average lifetime, the death of the 5 year old counts 7 times as much as the death of the 70 year old. Put another way, with the second scale, the death of one 5 year old would count as much of the death of seven 70 year olds.

A question related to this example is which individuals might die. Should a worker in the proposed nuclear power plant be counted as much as an individual living nearby? Using a measure such as total fatalities implies that this is the case. If individuals who have voluntarily accepted the associated risks (e.g., they work in the plant) are to count less, a value judgment is needed to determine how much less in constructing the scale. In any case, there is no denying the fact that such a value judgment will necessarily be made—explicitly or implicitly.

5.4.2 DEVELOPING CONSTRUCTED SCALES

Most constructed scales are meant to measure more than one facet of a complex problem. As a result, in addition to all the considerations necessary in the selection of natural scales, additional value judgments are made. As an illustration, consider the constructed scale which measured the objective "minimize the biological impact at the proposed site" in the WPPSS nuclear power plant siting study in Chapter 3. This scale, repeated in Table 5.1, was meant to describe the major possible site impacts

TABLE 5.1

CONSTRUCTED SCALE FOR BIOLOGICAL IMPACTS AT THE SITE

Scale value	Level of impact[a]
0	Complete loss of 1.0 sq mile of land which is entirely in agricultural use or is entirely urbanized; no loss of any "native" biological communities.
1	Complete loss of 1.0 sq mile of primarily (75%) agricultural habitat with loss of 25% of second growth; no measurable loss of wetland or endangered species habitat.
2	Complete loss of 1.0 sq mile of land which is 50% farmed and 50% disturbed in some other way (e.g., logged or new second growth); no measurable loss of wetland or endangered species habitat.
3	Complete loss of 1.0 sq mile of recently disturbed (e.g., logged, plowed) habitat plus disturbance to surrounding previously disturbed habitat within 1.0 mile of site border; or 15% loss of wetlands and/or endangered species.
4	Complete loss of 1.0 sq mile of land which is 50% farmed (or otherwise disturbed) and 50% mature second growth or other community; 15% loss of wetlands and/or endangered species.
5	Complete loss of 1.0 sq mile of land which is primarily (75%) undisturbed mature "desert" community; or 15% loss of wetlands and/or endangered species habitat.
6	Complete loss of 1.0 sq mile of mature second growth (but not virgin) forest community; or 50% loss of big game and upland game birds; or 50% loss of wetlands and endangered species habitat.
7	Complete loss of 1.0 sq mile of mature community or 90% loss of productive wetlands and endangered species habitat.
8	Complete loss of 1.0 sq mile of mature virgin forest and/or wetlands and/or endangered species habitat.

[a] Three main impacts captured by this scale are those on native timber or sagebrush communities, habitats of rare or endangered species, and productive wetlands.

on productive wetlands, migratory species, endangered species, and virgin and mature second-growth timber stands. Level 0 is defined as the least biological impact, since agricultural and urban land are already completely biologically disturbed. Greater levels indicate increasing disturbance.

Note the many value judgments necessarily built into the scale. With levels 3, 4, and 5, the loss of the land used by an endangered species and 15% loss of productive wetlands are implied to be equally important. For level 4, an area of native second-growth forest is deemed equally valuable as the same area of other (nonforest) undisturbed communities. Such value judgments are difficult to make and context dependent. In the case illustrated, the judgments were made by experienced biologists who had visited the proposed sites to be evaluated and were familiar with the biological resources of the area.

The use of a scale such as that illustrated in Table 5.1 does not eliminate the possibility of including other biological factors in the evaluation. For instance, some of the proposed Washington sites were in the eastern part of the state in semidesert areas. A major biological impact at these sites was the loss of 1 sq mile of mature sagebrush communities and their associated small animal communities. In estimating the biological impact at such a site, the biologists must specify the likelihood that the impact would be between levels 2 and 3, 3 and 4, and so on. This required professional value judgments about the value of the sagebrush habitat relative to the defined level 2, 3, and 4 impacts. Regardless of the difficulty of making such judgments, they are an inherent part of the problem and must be either implicitly or explicitly addressed.

To gain some insight into the informal processes that are considered in developing a constructed scale, let us examine a formal model of this informal process. Consider a simplification of the Washington State problem where one is concerned only with the productive wetland loss and with the presence of rare species. Suppose it is reasonable to use two natural scales of this biological impact: X_1, acres of productive wetland and X_2, the number of a rare species present. If we let these range from 0 to 640 acres and 0 to 50 members, respectively, the possible consequence space is as illustrated in Fig. 5.5.

Clearly, the best biological impact is none, corresponding to the point $(x_1 = 0, x_2 = 0)$, which is 0 acres of productive wetland and 0 rare species present. The worst impact is $(x_1 = 640, x_2 = 50)$. Other possible impacts fall between these bounds. Suppose, through questioning biologists and ecologists, the indifference curves illustrated in Fig. 5.5 are elicited. From this, the constructed scale called Y in Table 5.2 could be developed. Note that the amount of impact is defined to increase with increasing levels.

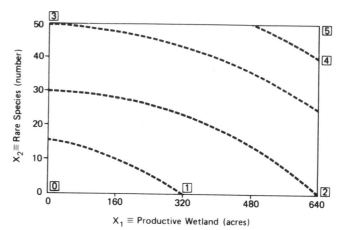

Fig. 5.5. Consequence space for biological impact.
Dashed lines indicate indifference curves and boxes indicate impact levels.

Several professional value judgments are used in the table. At level 2, the loss of 640 acres of productive wetland with no members of the rare species present is equated with no loss of productive wetland and 30 members of the rare species present. From Fig. 5.5, one can see that both of these are equivalent to the loss of 320 acres of productive wetland with 20 members of the rare species present. For simplicity, it is often convenient to use reference impact levels with only one kind of detrimental impact.

To develop more formally a constructed scale Y, one could assess a value function v over the (x_1, x_2) consequences. This would provide a

TABLE 5.2

ILLUSTRATION OF A CONSTRUCTED ATTRIBUTE Y OF SITE BIOLOGICAL IMPACT

Level	Description of equivalent impact
0	No loss of productive wetland and no members of the rare species present.
1	Loss of 320 acres of productive wetland and no members of the rare species present.
2	Loss of 640 acres of productive wetland and no members of the rare species present or 30 members of the rare species present and no productive wetland loss.
3	No loss of productive wetland and 50 members of the rare species present.
4	Loss of 640 acres of productive wetland and 40 members of the rare species present.
5	Loss of 640 acres of productive wetland and 50 members of the rare species present.

complete set of indifference curves over the consequence space in Fig.
5.5. Then a specific level of impact y could be defined as $v(x_1, x_2)$ which
would serve to define the constructed scale Y. This is in essence the
process used to generate all constructed scales. The distinction is the degree of formality or informality with which the value function is determined and whether the entire scale or only specific points on the scale are
clearly defined. When there are no clearly identifiable X_1 and X_2 scales,
the approach is necessarily done informally.

In using the constructed scales to evaluate alternatives, one usually as-
sesses a probability distribution for the likely impacts of each alternative
in terms of the possible levels of impact. The probabilities are combined
with preferences for the impacts for evaluating the alternatives. This
process was illustrated in Fig. 5.4. There is one exception to this, involving direct preference measures, which were introduced in Section
5.3.3. In this case, the probabilities and preferences are holistically combined for each alternative in one step to provide a direct preference indicator for a specific objective. This constructed scale provides a rank order
of all the alternatives with respect to the stated objective. All the value
judgments necessary for other constructed scales are also used in making
direct preference measurements.

5.4.3 DEVELOPING MORE INVOLVED SCALES

In many situations, there is no useful one-to-one relationship between
objectives and attributes or scales. In these cases, it is not necessary for
our discussion to differentiate whether the attributes are direct or proxy
or whether the scales are natural or constructed. The concepts are identical in all four possible situations.

The first case to consider is when more than one attribute is used to indicate achievement on one objective. For instance, suppose X_1, acres of
productive wetland, and X_2, number of a rare species present, were used
as the two attributes for the objective O, "minimize environmental impact." Then, by assessing a utility function $u(x_1, x_2)$ and probability distributions $p_j(x_1, x_2)$ for each alternative S_j, the analysis can proceed exactly as indicated in Section 5.3. Rather than consider that two attributes
are used to measure one objective, it is possible to consider two separate
objectives, "minimize loss of productive wetlands" and "minimize the
disruption of a rare species," each with one attribute. Then we have exactly the situation addressed earlier in this section.

Another case is when one attribute is used for more than one objective.
For example, with a coal power plant, one may have the following objectives: "minimize human health impacts," "minimize agricultural losses,"

and "minimize building degradation." For this problem, it may be that the main mechanism causing such problems is sulfur dioxide, so the single attribute tons of SO_2 emitted per year may be appropriate to measure all three objectives. In this case, one can consider that there is just one objective, "minimize SO_2 pollution" and this case is also identical to earlier one-to-one cases.

A more complex situation occurs when there is overlap among the attributes. Two environmental objectives may be to "minimize the loss of mature forest" and to "minimize the reduction of deer habitat." These objectives may be measured by acres of forest cut and number of deer living in the proposed site area. It may be that the value of part of the forest is due to its deer habitat. If so, then these two attributes overlap. The implication is that the probability distribution for impact measured on one of these attributes will be dependent on the other. Thus, one must be careful not to double-count the impact on deer by counting it within both attributes.

The final case concerns the omission of attributes for all or part of some objectives. For instance, with the objective O, "minimize health impacts," there may be concerns for both mortality and morbidity. Because of the difficulties in measuring morbidity for the alternatives, perhaps only the mortality aspect would be measured by an attribute such as "number of deaths." Then, of course, part of objective O would not be measured. To compensate for this, one can place a higher value on the mortality levels when assessing the overall utility function. This may be appropriate if mortality is highly positively correlated with morbidity for the problem being addressed.

5.5 Desirable Properties of the Final Set of Attributes

The important problem of concern here is what is it that makes a set of attributes appropriate? If one could specify the properties of a set of attributes, what would they be? In a way, this question focuses on a metadecision which must be made in any analysis, namely, the choice of objectives and attributes. Value judgments are of course needed and value tradeoffs will, alas, again have to be made. This decision will usually be made in an informal fashion by the siting team with feedback from the client.

There are five important properties for a set of attributes in a siting problem. It is important for the set to be

1. complete, to cover all aspects of the problem,
2. operational, to be useful for helping to make and justify a siting choice,

3. decomposable, to reduce the complexity and increase the penetration of the decision analysis,
4. nonredundant, to avoid double counting of possible impacts, and
5. minimal, to reduce the time and cost necessary for the study.

A bit oversimplified, completeness refers to what we would like to do, operationalness refers to whether we can do it, and the other three properties refer to how easy it is to do. We will briefly discuss each of these.

5.5.1 COMPLETENESS

A set of attributes is complete if knowing consequence $(x_1, x_2, ..., x_n)$ gives a clear picture of the degree to which the overall objective is met. Completeness then addresses a theoretical concern in the sense that questions about practicality in obtaining information about the $(x_1, x_2, ..., x_n)$ are not an issue. There are implications for each attribute individually which follow from completeness.

Each of the individual attributes should measure the lower-level objective to which it is associated as thoroughly and uniquely as possible. For instance, if an objective is to minimize the aesthetic disruption of a proposed transmission line, simply measuring the length of the line may be inadequate since sections through different areas may have greatly varying aesthetic impacts. A thorough attribute would include this possibility. The pumped storage siting study discussed in Chapter 9 does exactly this by weighting the different environments through which proposed transmission lines pass.

If one intends to evaluate various oil refinery sites, an objective might be "minimize health impacts via air pollution," as indicated in Fig. 5.2. An attribute involving pollution concentrations in the vicinity would not be specific to the refinery since other sources would also contribute to the concentrations. If one used a proxy attribute concerning pollution, rather than a direct attribute for health impact, an indicator of emissions from the refinery would be specific to the proposed facility.

It should be clear that obtaining a complete set of attributes is not in itself a difficult task. A set which would indicate in exhaustive detail what might happen to every individual in the vicinity, to every animal and tree, and so on would certainly be complete. However, it would be completely impractical to collect the necessary data and such data would be unmanageable to use. To some extent, this is the position we are in now in the United States with regard to environmental statements required for most energy facilities because of the National Environmental Policy Act. What is needed is a logical, systematic, and defendable system for aggregating the information into a useful form. This leads to the next property.

5.5.2 OPERATIONALNESS

A set of attributes is operational if it is possible to obtain the information necessary to proceed with the analysis and if the analysis would then provide insights for (1) selecting a best site and (2) justifying the choice to others. The information necessary to proceed is the probability distribution $p_j(x)$ and the utility function $u(x)$. There are several reasons for not being able to obtain this information. It may be too costly or time consuming to gather the data. For instance, analyzing all the possible impacts on a town due to a boom–bust cycle would be prohibitive for these reasons. Summary information may be more appropriate. In other cases, the extent of scientific knowledge limits our abilities to describe some possible consequences: the health effects of pollutant or radiation emissions, and the noneconomic costs (and some of the economic costs) of large-scale solar energy facilities.

Several methods of dealing with the difficulties of obtaining the $p_j(x)$ are found in Sections 5.3 and 5.4. The ideas included the use of proxy attributes, constructed scales, and direct preference measurements. Basically, all of these shift some of the problem of obtaining the $p_j(x)$ into the problem of obtaining $u(x)$.

The main reason for not being able to assess the utility function is usually politics. It may not be politic for a client to be clear and honest about preferences. This is a fault not only of the client, but also of the system by which the actions of the client are monitored and appraised. Some individuals believe that the less one discloses, the less one can be attacked by adversaries. Concomitant with this is the fact that there is less that can be supported by those in agreement. As a result of new laws and regulations, as well as a changing climate of society, the trend is to be more explicit about one's preferences.

5.5.3 DECOMPOSABILITY

It may be difficult to quantify the probability distribution $p_j(x)$ for each site S_j, $j = 1, ..., J$ and the utility function $u(x)$. The preferred solution is to break these assessments into parts which are easier to handle than the whole, and then appropriately integrate these parts. Mathematically, this means we wish to find simple forms of f_j and g such that

$$p_j(x) = f_j[p_j^1(x_1), p_j^2(x_2), ..., p_j^n(x_n)], \qquad j = 1, ..., J, \qquad (5.4)$$

and

$$u(x) = g[u_1(x_1), u_2(x_2), ..., u_n(x_n)], \qquad (5.5)$$

where p_j^i is a probability distribution for attribute X_i, given site S_j is selected, and u_i is a utility function over attribute X_i.

There are various sets of assumptions which lead to forms such as (5.4) and (5.5). These involve probabilistic independence in the case of (5.4) and utility independence and preferential independence in the case of (5.5). Details of the requisite conditions for different specific forms are found in Chapter 6 and 7, respectively, for p_j and u.

One refers to the set of attributes as being decomposable if forms such as (5.4) and (5.5) are appropriate. In some cases, the attributes can be only partially decomposed. Even this may be of considerable help in the analysis. For instance, a probability distribution such as

$$p_j(x) = f_j[p_j^1(x_1, x_2), p_j^3(x_3), \ldots, p_j^n(x_n)], \tag{5.6}$$

where there is a probabilistic dependency between attributes X_1 and X_2, may hold. A corresponding example of a partially decomposable utility function would be

$$u(x) = g[u_1(x_1), u_1'(x_1), u_2(x_2), \ldots, u_n(x_n)], \tag{5.7}$$

where two separate single-attribute utility functions over X_1 are needed because relative preferences over X_1 levels could depend on the levels of the other attributes.

5.5.4 NONREDUNDANCY

The set of attributes should avoid redundancy, that is double-counting, of impacts in the evaluation. This can be rather tricky because double-counting can occur in two ways. One involves double-counting of the impacts *per se* and the other involves double-counting of the values for the impacts. Some examples should clarify the distinction.

Suppose two attributes used to measure the environmental impact of an Arctic pipeline were the number of species whose habitat was disrupted by the pipeline and the acres of caribou habitat removed. Because the caribou is a species, this would involve double-counting of the impact. Such a problem could be corrected merely by changing the first attribute to the number of species, except caribou, whose habitat was disturbed.

Cases of value double-counting often result from including both means and ends objectives in the objectives hierarchy. Suppose for a refinery siting hierarchy there was an objective to reduce SO_2 emissions and another to minimize impact on the health of individuals in the neighborhood of the facility. If it were the case that the only adverse impact of the SO_2 pollution was to the health of individuals and also that the only adverse health impacts were due to SO_2 pollution, then any utility function which included both attributes would be a clear case of double-counting.

However, what is more likely to be the case is that only some of the impact of SO_2 pollution is related to health effects and that only some of the health effects are from SO_2 pollution. Then the utility function may be double-counting some of these impacts. The degree to which this actually occurs is difficult to ascertain, as can be illustrated by the refinery objectives hierarchy of Fig. 5.2. Note that the individual health effects of pollution are separated from the air pollution impacts to fauna and flora on land. The latter may be measured by an attribute of, say, tons of SO_2 emitted per year. In assessing the resulting utility function, if the levels of this attribute are evaluated only as they pertain to the flora and fauna impact, then there is no double-counting. If they are evaluated including health effects, there is double-counting. The analysts must both make sure the client understands the purpose of each attribute before assessment and carefully monitor the utility assessment process itself to avoid such problems.

5.5.5 MINIMALITY

Generally, the smaller the number of attributes in a set, the easier it will be to conduct the analysis. There will be fewer probability distributions over the single attributes and fewer single-attribute utility functions to assess if the attributes are decomposable. If they are not, the dimensionality of the problem is less with fewer attributes. However, one motivation for subdividing attributes—i.e., building lower levels of the objectives hierarchy—is to obtain operational attributes. Hence, operationalness and minimal size are often in conflict for a particular siting study.

At the extreme, one can have just one objective, the overall objective such as "maximize the well-being of society" in Fig. 5.2, and one associated attribute. This attribute could be defined to be complete, and, of course, it is nonredundant, of minimal size, and decomposable by definition. It is, unfortunately, far from being operational. Any resulting analysis would necessarily be carried out in one's head and that is what we wish to avoid. The analysis is meant to bolster the client's intuitive processes, not to force total reliance on them.

This subsection should clarify the necessary value tradeoffs which the siting team must make in selecting a set of attributes. It is invariably the case that improvement in some of the desired properties of the set can come only at the expense of others.

Once the set of attributes is defined, they and not the objectives *per se* are used in the ensuing analysis. For each candidate site S_j, a probability distribution $p_j(x)$ is specified, which describes the possible consequences and their likelihoods, and a utility function $u(x)$ is assessed, which quantifies the client's value structure for consequences. The methodology and procedures for completing these tasks are discussed in detail in Chapters 6 and 7, respectively.

CHAPTER 6

DESCRIBING POSSIBLE SITE IMPACTS

In this chapter we are concerned with how the candidate sites measure up in terms of the objectives. The setting is as follows. The candidate sites S_j, $j = 1, \ldots, J$ have been identified as well as an objectives hierarchy with attributes X_i, $i = 1, \ldots, n$ associated with the lowest-level objectives. Thus the consequence of a decision to locate an energy facility at site S_j can be described by the vector $x(j) = (x_1(j), \ldots, x_n(j))$, where $x_i(j)$ is a level of attribute X_i at site S_j.

If there are no uncertainties in the problem, meaning that the impacts for each site are known, the task of selecting a best site is reduced to selecting the best $x(j)$. First, however, the $x_i(j)$ must be determined for each site and attribute. The process of making these determinations is one topic addressed in this chapter.

More often it is not possible to estimate accurately the consequences associated with a particular site. Because of lack of information or the excessive costs of obtaining it, environmental impacts will not be precisely known. Uncertainties about the occurrence of both natural hazards (e.g., earthquakes) and human-caused accidents imply uncertainties about health and safety impacts. The complexity of social processes results in uncertainties about socioeconomics and public attitudes. And it is universal knowledge that costs can vary greatly from deterministic estimates. Because of their importance, it is appropriate to include uncertainties in siting evaluations. One can describe the uncertainty using probabilities. Each site S_j is characterized by several possible consequences x_j and their associated likelihoods of occurrence. Formally, the possible impacts at site S_j are quantified by a probability distribution $p_j(x)$, $j = 1, \ldots, J$.

As outlined in Section 1.4, the long time horizons often applicable to the siting of energy facilities complicate the description of possible impacts. This results from two distinct factors. One is that the same impact occurring at different time periods is not necessarily evaluated the same. This is clearly the case for economic costs: 1 million dollars cost today does not have the same value as 1 million dollars cost in 10 years. The second factor is that we usually have greater uncertainties about impacts years in the future than in the short term. As a result of these factors, it may be appropriate to have the probability distribution describe possible impacts over time.

Outline of the Chapter

Section 6.1 describes the methods for quantifying the possible impacts of any particular site. The purpose is to familiarize one with the different approaches and procedures. These are presented only in summarized fashion both to indicate the basic ideas as effectively as possible and because a voluminous literature exists on the subject for anyone wishing more details. Appropriate references are given for each of the approaches.

The rest of the chapter is concerned with illustrating the methods. Sections 6.2–6.6 concern the impacts with respect to the environment, economics, socioeconomics, health and safety, and public attitudes. General approaches to describing impacts specific to those concerns, as well as short descriptions of applications, are indicated. Sections 6.7 and 6.8 present more detailed applications that both combine the methodological approaches suggested in Section 6.1 and illustrate the procedures on important problems which are not commonly addressed in a formal manner.†

6.1 Methods for Quantifying Impacts

There are essentially three different approaches used to describe the possible impacts at any site. These are:

1. collecting and synthesizing data,
2. modeling the processes leading to site impacts, and
3. assessing professional judgments directly.

† Sections 6.2–6.8 can be skipped without a loss of conceptual continuity. These sections all provide detail on how to quantify impacts as well as case studies.

Often, combinations of these are used. For illustration, each of the approaches will be considered separately. Examples throughout the chapter illustrate their combined use. This section concludes with discussions of methodological and practical issues relevant to implementing the approaches.

6.1.1 COLLECTING AND SYNTHESIZING DATA

Sometimes there is very little or no uncertainty about a particular impact, given that site S_j is chosen. Examples in the WPPSS nuclear siting study of Chapter 3 were the site population factor attribute and the length of intertie line attribute. Usually for attributes with essentially no uncertainty, one can collect the data to indicate the impact. In the WPPSS case, the population at various distances up to 50 miles from the proposed plant was tabulated. Then a simple calculation using (3.1) gave us the site population factor level for the site. The intertie line distance was measured from a map indicating the candidate sites and the main electrical transmission grid in the Pacific Northwest.

When it is appropriate to include uncertainties, data collection alone is not usually sufficient to quantify the probability distributions. The data would probably be used in conjunction with the other approaches. For instance, in the WPPSS study, data was collected about the environmental conditions at the candidate sites, the number of salmon annually using nearby rivers for spawning, the size and fiscal condition of towns near sites, and so on. This information was used in assessing site biological impact, salmonid impact, and socioeconomic impact. Section 6.2 has more details on the first two of these. With the site biological impact and the socioeconomic impacts, the probabilistic assessments were made using direct professional judgments and the available data. A small model was built to incorporate the salmonid data in assessing impacts. A different illustration of socioeconomic impacts using a more detailed model is found in Ford [1979] and briefly discussed in Section 6.5.

6.1.2 MODELING THE PROCESSES LEADING
TO SITE IMPACTS

Modeling is used for quantifying both deterministic and probabilistic descriptions of impacts. In the pumped storage siting study discussed in Chapter 9, an economic model is used to specify a deterministic estimate of first-year equivalent cost for each site. The input to the model includes capital costs of constructing the facility and transmission lines, acquiring necessary land, engineering design, construction management, overhead, surveying and licensing, operation, and so on. The output of the model is one number representing first-year equivalent cost. A similar deterministic model of economics is used in the WPPSS study.

Because there are so many uncontrollable factors influencing the costs of energy facilities, we feel it is more appropriate to build a probabilistic model of costs. Such a model would have, as inputs, information about land costs, geology and seismology, prices of construction materials, possible delays in licensing various aspects of the facility, productivity of construction workers, the weather (e.g., heavy rains or snow could slow construction), interest rates, and several other factors. Clearly, each of these factors could have a significant effect on the cost of a proposed facility, and yet each might have large uncertainties. The uncertainties can be quantified with either available or collectable data and/or the professional judgments of experts.

The probabilistic model may be either analytic or simulation. It would involve describing the relationship between the set of input variables and the output variables which would be the economic attributes for the problem. This may include both the construction and operation costs of the facility year by year over the lifetime of the facility. Given that the inputs are probabilistically described, the probability distributions for the outputs will come from the model.

A graphical illustration of a probabilistic impact model is shown in Fig. 6.1. This figure is a summarized view of step 3 of decision analysis (i.e., describing possible impacts of each site) illustrated in Fig. 2.1. Professional judgments must be made about what input variables to include, how they relate to attributes, and the probability distributions over the input variables. There are literally hundreds of books and articles devoted exclusively to the construction of impact models. These include most texts on simulation, probabilistic methods, and statistics. Throughout this chapter, reference will be made to specific works relevant to the modeling of site impacts.

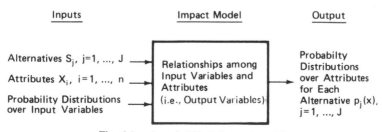

Inputs	Impact Model	Output

Alternatives S_j, $j=1, ..., J$ → | Relationships among Input Variables and Attributes (i.e., Output Variables) | Probabilty Distributions over Attributes for Each Alternative $p_j(x)$, $j=1, ..., J$

Attributes X_i, $i=1, ..., n$ →

Probability Distributions over Input Variables →

Fig. 6.1.   A probabilistic impact model.

6.1.3   ASSESSING PROFESSIONAL JUDGMENTS DIRECTLY

Usually, if direct assessment of professional judgments is required, there is uncertainty surrounding the assessments. Thus, the judgment would be quantified using a probability distribution. There are basically two ways to proceed. One is to assess the entire distribution directly and the other is to select a standard probability distribution and directly assess parameters for this form. Because one of the major distinctions of decision analysis *vis-à-vis* other approaches is the formalization of professional judgments, and because such judgments are necessary for handling several crucial aspects of siting studies, we shall spend more time on this technique than the previous two.

Directly Assessing a Discrete Probability Distribution

The procedures used for variables with discrete and with continuous possible levels are slightly different. The discrete case is simpler so we will begin with it.

Let us suppose the attribute (or input variable) is denoted by Y with possible discrete levels y_k, $k = 1, ..., K$. In some cases the y levels may be ranges. However, the y_k are such that one must occur and only one can occur. That is, they are mutually exclusive and collectively exhaustive. The task is to determine the probability P_k, $k = 1, ..., K$ that level y_k actually occurs. Theoretically, the problem is straightforward. We simply ask the expert whose judgment is being quantified what the P_k are. However, in practice, the expert may require help in articulating these judgments and in insuring their internal consistency. For the typical case where significant help is needed, the assessment process may involve the following.

First, the likelihood of the various individual y_k levels is ordered from most to least likely. Let us suppose the order is $y_1, y_2, ..., y_k$. Then of course, barring ties,

$$P_1 > P_2 > \cdots > P_K. \tag{6.1}$$

If the likelihood of y_k and y_{k+1} is the same, clearly

$$P_k = P_{k+1}. \tag{6.2}$$

The next step may include some joint comparisons. For instance, the expert may feel that y_1 is more likely to occur than either y_2 or y_3. If so, then

$$P_1 > P_2 + P_3. \tag{6.3}$$

There is, in addition, the fact that the sum of these probabilities must be one since the y_k are mutually exclusive and collectively exhaustive, so

$$P_1 + P_2 + \cdots + P_K = 1. \tag{6.4}$$

Collectively, a set of equations such as (6.1)–(6.3) in conjunction with (6.4) can significantly restrict the ranges of the possible probabilities. And this is accomplished using only qualitative comparisons about which outcome is more likely.

To pinpoint the probabilities usually would require quantitative estimates about how much more likely one outcome is than another. For instance, the expert may feel that y_2 is twice as likely as y_4. Then

$$P_2 = 2P_4. \tag{6.5}$$

It is more often the case that an external characterization is used to calibrate or assess the P_k. An illustration of such a device is shown in Fig. 6.2. The expert is offered a choice of two lotteries, one related to the problem at hand and the other completely unrelated. If the expert chooses Lottery I, he or she wins a desirable prize if y_k occurs and otherwise nothing. If Lottery II is chosen, the probability of winning the prize is q and of not winning, $1 - q$. The probability q can be set externally by using an urn full of red and white balls. For example, if 40 of 100 balls are red, the probability q can be 0.4 and equal to the chance of drawing a red ball in a fair draw. By varying the number of red and white balls, the probability q is varied until the expert is indifferent to the two lotteries. Let us define this indifference probability as q_k. Because indifference should occur when the likelihoods of getting the prize with either lottery are equal, the probability of y_k occurring should be q_k, so

$$P_k = q_k. \tag{6.6}$$

Rather than set q very close to what might be expected for P_k, it is usually better to start far away. If we thought P_k might be approximately 0.2, it may be appropriate to set $q = 0.9$ in Fig. 6.2 and ask the expert to

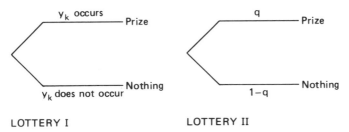

LOTTERY I   LOTTERY II

Fig. 6.2.   Assessing the probability of outcome y_k.

choose. Of course, in this situation Lottery II should be chosen. This would both help to insure we were communicating correctly and allow the expert to begin with an easy question. This aids the expert in thinking about the difficulty of the task ahead. The next level of q for comparison might be 0.6. Again, Lottery II should be preferred. This directly implies that $P_k < 0.6$. Then we may jump to $q = 0.05$ and if P_k is actually near 0.2, then the preference changes to Lottery I, and of course, it follows that $P_k > 0.05$. The probability q is changed upward when Lottery I is preferred and downward when Lottery II is preferred. Using this converging technique, the expert must eventually be indifferent. If this occurs at $q = 0.25$, then it follows that $P_k = 0.25$.

As a result of some equations such as (6.1)–(6.6), very tight bounds or exact estimates may be specified for P_1, ..., P_K. It would then be appropriate to conduct some consistency checks. This basically consists of asking more questions similar to those indicated above and verifying whether the assessed P_k conform to the responses to the questions. If the expert says that y_1 is at least three times as likely to occur than y_2, then $P_1 > 3P_2$. If either y_1 or y_2 is more likely to occur than all the other possible levels, then $P_1 + P_2 > 0.5$. And if y_6 is 10% more likely than y_7, clearly $P_6 = 1.1P_7$. When inconsistencies occur, adjustments must be made to arrive at a consistent set of assessments.

It is not unlikely that inconsistencies will be found. In fact, a major reason for formalizing such judgments is to identify and eliminate inconsistencies. Basically, most siting problems are too involved to expect individuals to keep all the necessary information consistently in their heads. Just as most people would prefer to multiply eight numbers on paper rather than mentally for fear of mistakes, it seems reasonable to write eight probabilities (i.e., a case with $K = 8$) down on paper to appraise, update, and eliminate any inconsistencies.

Directly Assessing a Continuous Probability Distribution

For the continuous case, let us consider the attribute (or input variable) denoted by Y with levels y. We want to determine the probability that the level of Y is less than or equal to y for every possible y.

To provide a concrete example, we shall use a recent study of oil terminal and marine service base sites on Kodiak Island, Alaska.† These facilities would support proposed offshore oil development in the Western Gulf of Alaska. One environmental concern was the disturbance of sea birds with a habitat in the vicinity of the proposed terminal. The attribute was

† See Section 7.7.3 for a brief description of this study.

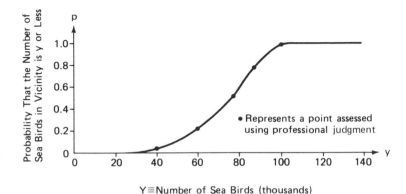

Y≡Number of Sea Birds (thousands)
Fig. 6.3. A judgmentally assessed cumulative probability distribution.

the number of sea birds in the vicinity. If it were necessary to obtain a probability distribution for this attribute, we would want to obtain a curve like the one illustrated in Fig. 6.3. The question is, how does one obtain the curve.

For illustration, suppose the local game warden is very familiar with the area and the birds making their habitats there. And to simplify gender references in the next few paragraphs, suppose the game warden is female. We, as analysts, wish to quantify her professional judgments about the number of sea birds present.

To begin, we might ask for an estimate of the number of birds such that she would be sure the actual number would exceed this estimate. Clearly, zero birds is one "correct" answer, but we want to raise the estimate as high as possible. Suppose she responds 25 thousand. This becomes our first-cut lower bound. Next, we ask for an upper estimate which she believes is higher than the actual number. This estimate might be 120 thousand. Each of these estimates is subject to change as we carry out the assessment process.

Next we ask for a median estimate, denoted by $\hat{y}$, such that she believes the probability that y is less than $\hat{y}$ is 0.5. We could ask for such an estimate directly or use a device analogous to that in Fig. 6.2 as an aid. This is shown in Fig. 6.4. In Lottery III, the game warden obtains a desirable prize if the true number of birds y is less than or equal to an arbitrarily chosen y'; otherwise, she will receive nothing. With Lottery IV, she has a 0.5 chance of either winning the prize or not winning it. This probability can be supplied by the flip of a fair coin.

The idea is to begin with easy choice questions using estimates of y' and to proceed to more difficult ones. The two lotteries must be indifferent when the probability that $y \leq y'$ is 0.5, and by definition this implies

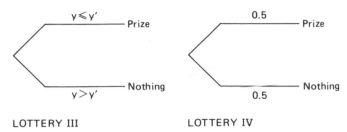

LOTTERY III   LOTTERY IV

Fig. 6.4.   **Assessing the median of a continuous probability distribution.**

$\hat{y} = y'$. To begin, y' may be set at 30 thousand. The game warden would probably choose Lottery IV since 30 thousand is only slightly over her minimum estimate of 25 thousand. When Lottery IV is chosen, the estimate of y' is increased, say to 50 thousand. Still, Lottery IV is preferred. Now y' is set at 90 thousand and the game warden prefers Lottery III. Then y' is dropped, perhaps to 70 thousand. Here Lottery IV may be preferred, but at 75 thousand there is no preference. This means

$$P(y \leq 75) = 0.5 \qquad (6.7)$$

so $\hat{y} = 75$ thousand. The point corresponding to $y = 75$ and $P = 0.5$ is then plotted on a blank graph with the axes of Fig. 6.3.

In technical terminology, the level $\hat{y}$ is referred to as the 0.5 fractile. We wish to obtain other fractiles at 0.01, 0.25, 0.75, and 0.99 and plot these on the graph. This can be done exactly as above by changing the 0.5 in Lottery IV of Fig. 6.4 to 0.01 for the 0.01 fractile and so on. Suppose the 0.01, 0.25, 0.75, and 0.99 fractiles are assessed to be 40, 60, 85, and 100 thousand, respectively. Then

$$P(y \leq 40) = 0.01, \qquad (6.8)$$

$$P(y \leq 60) = 0.25, \qquad (6.9)$$

$$P(y \leq 85) = 0.75, \qquad (6.10)$$

and

$$P(y \leq 100) = 0.99. \qquad (6.11)$$

The points corresponding to these are also plotted in Fig. 6.3. From these, the curve illustrated in the figure is drawn. This curve is the desired cumulative probability distribution for the number of sea birds inhabiting a specified area, as viewed by the game warden. From it, one can read the probability that the number of sea birds is less than any specified amount.

As in the discrete case, care should be taken to insure that the judgments are consistent. For instance, from Fig. 6.3, it follows that the likelihood that the true number of sea birds is less than 60 thousand equals the likelihood that the true number is between 60 and 75 thousand. Another implication is that, given the number of birds is greater than 75 thousand, it is equally likely that it is above or below 85 thousand. Also, reading from the curve, the likelihood that less than 50 thousand birds inhabit the area should be about 10% (i.e., a probability of 0.1). When inconsistencies occur, as in the discrete case, changes in some assessments should be made, the new implications appraised and, if necessary, more alterations made. The process stops when all the judgments are consistent.

Assessing Parameters for a Probability Distribution

There is often reason to believe that the attribute (or variable) of interest might be adequately described by a standard probability distribution such as the normal or Bernoulli probability distribution. Then to specify the entire distribution, one need only determine the parameters of the chosen distribution.

Let us illustrate the idea with an example. Suppose for the previous assessments of the sea bird populations, it was decided that a normal distribution would be appropriate. The cumulative probability distribution in this case is

$$P(y \leqslant y') = \int_{-\infty}^{y'} \frac{1}{\sigma\sqrt{2\pi}} \exp -\frac{(y - \bar{y})^2}{2\sigma^2}, \tag{6.12}$$

where $\bar{y}$ and σ are the mean (i.e., average) and standard deviation of Y. These can be assessed in a variety of ways. For example, given the 0.5 and 0.25 fractiles from (6.7) and (6.9), substituting into (6.12) gives two equations with unknowns $\bar{y}$ and σ. These can be solved using standard tables for $\bar{y}$ and σ.

Alternatively, it may be possible to ask for assessments of $\bar{y}$ and σ directly. Because the normal probability distribution is symmetrical, the mean and median are the same. Because, in addition, the distribution has only one mode (i.e., most likely level of Y), this is also equal to the mean. Hence, one could ask the game warden for the most likely level of Y and this would become $\bar{y}$ in (6.12). The standard deviation has no similar intuitive interpretation other than that the likelihood is approximately 0.64 that the true number of birds falls within one standard deviation of the mean. It is possible to ask directly for such an estimate, but it is probably easier for the professional to respond to a fractile question from which the standard deviation can be easily computed. The assessments for the salmonid impacts for the WPPSS study contained in Section 3.3.2 were done by specifying the mean and standard deviation for a normal probability distribution. These assessments are discussed in more detail in Section 6.2.3.

In some problems, it is a probability itself that is the desired parameter of a distribution. This is the case, for example, with the Bernoulli probability distribution. Here, a situation occurs where an event either does or does not happen. We want to know the probability that it will happen. As an illustration, the risk analysis of a liquefied natural gas (LNG) import terminal, discussed in Section 6.8, includes the possibility of ship collisions leading to LNG spills. Such a spill can either be ignited immediately or not. This probability can be calibrated using the device illustrated in Fig. 6.2 with y_k defined as "ignition of the spill."

Summarizing a Probability Distribution

In some cases where it is either extremely difficult or impossible to collect data and/or develop a model of the relevant process, it may also be difficult for the professional to express complete probability distribution functions. It may be sufficient simply to summarize such information with a "best estimate." Then with sensitivity analysis, we can examine whether it appears worthwhile to address the uncertainties formally later in the analysis.

One example of a best estimate of a professional judgment, generalized from the sea bird illustration, might be the number of a particular species living in the vicinity, say within 5 miles, of a proposed energy facility. Another example used in the appraisal of potential attributes for the WPPSS study involved describing the number of households near proposed sites which would experience a certain number of days of fog due to the facility. A meteorologist utilized professional knowledge of the local weather plus geographical information and data about the proposed cooling system to draw contours of areas expected to have 1, 2, ... days yearly of additional fog. Superimposing these areas on a map of the area, we simply counted up the number of houses involved to specify a best estimate. No formal meteorological model was developed; the integration of information was done mentally with the meteorologist using a mental model.

Because impacts involving no uncertainties are a special case of impacts involving uncertainties, much of what was said about consistency and assessment technique in the general case is also relevant to the special. There are several excellent references about how to assess information, and particularly probability distributions, from knowledgeable individuals. Winkler [1967a,b], Raiffa [1968], Schlaifer [1969], Savage [1971], Tversky and Kahneman [1974], Brown, Kahr, and Peterson [1974], Spetzler and Stahl von Holstein [1975], and Seaver, von Winterfeldt, and

Edwards [1978] all have detailed discussions of various procedures and practical insights helpful for implementation.

6.1.4 METHODOLOGICAL ISSUES RELEVANT TO IMPACTS

There are two main concerns which we wish to address here, probabilistic dependencies and biases or errors in judgments. Both are important to consider in an analysis.

Probabilistic Dependencies

The complexity of probabilistic dependencies was introduced in Section 2.3.3. The term means, given a particular site, the knowledge of the level of one attribute will influence the judgments about the level of another attribute. It is important to recognize that this dependency must be conditioned on a particular site. It may be that two (or more) attributes are dependent at one site and not at another. Let us suggest two examples where dependencies might be expected to occur, and two different ways of dealing with them.

One example can be drawn from the WPPSS study of Chapter 3. Recall that the salmonid impact for each site was indicated by two attributes: the number of salmonid in the stream and the percent lost annually due to entrainment, entrapment, etc. For the WPPSS case, the number of salmonid in the stream was deterministically estimated from historical escapement data, whereas the percentage loss was probabilistically assessed using professional judgments. Suppose that it had been necessary to assess both probabilistically.

For illustration, let us define attributes Y, number of fish in the stream, and Z, percent lost. The levels will be denoted by y and z. It may be that as estimates of the number of fish increase, the estimates of the percentage loss would decrease. (Note that this could occur and yet the number of fish lost could increase as the number of fish in the stream increases.) This dependency might occur, for instance, because more fish implies more water in the stream which implies that a smaller percentage is used for cooling.† It follows that thermal pollution per unit of water would decrease, reducing the likelihood that any particular fish is negatively impacted. This reduces the percentage loss of fish. To handle such a dependency, we could first assess a probability density function $f_Y(y)$ for the

---

† The data collected for the WPPSS study in Table 3.5 for the several candidate sites indicates such a dependency is plausible. The larger the number of fish in a stream, the smaller the percentage of fish lost.

number of fish in the stream. Then for any particular y in that stream, we would need to assess a conditional probability density function $f_{Z/Y}(z/y)$ for the percentage lost. The joint probability density used in the analysis is

$$f(y, z) = f_Y(y)f_{Z/Y}(z/y). \qquad (6.13)$$

From $f(y, z)$, we can calculate the joint probabilities of y and z being in any specified range.

A second example indicates that dependencies often result from impacts on different attributes being caused by a common means. Consider the siting of a large coal-fired power plant. Suppose two attributes in the problem are Q and R, measuring the number of days of sickness to individuals with respiratory ailments and the number of individuals in the vicinity irritated by the pollution, respectively. We want to assess the joint probability density $h(q, r)$, but this is difficult to do directly because of the probabilistic dependencies. Without these, we could determine the single-attribute densities h_Q and h_R from cumulative distribution assessments conducted as discussed in Section 6.1.3.

The dependency occurs because both Q and R depend on the sulfur dioxide pollution, which will be denoted S with levels s. The solution is to directly assess a probability density h_S over pollution levels and then conditionally assess probability densities $h_{Q/S}$ and $h_{R/S}$ for Q and R conditional on S. This does assume that Q and R are conditionally probabilistically independent given S, but this is a less stringent assumption than outright independence. Given the assumption, then

$$h(q, r) = \int_s h_{Q/S}(q/s)h_{R/S}(r/s)h_S(s) \, ds. \qquad (6.14)$$

This expression† simply weights the various probabilities of q and r pairs, given s, by the probability of s. Standard probability texts, such as Feller [1950], Parzen [1960], and Drake [1967], address probabilistic dependencies in detail.

These examples mainly concern the situation where elicitation of professional judgments is used to provide the probability distributions. When adequate data are available or a model of the process of concern is developed, no special techniques need be used to quantify the dependencies. With data collection, an appropriate tabulation of the data should indicate the dependencies. The data collected must indicate the occurrence of

---

† A major advantage of an expression such as (6.14) is that it provides a mechanism for encoding and logically combining information from different professionals. In this example, an engineer may be best informed to provide h_S about sulfur dioxide emissions, a medical doctor might provide $h_{Q/S}$ about sickness, and a sociologist might provide $h_{R/S}$ about public irritation.

various combinations of levels of attributes (or variables) rather than iso-late one attribute at a time. An impact model should include the mecha-nisms which lead to the resulting dependencies. If this is properly han-dled, the model output will indicate the probabilistic dependencies.

Biases in Judgments

It is well documented that there are biases that often influence profes-sionals in articulating their judgments. Tversky and Kahneman [1974] provide an excellent discussion of many of these biases. Here we will summarize some of the important ones to illustrate the potential problems. It is helpful for discussion purposes to divide the biases into those that are recognized by the professional whose judgment is being elicited and those that are not.

Sometimes a professional deliberately gives biased judgments. These invariably stem from value judgments of the professional. For example, a professional may prefer site S_1 to site S_2 and thus bias the cost estimates for S_1 downward and/or those for S_2 upward. Another reason for biasing a judgment is to sabotage an analysis or, perhaps, a team of analysts. If "odd" responses are given for judgmental data, then the analysis may re-sult in unbelievable or unjustifiable implications. The root of both these difficulties is the professional. In either case, such biased information could be provided regardless of the methodology used. Having it clearly articulated perhaps increases the chance that another professional or the client may recognize its inappropriateness.

Another source of deliberate bias results from reward systems which promote dishonesty. If an architectural engineering firm says it can build a facility for 100 million dollars, rather than the 125 million they really be-lieve is required, the firm has a greater chance of being awarded the con-struction contract. On the other hand, they do not get much of a reward for building a facility for 100 million dollars when the estimate was 125 million. Because such reward systems *de facto* promote initial low cost estimates, it is rational that professionals bias their initial estimates toward the low side. Reward structures have been developed which do promote honesty in providing assessments (see, for example, Winkler [1969]). These systems, however, seem more appropriate for cases where professional judgments are repeatedly assessed, such as weather fore-casts. With siting studies, which are unique events, the mechanism which will reduce this particular bias is a management enlightened in probabi-listic reasoning. Professionals and analysts could tremendously influence this process by making clear the assumptions under which judgments are made and by making judgments for alternative sets of assumptions.

A common bias which occurs in several contexts but which is not rec-

ognized by professionals providing judgments, is referred to as an-
choring. It pertains to situations where a series of estimates is made. It
happens that estimates later in the series are closer to the earlier estimates
than they should be. These estimates may be fractiles, a parameter in a
model, a probability, and so on.

As an example, refer to Fig. 6.3 where we discussed assessing the
number of sea birds in the vicinity of a proposed oil terminal and marine
service base. The first assessed point was the 0.5 fractile of 75 thousand
birds given by (6.7). Next, the 0.25 fractile was determined. The usual
mental process is to start at 75 thousand and reduce this until the 0.25
fractile is reached. A large amount of data show that individuals do not
move far enough from the starting point, referred to as an anchor, in this
case the 0.5 fractile of 75 thousand.

Another example involves estimating water acquisition costs for the
various sites. Since the same process will probably be used for other sites,
once such an estimate is calculated for one site, the other estimates are
likely to be biased toward the original anchor estimate.

Availability is another mechanism which biases professional judgments. In an analysis of potential oil import terminals, an estimate of the
amount of oil spilled in a possible collision may be needed. If there had recently been a large, well-publicized oil spill, such an estimate would probably be larger than otherwise. It is the availability of the event in the memory when making the assessments which biases the estimates. This is in
addition to a possible increase in the estimate simply due to data from one
additional spill.

There are other types of biases discussed in Tversky and Kahneman
[1974] which influence the articulated judgments of professionals. By
understanding the heuristic procedures which professionals use in making
their estimates, it is easier to ascertain when and how the procedures
influence judgments. By forcing the professional to consider the problem
from another viewpoint, it may be possible to remove the biases. An example of this is the convergence technique illustrated in Section 6.1.3 for
arriving at estimates by approaching them from both sides.

6.1.5 PRACTICAL ISSUES RELEVANT TO QUANTIFYING IMPACTS

For almost all problems, there is a combination of data, models, and
professional judgments used to determine the possible impacts of siting at
each candidate site. These may be used separately for some attributes and
in conjunction for other attributes.

In every siting problem, and indeed every decision problem, professional judgments must be used either indirectly or directly. If the data are available for quantifying impacts, then professional judgments must be used in selecting what data to use and how much to collect. If a model is built to indicate possible impacts, then professional judgments are used both to construct the model and to provide the input information. And, of course, if direct assessment of information is used on the problem, it is obvious where the professional judgments are utilized.

Judgments must also be made concerning the appropriate balance of these three methods of quantifying impacts. The proper balance will depend on the time and cost required to do each, as well as the probable effect various parts of the problem will have on the eventual choice of a site. Since such judgments must be made before the results of the analysis are known, it is appropriate to reappraise these decisions made by the siting team after the first round of results is available. As a consequence, some parts of the analysis should sometimes be repeated with more data, a better model, or more careful direct judgmental input.

Sensitivity analysis should help considerably in deciding if and where to improve the overall quantification of impacts. For instance, it may be readily apparent that three of nine sites simply do not measure up to the other six, even making allowances for those which appear inferior. Then, of course, no further effort should be expended on the three. If two professionals disagree on some judgments, the evaluation of sites, which follows the quantification of the value structure for impacts (see Chapter 7), will help to indicate whether or not it is worthwhile attempting to reconcile these judgments.

Every method for evaluating sites for energy facilities uses some data, some models, and considerable professional judgment. When they are used properly, all the methods, including decision analysis, make it clear where the data come from and how and why the model was constructed. The main distinction in these methods for specifying impacts is that decision analysis alone clearly articulates the professional judgments which must be made in any siting study. With these judgments made explicit,

1. it is more likely that biases, deliberate or unintentional, will be detected and corrected;
2. it is more likely that the siting model will help identify the best site for the client; and
3. it is more likely that the justification of the site selection process and the site itself will be accepted by regulatory authorities.

These circumstances lend credibility to the claim that decision analysis provides a much better approach for the selection of sites for energy facilities.

In Sections 6.2–6.6 we will focus on the five general concerns, environ-

ment, economics, socioeconomics, health and safety, and public attitudes. The general approaches outlined in Section 6.1 for assessing impacts are used for all five concerns. However, some approaches are more
appropriate in some cases than in others. The class of models developed
for each of the general concerns varies in detail and sophistication, the
availability of natural indices varies greatly, and the ease with which
appropriate data can be collected are different for the different concerns.
The purpose of Sections 6.2–6.6 is to indicate the main distinctions and to
summarize some applications. Sections 6.7 and 6.8 then present in detail
two important applications.

6.2 Quantifying Environmental Impacts

There are several ways in which the environmental impact of a particular site might be measured. For example, the environmental impact might
be categorized as floral and faunal, aquatic and terrestrial, during construction and during operation of the facility, or via air, water, and land
pollution. Depending on the choice, the attributes selected and hence the
procedure for assessing impacts may vary.

6.2.1 MEASURING ENVIRONMENTAL IMPACT

Basically, the environmental impacts of concern are either on the facility site itself or nearby. The mechanisms for these impacts are different.
Site impacts are due to construction and existence of the facility. Nearby
impacts are often, but not always, due to operation of the facility and are
caused by pollution or the existence of more people in the vicinity.

The site impacts can be measured either by the proxy attributes
describing the amount and type of land and water resources disturbed or
by the direct attributes describing the amount and types of particular biological species disturbed or eliminated. It is, of course, easier to gather data
for the proxy attribute, but the assessment of preferences (covered in
Chapter 7) may be more difficult in this case. The choice of which to use
depends on the problem and must be made by the siting team.

There would appear to be at least three approaches for measuring the
nearby environmental impacts. For instance, if these result from pollution, one could define attributes concerning either pollution emissions,
ambient pollution concentrations, or the number and type of species disturbed. For example, for a coal-fired power plant, one is obviously concerned about possible impact due to sulfur dioxide. This may be the case

whether or not scrubbers are used. Anyway, one proxy attribute may be the amount of sulfur dioxide emitted, measured in tons per year. Another may be the ambient sulfur dioxide level, measured in parts per million. The natural attributes may be the pine forest disturbed, measured in acres, and the number of game birds in the region of increased pollution. Any one of these three approaches for measuring nearby environmental impacts is sufficient. To include more than one approach might result in double-counting.†

6.2.2   Models of Environmental Impact

To elaborate on this case, refer to Fig. 6.5 which illustrates a generic pollution impact model. There are pollutant emissions that, as a result of weather conditions, cause pollutant concentrations. These concentrations and the biological processes result in the environmental disturbance. Let us generically define attributes for emissions, concentrations, and disturbance as E, C, and D, respectively. Levels of these will be denoted by e, c, and d. We have indicated that environmental impact can be measured by any of these three attributes. Let us consider each.

If emissions are used as the attribute, then an engineer familiar with the proposed plant might describe impacts by the probability distribution $f_E(e)$. The uncertainties, which may be rather small relative to the average emission, would result from uncertainties about fuel purity and mitigation (e.g., scrubbers) effectiveness. The greatest difficulty would be in assessing a utility function $u_E(e)$ over different e levels. The individual being assessed would need to remember the possible weather conditions, the biological processes resulting in environmental disturbance, and the relative desirability of the possible disturbance levels. This is, needless to say, a great deal of information to know and keep straight in one's head.

If concentrations are used as the attribute, several questions automatically arise. Where should concentrations be measured, how should concentrations at different locations be integrated, and how should concentrations over time be addressed. Let us assume these questions are answered and the result is the single attribute C in Fig. 6.5. We need a probability distribution $f_C(c)$ and a utility function $u_C(c)$. The probability distribution might be directly assessed, but it may be more appropriate to

---

† "Might result" is used rather than "would result" in double-counting for the following reason. Sulfur dioxide emission could be used as a proxy attribute for the impact on the game birds only, and acres of pine forest could be used for the forest. This does require careful consideration during the evaluation phase to avoid "overweighting" sulfur dioxide emission because of the forest impact.

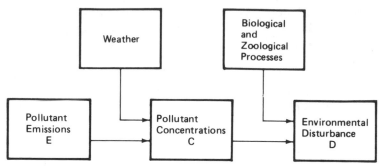

Fig. 6.5. A generic pollution impact model.

model it using as input the $f_E(e)$ for emission and the weather. A meteorologist might use available data to build a weather model which would provide the conditional probability distribution $f_{C/E}(c/e)$ for concentrations, given emissions for each possible e. Then the marginal (i.e., unconditional) probability distribution for concentrations is calculated from

$$f_C(c) = \int_e f_{C/E}(c/e)f_E(e) \, de. \qquad (6.15)$$

In assessing the utility function $u_C(c)$, the individual assessed would have to remember the implications of various pollution concentrations for biological processes.

If environmental disturbance is used as an attribute, there are also a number of choices to be made. The probability distribution $f_D(d)$ for disturbances might be directly assessed. This would require a good deal of knowledge about how much pollution might be emitted, how it would interact with the weather to provide concentrations, and how these would combine with the natural processes to cause disturbances. Alternatively, the impacts might be modeled using (6.15) and a model of the relevant biological processes. The latter would essentially be a dose–response model developed for the different species by biologists. This would result in a conditional probability distribution $f_{D/C}(d/c)$ for environmental disturbance, given concentrations. The marginal probability distribution $f_D(d)$ could then be calculated from

$$f_D(d) = \int_c f_{D/C}(d/c)f_C(d) \, dc. \qquad (6.16)$$

The utility function $u_D(d)$ could be assessed considering only the relative values associated with different levels of disturbance.

With an overview of the three options for attributes, that is E, C, and D, it should be clear that the same information is required in all three cases.

The basic tradeoffs concern whether this is used in determining the probability distributions or the utility function. If E is used, the probability distribution is relatively easy to obtain but the utility function requires a knowledge of the entire process represented by Fig. 6.5. If C is used, the probability distribution is more difficult, but the utility function does not require as much knowledge. And if D is used, the probability distribution integrates all the knowledge, and the preference assessment is straightforward.

Usually it is advantageous to use the direct attribute, which is D in this case. Then the preferences can be assessed by the client or others (e.g., regulators or the public) with no special understanding of, for example, the biological processes. If either E or C is used, the preferences of a professional with such an understanding should be used. With attribute D, it is probably best to model the environmental disturbance using (6.15) and (6.16). A great advantage of this modeling is that professionals in various disciplines need to provide information only within their domain of expertise. The model ((6.15) and (6.16)) then systematically and logically integrates this information. In this case, f_E might be provided by engineers, $f_{C/E}$ by meteorologists, $f_{D/C}$ by biologists, and u_D by the client.

6.2.3 CASES OF ASSESSED ENVIRONMENTAL IMPACT

As a result of the National Environmental Policy Act, the construction of any energy facility requiring federal government approval requires an environmental impact statement (EIS) to identify, describe, and evaluate the significance of the environmental impact of the project. The typical EIS is very lengthy with elaborate detail listing a large number of possible adverse and beneficial environmental impacts.† The EIS is compiled once a site has been selected as the best for a particular energy facility. Rarely does the EIS contain an indication of the likelihoods of the various impacts included, and yet the uncertainties are large. At the site selection stage of the project, the uncertainties about possible impacts at each candidate site are even larger since it is too expensive to conduct a full-scale baseline monitoring at each of the candidate sites. Thus it is very important to include uncertainties at this stage.

Generic models are useful in providing indications of the local resources to be used or the pollution levels. However, the local environmental impact caused by energy facilities is specific to the site. Thus,

---

† Many environmental impact statements have been so voluminous that they were of little help in ensuing decision making processes (Fairfax [1978]). Recently, the Council on Environmental Quality [1978] has taken steps to correct this.

generic models of the environmental impact conditioned on resource use or pollutant levels are often of little use. Detailed site-specific studies are required. The following cases illustrate both the generic and site-specific models.

Pollution Emissions from a Coal-Fired Power Plant

Most of the environmental consequences at the sites of energy facilities occur either from the use or the pollution of local resources such as air, water, or land. It is often unfortunately the case that less pollution of one type can be achieved only if one is willing to accept an increase in another type of pollution. For instance, the use of scrubbers in coal power plants results in a reduction of SO_2 emitted but also in an increase in solid wastes at the site. Lincoln and Rubin [1979] examine such implications, which they refer to as cross-media environmental impacts, for coal power plants.

Gruhl [1978] has developed a model to characterize pollution emissions of coal-fired power plants as a function of the size and operations efficiency of the plant, its emission stack height, the source of coal utilized, and the local meteorological conditions. The stack height and meteorology affect the dispersion of pollutants and the coal type determines the presence of contaminants, heat content, and so on.

The model provides as output five points on the cumulative probability distribution for each of several air pollutants, including SO_x, NO_x, particulates, CO, CO_2, hydrocarbons, and trace elements, as well as water pollutants and solid wastes. Typical outputs of the model are illustrated in Table 6.1 for a 1900 MWe unit, operating at 70% capacity, with a 235 m

TABLE 6.1

Characteristic Information from a Probabilistic Pollution
Emission Model for Specific Power Plant Sites

	Minimum	1 Standard deviation low	Median	1 Standard deviation high	Maximum
Fractile	0.0	0.16	0.5	0.84	1.0
Pollutant					
SO_x (grams/minute)	102	1287	42,475	129,900	485,230
Radioactivity to air (curies/year)	0.0003	0.029	0.110	0.237	0.585
Thermal pollution into water (10^{12} Btu/year)	25.51	28.12	30.15	33.02	36.94
Solid wastes (10^6 tons/year)	0.608	2.66	3.85	5.32	11.08

stack, using southern West Virginia bituminous coal, in a region with "average" U.S. meteorology. It is obvious from the table that the range of uncertainty is significant.

Site Impacts from a Nuclear Power Plant†

Chapter 3 presented a siting study conducted for the Washington Public Power Supply System to identify a site for a 3000 MWe nuclear power plant. Two environmental impacts were chosen to characterize the impacts at the power plant site. These were impacts on salmon and on other biologically important features. They are illustrated in more detail here since the former involves impacts measured with a natural attribute and the latter with a constructed attribute. Both attributes required some "creative development" in order to be appropriate for evaluating the set of candidate sites. This requirement is common for the environmental impacts in many studies.

We chose to separate salmon from other biologically important features because (1) their commercial, recreational, and aesthetic value is an extremely important resource to the people of the Pacific Northwest and (2) experience in the Pacific Northwest suggested that salmon would be at least as important as all other biotic factors together. The public, government agencies, environmental groups, commercial fishing interests, sports clubs, native Indians, and academia will all rise to the defense of the fish.

In addition, the egg, fry, and juvenile stages of salmon are generally considered more sensitive to environmental perturbations than are many other common or important aquatic species, and probably serve as a fair indicator of water quality (see Kemp *et al.* [1973], Bell [1973], and Fulton [1970]). Thus, if the impacts on salmonids are minimized, most of the other aquatic resources such as trout, shad, sturgeon, and plankton will experience at least a degree of protection.

Impacts were assessed, based on the professional judgment and experience of two biologists‡ on the siting team. For salmon impact, they utilized data on the total river flow, annual average spawning escapement, distribution of fish in the cross section of the stream (i.e., juvenile fish are often concentrated on the edges rather than in the middle), likelihood of disturbing spawning grounds, and other related factors. Information from site visits were used in making site biological impact assessments. There was not always complete agreement, at least initially, between the biologists but the model provided a formal framework within which their differences could be resolved (as was the case in this project). If agreement could not

† The discussion of this section is liberally adapted from Keeney and Robilliard [1977].

‡ The biologists were Gordon A. Robilliard and David White.

have been reached, the model could have given a measure of the type and magnitude of disagreement.

Salmon impact.   One main ecological objective of the siting study was to minimize the adverse impacts on salmonids (i.e., salmon). Salmonids are defined as the five species of salmon (silver, chinook, chum, humpback, sockeye) and the steelhead trout which occur in Washington/Oregon waters. These salmonids are all anadromous fish, that is, they spawn in gravel beds in cool, clean, freshwater streams and lakes, and the eggs incubate for several months. The fry emerge to spend some time (from a month to two years depending upon the species) in fresh water before heading downstream to the ocean as juveniles. They mature for two or more years in the ocean before returning to the fresh water to spawn, thus completing their life cycle.

Adverse impacts were defined as those which result in an immediate and/or long-term decrease in population size in the affected water bodies. The decrease could result from entrainment in the power plant cooling system, entrapment in the discharge plume, impingement at the intake structure, destruction or alteration of spawning beds or juvenile maturation areas, or sublethal effects which result in lower reproductive success.

It is desirable to identify a practical measure of adverse impact which has a historical record, is widely used and interpreted, and can be applied in almost all situations. Two measurements seemed to satisfy these conditions: average annual number of spawning escapement lost and average annual percentage of spawning escapement lost. Spawning escapement is the number of adult fish that return to a particular stream to spawn. There are good historical records of the escapement of adult fish for most major salmon streams (Wright [1974]).

Numbers alone are misleading. A loss of 10,000 fish in the main stream of the Columbia River would represent 1–5% of the annual escapement in that river, depending on when and where the loss occurred. Such losses, although important, would probably not seriously affect the population dynamics of salmon in any particular tributary river because only a small portion of the 10,000 would probably be destined for that particular tributary. On the other hand, a loss of 1000 fish in the North Santiam River might represent 25–50% of the total escapement to that river and could cause a major change in the population dynamics and genetic makeup. Furthermore, there is considerable variation in escapement from year to year. In smaller streams, it is conceivable that the loss of 1000 fish might represent the total population in a low year, thus effectively eliminating the run in the ensuing cycle year.

Two factors are important in measuring adverse impacts on salmonid.

First, commercial, recreational, and aesthetic losses occur, and these can be better indicated by the number of fish lost. Second, the genetic history and composition of the salmonid population from each stream are somewhat distinct and cannot be "replaced" by restocking with fish from other streams or hatcheries. Impacts on this second factor can be better estimated by the percentage of fish lost in a given stream. The second factor is not considered significant in the Columbia River because most of the salmonids there are destined for one of the tributary streams and use the Columbia River simply as a passageway; i.e., few salmonids actually spawn in the Columbia River, especially in the lower reaches. Also, salmonid escapement in the Columbia River usually exceeds 300,000, whereas the next largest escapement is under 100,000.

For streams with less than 100,000 escapement, two measures (attributes) of adverse impact on salmonids are used:

Y = number of salmonids in the stream,
Z = percentage of salmonid escapement lost in a year.

Attribute Y was chosen as number of fish in the stream rather than number of fish lost, because one implies the other when interpreted in conjunction with attribute Z, and the preference assessments were easier using number of fish in the stream. For the Columbia River, the only concern was the number of fish lost which, of course, is equal to Y times Z.

The main hazard to salmonids will probably be entrainment and impingement, even though the water intake structure for the power is designed to minimize these dangers. Construction on the Columbia River or the presence of thermal plumes in the river will cause essentially no disturbance to spawning and rearing areas, since few exist in the Columbia River mainstream. But on smaller rivers, spawning and rearing areas immediately downstream from the site are likely to be eliminated. In these smaller rivers, adult fish may be blocked from reaching upstream spawning areas by construction activities or by the thermal plume.

The biologists felt that possible impacts could be qualitatively described as follows. There is a small chance of very little loss of salmon; this chance increases up to a most likely level of 1–15% loss, depending on the size and salmon-spawning potential of the river, and then decreases. There is a very small likelihood of a large—greater than 50% or 100,000 fish—loss. Hence the probability distribution is skewed. One could assess skewed probability distributions to describe such impacts, but, after checking, it appeared that a normal distribution could adequately approximate the likely impacts. We used the normal distribution for convenience. The assessed parameters of the distributions are given for the nine candidate sites in Table 6.2.

TABLE 6.2

SALMONID IMPACTS

Site	River affected	Annual average escapement y (in 1000s)	Mean percent lost $\bar{z}$	Standard deviation of percent loss σ_z
Benton	Columbia	430	1	0.5
Umatilla	Columbia	365	1	0.5
Clatsop	Blind Slough on Columbia	5	15	7.5
Grays Harbor	Wynoochee	5.5	15	7.5
Wahkiakum	Elochoman	17	15	7.5
Lewis 1	Cowlitz	55	8	4
Lewis 2	Cowlitz	55	8	4
Lewis 3	Cowlitz	55	8	4
Linn	North Santiam	3	15	7.5

Site Biological Impact.   During the construction and operation of the power plant, it is important to minimize the biological disturbance. Many features are included under this heading. For the sites under consideration, the main biological concerns (other than salmonids) are preservation of threatened and endangered species, protection of habitat of migratory species (especially waterfowl and game birds), maintenance of productive wetlands, and preservation of virgin or mature second-growth stands of timber or "undisturbed" sagebrush communities.

There did not seem to be any convenient measures for indicating the degree to which a power plant would cause biological disturbance as defined above. One possibility was to estimate the land area involved in each of the categories mentioned, but we felt that it was too difficult to relate areas *per se* to impact. As an alternative, we chose to construct an index of potential short-term and long-term impacts. This scale, illustrated in Table 3.3, was defined after site visits by the client and the project team members, including the two biologists. The scale goes from 0 to 8; larger numbers are associated with greater biological impact. The scale is defined to include the important features which distinguish the sites as well as to illustrate and communicate in realistic terms the degree of biological impact.

The likely biological impact at each site was directly assessed by the biologists after the site visits and a review of available publications concerning biological activity in the vicinity of the sites. For each site, the probability that an impact fell in the range of 0 to 1, 1 to 2, ..., 7 to 8 was assessed. Several internal consistency checks were used in this activity. For instance, refer to the Lewis 2 and Lewis 3 data in Table 6.3. We

TABLE 6.3

POSSIBLE BIOLOGICAL IMPACT[a] AND EXPECTED UTILITY

Site	Range of impact[b]							
	0–1	1–2	2–3	3–4	4–5	5–6	6–7	7–8
Benton	0.1	0.5	0.4					
Umatilla	0.7	0.3						
Clatsop				0.2	0.5	0.3		
Grays Harbor			0.2	0.8				
Wakhiakum				0.2	0.5	0.3		
Lewis 1		0.9	0.1					
Lewis 2		0.9	0.1					
Lewis 3		0.8	0.2					
Linn		0.3	0.6	0.1				

[a] Data represent the probability that the impact at each site will be in the range indicated.
[b] Based on Table 3.3.

asked "Is the likelihood of a 2–3 impact twice as great at the latter site as at the former?" A yes response was consistent with the original assessment. The data in Table 6.3 represent the final adjusted numbers and are meant to quantify and thus complement the qualitative descriptions of possible site impacts, two of which are briefly summarized below:

Benton. This is used mostly for wheat farming and some grazing. There is relatively little undisturbed sagebrush habitat and there are no wetlands or known endangered species habitats. The proportion of agricultural area to undisturbed habitat will vary depending upon exactly where the site is located; hence, the distribution is 0–3.

Clatsop. The site region is made up of varying proportions of mature second-growth forest, logged areas, and some small agricultural areas. There are some small swampy areas and nearby wetlands. There is a strong possibility that Columbia white-tailed deer, an endangered species, may occupy the site or environs. The distribution range is 3–6.

6.3 Quantifying Economic Impacts

At first glance, economic impacts should be the easiest of the siting concerns to handle. One simply wants to know how much it costs to build and operate an energy facility. That is true, but the problem of determining these costs is very difficult for three principal reasons:

1. there are many uncertainties, which are not completely under the control of the client, affecting costs (see *Business Week* [1979a]);

2. a dollar cost for construction is not necessarily valued the same as a dollar cost of operation or maintenance; and

3. a dollar cost at one time is not valued the same as a dollar cost at another time.

Let us elaborate.

6.3.1   MEASURING THE ECONOMIC IMPACTS

The uncertainties stem from many sources. Some of these are due to natural causes. The geology in the vicinity of a proposed pumped storage unit can have significant effects on the construction costs. Natural occurrences such as a landslide or an earthquake can cause construction delays, damage a facility, or force it to shut down for a complete inspection. Other crucial uncertainties are directly attributable to human actions. Such factors as worker strikes, possible court actions, and delays in licensing can greatly affect the cost of an energy facility. Clearly, these possibilities should be included in quantifying economic impacts. The capital costs of building energy facilities are also uncertain. Furthermore, in the recent past, these capital costs have escalated faster than inflation (see, for example, Shaw [1979]).

Because of various laws affecting the manner in which dollars are treated for tax or revenue purposes, every dollar of cost is not equal from the viewpoint of a client proposing an energy facility (see, for example, Gándara [1977]). The government may grant tax credits to firms building capital assets, but maintenance costs may not get this favorable treatment. This would make a construction dollar less expensive to the client than a maintenance dollar. Operating and maintenance costs can be passed on to customers of utility companies, whereas capital construction costs sometimes cannot be passed on. This makes a dollar cost of the former preferred to a construction dollar. A regulatory agency charged with setting rates for energy usage causes the net impact of different costs to vary with time and circumstances. All of these situations imply that different costs should be added up separately in determining economic impacts.

Due to inflation and interest rates, a dollar cost today is more important than a future dollar cost. Consequently, the economic impacts also need to be separated according to when they occur. Thus, we might break down the time horizon for a facility into years, where X^t designates a cost attribute in year t. The overall impact on this attribute requires the consequence $(x^1, ..., x^t, ..., x^T)$, where x^t is the cost in year t and T the lifetime in years of the facility.

6.3.2   MODELS OF ECONOMIC IMPACTS

The most appropriate way to handle the economic impact would be to model it. Most firms which build and operate energy facilities have elaborate economic models to predict construction, operating, and maintenance costs over the years. Several books and articles are concerned almost exclusively with developing such models (see, for example, Miller [1976] and Doane et al. [1976]). The main shortcoming often present in these models is that uncertainty is not explicitly considered. They will, for instance, include construction costs, personnel costs, fuel cost, scheduled maintenance, and all other auxiliary costs. But these are treated as if they were known for certain, many years into the future. This is simply not the case. Also not included in these standard cost models is the possibility that accidents, such as earthquakes, either produce damage or require unscheduled maintenance.

Historically, the actual construction costs have deviated greatly— usually upward—from the price estimate (Olds [1974] and Finch et al. [1978]). There is really no valid reason for deterministic forecasting since the methodology is available to do a better job. What is required is judgmental assessments of the likelihoods that various circumstances will occur, based on the best data we can collect and a model of the circumstances. Some energy companies, such as Florida Power and Light, have taken major steps in this direction (see *Electric Light and Power* [1977]).

The geology of an area is never known with complete certainty. Borings and soundings indicate structure, but local irregularities could significantly affect costs. In the sensitivity analysis of the pumped storage study discussed in Chapter 9, probabilities of alternative geological structures were assessed and the effect these uncertainties might have in the ranking of sites was appraised.

As another example, there is some historical record of the time required for hearings associated with different regulatory processes. Using this, plus experts' judgments on the particular site and the "mood" of the regulatory body, assessments for the distribution of time required could be obtained. The probability of rejection of the site could, and should, also be included.

It is also possible to make assessments of the likelihood of natural disasters. Models for the occurrence of earthquakes are well developed for predicting the probability that an earthquake of certain intensity will occur in a specified time interval (see Gutenberg and Richter [1954], Greensfelder [1977], Ryall *et al.* [1966]). Other models concern tornados, hurricanes, tsunamis, floods, and so on. In Section 6.7, a model of the landslide potential at a proposed geothermal power plant is presented in detail. Such a model, which incorporates data with the experience of geologists, seismologists, and engineers, can provide important input information for an economic model.

6.3.3   CASES OF ASSESSED ECONOMIC IMPACT

Prior to the National Environmental Policy Act (NEPA) of 1970, most, if not all, siting studies of energy facilities were based primarily on economic grounds. The economic studies were conducted by the energy company and as a general rule, the results were not publicized in any detail. Since NEPA, disclosure of possible environmental impacts of alternative sites is required. However, since the energy companies continue to conduct the economic aspects of siting studies and since governmental licensing regulations do not require the public disclosure of the economic analyses of alternative sites, such figures still rarely appear. Consequently, complete cases discussing economics in the context of the uncertainties inherent in the problem simply are not publicly available.

Typically, the economics are included in siting studies by having the energy company provide summary data concerning, for instance, annual cost of operation, assuming its chosen interest rate. This is what occurred in the nuclear and pumped storage siting studies of Chapters 3 and 9. There is obviously a great amount of thinking, modeling, and calculating done to arrive at such numbers, but without knowing the assumptions made in the process, it is difficult to conduct reasonable sensitivity analyses of the results. The biggest weakness of these studies is the cursory manner in which uncertainties and financial impacts over time are considered. The case briefly summarized below indicates the type of economic costs for any energy facility. This provides a basis for a more realistic treatment of uncertainty and impacts over time.

Coal-Fired Power Plants in Florida

Eight candidate sites for two 640 MWe coal-fired plants were evaluated in a recent study of Woodward-Clyde Consultants [1978] for the Florida Power Corporation. Much of the cost data, summarized in Table 6.4, was provided by the Florida Power Corporation. It is perhaps tempting to conclude from the table that, since the plant costs represent in the neighborhood of 90% of the capital costs, and since the fuel represents the same

TABLE 6.4

Cost Summary in 1978 Dollars for Two 640 MWe Units

Site	Astor	East of Orlando	Gulf County Canal	Lake Jessup	Lake Kissimmee	Phosphate Zone	Shingle Creek	Suwannee River
Capital costs ($ millions)								
Land acquisition	1	3	2	3	3	3	3	2
Site preparation & foundations	30	28	28	32	28	40	28	28
Plant	950	950	950	950	950	950	950	950
Coal delivery facilities	2	11	12	10	8	1	3	2
Water supply	1	16	7	7	1	20	2	1
Transmission system	24	18	190	18	13	7	9	111
Totals	1008	1026	1189	1020	1003	1021	995	1094
Annual costs[a] ($ millions/yr)								
Fuel	142	144	121	144	143	142	144	139
Transportation system losses (differentials)	3	3	4	3	3	0	1	3
County taxes	10	12	2	9	6	12	12	2

[a] It is assumed that except for fuel costs, operating and maintenance costs are equal for all sites. Therefore, they are excluded.

percentage of annual costs, that uncertainties and impacts over time do not matter. However, if capital costs are annualized at a typical rate, the annualized capital costs are only about one-half as large as annual costs. Thus, even if we know the (short-term) capital costs precisely, the uncertainties in future fuel prices could greatly influence the choice of a best site. Also, the manner in which the financial impacts over time are evaluated (see Section 7.3) is critical to the site ranking.

With each of the capital costs included in Table 6.4, one has a model of the components. For instance, the coal delivery facilities included rail spurs which required obtaining rights-of-way, clearing and grading, track construction, signal and communication systems, and so on. With annual fuel costs, assumptions are made about the location of the fuel source (southern Illinois in this case), its Btu content, the means of transportation to each site, and projections for future costs.

If the same coal source were being used by all facilities, the forecast annual costs of each would increase as coal prices rise. However, there may be a differential rise because the cost of coal at the source and the cost of coal transport may not rise at the same rate. More importantly, with higher annual costs and/or with greater value† placed on these costs, the plant sites with larger initial capital investment and lower annual expenses will improve in the ranking. It is this circumstance which one wishes to model adequately by including uncertainties and impacts over time explicitly in an economic analysis of sites. The importance of this task can be highlighted by quoting from a recent paper‡ of Nagel [1978], an executive in the utility industry.

> Electric utilities are without doubt the most capital-intensive of all economic enterprises. The power industry today requires approximately $5 of investment per dollar of annual revenue. This is in sharp contrast to most manufacturing industries, which require $1 or less of investment per dollar of annual revenue. In addition to the high costs of facilities is the need to raise much of the investment in the marketplace. Thus, the power industry is inherently sensitive to the cost of money. It is also sensitive to the cost of the primary fuels used in the generation (conversion) process. Historically, the cost of operation —mainly that of fuel—represented approximately 50 percent of the annualized cost of electricity produced and delivered to the generating plant's terminals. The percentage is even higher today.

---

† Greater value here refers to weighting the future costs more relative to the current costs than in the base case evaluation. This may occur, for example, if fuel costs continually increase faster than general inflation.

‡ Copyright 1978 by the American Association for the Advancement of Science.

He also stated "Uncertainty is the very essence of the problems facing utility management today." It certainly seems appropriate to include these economic uncertainties in evaluating candidate sites for any proposed energy facility.

6.4 Quantifying Socioeconomic Impacts

Broadly speaking, the socioeconomic impacts are meant to encompass two types of impacts. The first is the impact on the communities in the vicinity of the proposed energy facility. The main individuals influenced are those living in these communities. The second is the aesthetic impact of the facility. This influences those living or traveling near the facility.

The following recommendation, an excerpt from the U.S. Nuclear Regulatory Commission [1979a], indicates how important socioeconomic impacts are in siting.

> The Nuclear Regulatory Commission staff has concluded, after balancing environmental impacts and costs, that a proposed nuclear power plant should not be built on a site at Cementon, New York.
>
> Among the most important impacts and costs which are the basis for the staff's conclusion are:
>
> 1. As proposed, the Greene County Nuclear Power Plant would use a natural draft cooling tower which would entail an unacceptable aesthetic impact on local, regional, and national historic, scenic, and cultural resources, particularly the Frederick E. Church house (Olana), a national historic landmark.
>
> 2. The impacts on the community of Cementon itself. Construction of the proposed Greene County facility would require 280 to 303 acres of land currently owned by the Lehigh Portland Cement Company and could result in the closing of those facilities; this would result in an annual tax loss of $212,000. Use of the site by the Power Authority of the State of New York would result in in-lieu-of-tax payments estimated at $10,400 a year. Other activities might generate much larger tax payments.

6.4.1 MEASURING THE SOCIOECONOMIC IMPACTS

The local community may be defined as those people directly influenced by the building or operation of the plant at a particular site. Essentially, depending on the tax laws which prevail and the size of the political jurisdictions, the main impact falls on the nearest town where construction workers live and the county or town receiving the property tax from the facility. The main impact during the construction phase is due to the boom–bust cycle associated with a rapid influx of workers at the beginning of construction and the replacement of these workers with a much smaller group of permanent employees when the facility is ready for operation. There is much literature† about the impacts of boom–bust cycles on public debt, social and cultural institutions, schools, municipal services, housing, and quality of life in general. There are also reported case histories of energy companies taking effective mitigation procedures to minimize detrimental local impact and create local socioeconomic benefits. Examples are in Washington (Myhra [1975]) and Montana (Myhra [1976]).

There appear to be several ways to measure the local socioeconomic objectives. One is to categorize the impact by the population of the local community and the number of workers moving into it. This, of course, assumes, for instance, that 2000 workers‡ moving into a town of 25,000 would have the same impact regardless of the town. Such an assumption may be a reasonable approximation for the real situation in some cases. The constructed scale of the WPPSS socioeconomic impact in Table 3.4 was developed using this idea.

Another way to measure the local impact is by indices over time of the socioeconomic "health" of the community. These include the population, public service capital per capita, retail and service capital, housing shortage, mobile homes, property tax rate, construction jobs, and so on. This was the approach taken by Ford [1977] and is briefly described in Section 6.4.2.

A third approach to measuring socioeconomics is by the direct effects on local laws, institutions, and lifestyles. An example of one such case concerned the evaluation of sites for oil terminal and marine service base sites on Kodiak Island, Alaska. Attributes included zoning and ownership conflicts, loss of recreational facilities, possible accidental destruction or disturbance of archaeological or historical resources, municipal services, disruption of the local fishing industry, and impacts on native lifestyles. The manner in which these possible impacts were evaluated is discussed briefly in Section 7.7.3.

The aesthetic impacts of energy facilities are mainly due to visual degradation, odors, or noise. In each case we are concerned with how many

---

† See, for example, references in Ford [1977] and Harbridge House [1974].

‡ It is generally assumed that with families and support staff to provide necessities such as food and shelter, the total population increase to a community is four to five times the number of workers.

people are affected and how strongly they are affected (the intensity and time duration of the situation). Depending on the problem, it may only be necessary to address one of these points if the other is fairly constant. For example, in the siting of a transmission line, if it can be assumed that the number of people influenced would be about the same for all routes, an index of visual degradation would be appropriate to capture aesthetic impact.

For visual impacts, the use of pictures to define points on a scale may be an effective way to characterize intensity. A summary of interesting work in this regard is found in Craik and Zube [1976]. A related issue with transmission corridors concerns the integration of impacts for different locations of the corridor. The question is whether, for instance, 50 miles of line disrupting a beautiful area coupled with 50 miles in an already disrupted area is worse than 100 miles of line in a reasonably nice area. Value judgments are required to address this question.

6.4.2 MODELS OF SOCIOECONOMIC IMPACTS

The dynamics describing what happens to a town that receives a relatively large influx of individuals is complex. We would like to model these dynamics in order to evaluate various mitigation measures which might be taken in conjunction with the siting decision. For instance, if the client planned on front-end grant or loan money to the communities impacted, then this could influence the siting decision if there was a differential impact between candidate sites.

The boom town model for evaluating mitigation policies, as described by Ford [1977], requires information about potential mitigation policies and data on the town and the energy project. The model itself uses this information to provide outputs—a description of how the town would be changed, given that the project is carried forward and a particular mitigation policy is followed. The outputs could describe community impact either by the effect on a number of socioeconomic indices or by the direct effects on citizens.

Ford's model reported descriptions of six indices over a twenty-year time period before, during, and after construction. These were the public service capital per person, property tax rate, retail and service capital, construction jobs, houses, and population. From these time indices, data relating to the percentage changes (e.g., the rise in the property tax rate, the percent of housing shortage) and durations were calculated. These were used as attributes for a value model concerning possible mitigation policies for boom effects in Farmington, New Mexico.

In some situations, it may not be clear which towns will be primarily impacted by a proposed energy facility. Perhaps the workers will settle in two or three different nearby towns, dispersing and reducing the overall impact. This, however, complicates the problem by requiring a model to predict how many workers will settle in each location.

Models of aesthetic impacts are different, depending on the sense (sight, smell, or hearing) involved. In all cases, the duration of exposure can be measured in time and the number exposed by numbers. However, there do not seem to be any mathematical models to help determine visual impact. There are meteorological models which would help to describe where odors would travel and how much they would be diluted before reaching people. Similar considerations apply to noise transmission. For all the aesthetic impacts, the following question is relevant. How should multiple exposures be considered? For example, is one person exposed to 20 min of a certain noise level equivalent to two people exposed to this level for 10 min each? The answer requires a value judgment which is problem dependent. If such a value judgment is not made explicitly, it will necessarily be made implicitly.

6.4.3   CASES OF ASSESSED SOCIOECONOMIC IMPACTS

This section summarizes two distinctive approaches for assessing the possible socioeconomic impact of a major energy facility on the surrounding area. The first discusses the qualitative evaluation of the possible impact of two 1150 MWe nuclear power plants proposed for Montague, Massachusetts. The second describes a general model developed to predict socioeconomic impacts of major energy developments in local areas of North Dakota.

A few other cases involving socioeconomic impact are described elsewhere in this book. Section 6.6.3 summarizes the work of Ford and Gardiner on quantifying local public attitudes on energy development near two small western towns. The actual local socioeconomic impacts were predicted using a model discussed in Ford [1977]. Section 7.7.3 includes a brief summary of the socioeconomic impacts to native Indians of proposed energy facilities on Kodiak Island, Alaska.

The Montague Case

Northeast Utilities has proposed to the Nuclear Regulatory Commission that it build a major nuclear power plant near Montague, Massachusetts (U.S. Nuclear Regulatory Commission [1975a]). As part of the overall collection of studies involving this site, Northeast Utilities gave funds to the town of Montague to conduct a study of the potential social and economic impacts on the town. Montague, in collaboration with the Franklin County Planning Department (Montague is in Franklin County), defined the scope of the study and presided over its conduct. The study was conducted by Harbridge House, Inc. [1974].

The study attempted to contrast the future of Montague with and without the nuclear power plant. The past and current status of the town with regard to population, economy, land use, and government was first documented. This provided the *status quo* situation from which little change was expected in the absence of the proposed facility.

To forecast the socioeconomic impacts with the facility, case histories of four other towns (Plymouth, Massachusetts; Waterford, Connecticut; Wiscasset, Maine; and Scriba, New York) near recently built nuclear power plants were studied. The analyses concentrated on land development, taxes, town–utility relationships, political impacts, community life, and any unexpected impacts. An attempt was made to identify any control measures which the local community did take or might have taken to affect the resultant impacts. As a result of these analyses, forecasts of the possible impacts at Montague were made.

Because of the complexity involved and because of the major effect on impacts that exogenous factors may have, several basic assumptions were necessary in the forecast. These included an assumption that commencement of construction would be delayed at most one year, that the nearby University of Massachusetts at Amherst would not expand at a rapid pace, and that the local property tax structure would remain the means of raising local revenue. Subject to these assumptions, it was concluded that Montague would grow 25–40% in the 1974–1984 decade with 20% of the impact by 1977 and 70% by 1981. This would result in proportionally more career people and young children than at present.

The net difference in the estimated tax rate with and without the nuclear plant was estimated at about $2000 per year on a house assessed at $20,000.

Harbridge House felt that, depending on the population influx, the effects on community stability and the local political processes could be substantial. However, it was felt that with effective policy action taken by the community, most of the detrimental impacts could be avoided. These options included land-use controls, limiting local expansion of the road network, hiring professionals who would report to the Board of Selectmen to help manage town affairs, and sharing fiscal resources with other local communities to reduce local flow of residents to Montague.

As noted in the Harbridge study, more work of many kinds is needed to make better decisions concerning the socioeconomic impacts of major

developments near small communities. This particular study, in addition
to its local relevance, contributes to this more general goal.

The North Dakota Model

The North Dakota model described in Leistritz, Murdock, and Jones
[1978] is composed of several modules as shown in Fig. 6.6. It has been

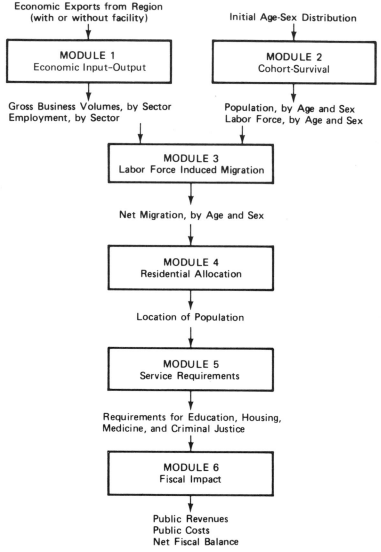

Fig. 6.6. The North Dakota socioeconomic model.
(Adapted from Leistritz *et al.* [1978].)

used to forecast socioeconomic impacts on a variety of local regions such as a community, county, or regional district.

The economic input–output module 1 relates various industrial segments within the economy being examined. The model coefficients were collected and validated using different North Dakota data. This module output estimates the gross business volume for each economic sector. Using local data on the gross business volume per worker for each sector, employment needs are tabulated.

Module 2 is a cohort-survival model which projects local population characteristics over time, given baseline population data, birth rates, migration rates, and survival rates. By comparing the required labor force from module 1 with the available labor force from module 2, the in and out migration of workers with particular skills is forecast in module 3.

Based on the migration, a gravity model for residential settlement of these workers is utilized in module 4. The model accounts for the different settlement patterns of worker types. These types are energy facility construction, energy facility operations, nonenergy-related operations, jobs indirectly created by the energy and other projects, and baseline economic activity. Based on population location, the service requirements module 5 projects needs for educational services, housing, medical services, and criminal justice services.

The fiscal impact module 6 forecasts the public sector costs and revenues for any region for which the alternatives are examined. Comparison with the baseline case indicates the changes induced by the proposed energy development. The fiscal model accounts for all tax revenues (sales, income, use, severance, property), state and federal revenue sharing, capital expenses for expanding and building facilities, and so on.

If probabilistic inputs and/or probabilistic descriptions of relationships in the modules are used, the output will be probabilistic. For instance, if there are major uncertainties about sales of the energy resource to outside regions (input for module 1), local migration factors, residential settlement, or a proposed new energy-related tax, the overall model can easily be used to examine the impacts of these uncertainties on local population growth, service needs, and fiscal effects.

Leistritz *et al.* [1978] describe the manner in which this model could be integrated with an environmental model. Currently, most environmental models of impacts of proposed energy facilities address, mainly, the direct effects of construction and operation of the facility itself. In some cases, especially when the development is in a sparsely populated area, the indirect environmental impacts induced by the socioeconomic changes due to the facility could exceed the direct impacts. An integrated socioeconomic environmental model could be used to examine this situation.

6.5   Quantifying Health and Safety

The construction and operation of an energy facility can result in sickness, injuries, and fatalities to individuals. These individuals may be workers associated with building or operating the facility, citizens living in the vicinity of the facility, or visitors near the facility during an accident (e.g., a dam break). Since there is no way to reduce the possibility of health and safety impacts to zero, they should be included in the evaluation of candidate sites. Carter [1977] discusses a case where safety considerations may result in the termination of a billion dollar dam project, even though over 150 million dollars have already been spent.

For reasons other than lack of data or time, health and safety is often handled using proxy attributes. The likelihood of coal train accidents with automobiles is reported rather than the number of people who may be hospitalized from injuries. Emissions are provided rather than a suggestion of how many sicknesses to expect from pollution. The likelihood of accidents is indicated rather than a specific number of deaths due to the accidents. This occurs because the public, the government, and many companies prefer to state such possibilities in a euphemistic fashion. They are an important part of siting problems and should be explicitly addressed. Such an approach should not only facilitate better decisions, but it may actually result in fewer fatalities (see Okrent [1980]). The willingness of society to deal conceptually with possible fatalities is currently the main factor limiting the quality of analysis being done on this problem. I personally expect this willingness to increase greatly over the next decade.

6.5.1   MEASURING HEALTH AND SAFETY

One aspect is clear about health and safety impacts: the concern is always with human injuries, sickness, and fatalities. This is invariant from site to site and even from siting study to siting study. The mechanisms which cause the impacts may vary from site to site and do vary from study to study, however.

The set of measures for health and safety may deal with either the fundamental concerns or the mechanisms which lead to them. For instance, it may be possible to use three attributes: the number of sicknesses, the number of injuries, and the number of fatalities due to construction and operation of the facility. The use of these requires the acceptance of several assumptions such as (1) sickness and injury can be defined and (2) all sicknesses are equal. Do sore eyes count as sickness? It is likely that a day of sore eyes is not equivalent to a long stay in a hospital with a respiratory disease. Should the intensity of the illness be included? This might be done using the number of sick days as an attribute rather than the number of cases of sickness. In any case, several value judgments must be made in selecting the index (or indices) for sickness. A similar set of value judgments must be made with regard to injuries.

Fatalities propose related difficulties, but the issue is conceptually simpler because death does not come in degrees. With sickness, the intensity can vary as well as the client's (or society's) value for that intensity of sickness. With death, only the client's value for the different fatalities is relevant. It is likely that a work-related fatality is not valued as much as a fatality in the public at large. Perhaps the death of an elderly bedridden individual is not valued as much as the death of a healthy child.

By using the attributes of fatalities, injuries, and sicknesses, it is not assumed that one knows for certain the number of these impacts caused by the facility. Uncertainties are inherent in the impacts for several reasons. First, the site selection is done prior to the construction (or operation) of the facility, so possible impacts must be specified beforehand. There are major shortcomings in data available for estimating the health effects of pollutant levels (see Neyman [1977]). Finally, it is difficult to ascertain the cause of death in circumstances where pollution levels may have been a factor. As a result, it is not possible to accurately forecast the fatalities which may result from pollution emitted by the energy facility.

If the attributes used for health and safety deal with the mechanisms by which the impacts occur, they will usually be proxy attributes. For a coal-fired power plant, the sulfur dioxide emitted may be one such attribute. This could be used to account for sickness and deaths due to the pollution. The miles traveled by coal trains might be another proxy attribute. This would address injuries and fatalities resulting from automobile collisions with trains. In these cases, the relationship between the proxy attributes and the fundamental concerns must be kept in mind when the utility function is assessed.

6.5.2 MODELS OF HEALTH AND SAFETY IMPACTS

A generic model of pollution impacts on health and safety is identical to that in Fig. 6.5 concerning environmental impacts of pollution. There are pollutants which interact with weather conditions to produce concentrations of pollutants. These interact with the physiological systems of humans causing sickness and possibly fatalities. What is needed are dose–response relationships for the various types of pollutants. One can either formally model these relationships and assess preferences over fatalities or assess preferences over emission levels and require the respondent being assessed to do the modeling mentally during the assessment process.

In most siting studies, the latter method is used, whereas the former seems to have several advantages. It provides the information of interest directly; it does not require a difficult information processing task to be done mentally; it allows individuals in different disciplines (e.g., engineering, meteorology, physiology, medicine) to contribute expertise only in their field and it logically integrates this information. The difficult part of the method is determining the dose–response relationships, although relationships of some air pollutants to health are fairly well known (see, for example, Comar and Sagan [1976]). Dose–response relationships for radiation effects are also well studied (see, for example, McBride *et al.* [1978]). An example, Fig. 6.7 taken from U.S. Nuclear Regulatory Commission [1975b], illustrates the relationships between a single dose of radiation, the degree of treatment, and the likelihood that it is fatal. An advantage of modeling dose–response relationships for siting studies is that the same models can be used on all studies, subject of course to updating with the acquisition of new knowledge.

One advantage of using proxy attributes for health and safety is that it saves some time and effort in the site selection. The amount saved appears insignificant, however, especially since cumulative experience from previous studies can be brought to bear to reduce the effort needed. A second advantage is that it is politically imprudent to be so explicit about potential fatalities. However, fatalities are a critical part of siting

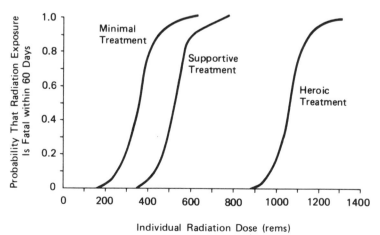

Fig. 6.7. **A dose–response relationship for a single exposure to radiation.**

problems and I would prefer to have them explicitly considered. This
point is discussed in more detail in Section 7.5.

For many of the normal operations related to energy facilities, there is a
wealth of historical data describing health and safety impacts. There are
data on mining accidents of all kinds, on shipping accidents for sea and
land transport, on incidences of black-lung disease, on drownings, and so
on. As an example, some comparative data on expected fatalities associated with coal and nuclear power plants are illustrated in Table 6.5. To
the extent that it is available and appropriate, such information should be
utilized.

There is another class of accidents for which available data may not be
appropriate because each situation is unique. For instance, the average
number of fatalities associated with a particular size dam break may not
have a large bearing on the expected fatalities if a similar sized dam imme-

TABLE 6.5

EXPECTED ANNUAL FATALITIES FROM COAL AND NUCLEAR PLANTS PRODUCING
800 MWE PER YEAR[a]

Fuel cycle component	Occupational		General public		
	Accident	Disease	Accident	Disease	Total
Coal-fired plant					
Resource recovery (mining, drilling, etc.)	0.3–0.6	0–7	b	b	
Processing	0.04	b	b	10	
Power generation	0.01	b	b	3–100	
Fuel storage	b	b	b	b	
Transportation	b	b	1.2	b	
Waste management	b	b	b	b	
Total for coal	0.35–0.65	0–7	1.2	13–110	15–120
Nuclear plant					
Resource recovery (mining, drilling, etc.)	0.2	0.038	~0	0.023	
Processing	0.005	0.042	b	0.002	
Power generation	0.01	0.061	0.04	0.011	
Fuel storage	b	~0	b	~0	
Transportation	~0	~0	0.01	~0	
Reprocessing	b	0.003	b	0.050	
Waste management	b	~0	b	0.001	
Total for nuclear	0.22	0.14	0.05	0.087	0.50

[a] From Gotchy [1977].

[b] The effects associated with these activities are not known at this time. Such effects are
generally believed to be small.

diately upstream from a town failed. Even data on the statistical likelihood of such a failure may be inappropriate since local features may be a large determinant in the dam's safety. For appraising the likelihood of accidents in rather new types of energy facilities (e.g., nuclear or solar power facilities), there may not exist any historical data.

In such situations, a specific model of accidents of concern can be constructed. Mark and Stuart-Alexander [1977] argue that possible disasters should be included in evaluating water resource projects. Because of differential impacts over sites, they should definitely be included in siting studies. Some of the models can include information about naturally occurring initiating events for disasters. An example is the model of earthquake frequency and magnitude (see Gutenberg and Richter [1954]) used in the landslide analysis discussed in Section 6.7.†

When there is a large series of events which must occur for an accident to result, an event tree may be a reasonable tool to relate events to the accident. An event tree (1) shows the logical relationships between individual events and the accident and (2) includes probabilities for calculating the likelihood of the accident. Perhaps the most widely known example of an event tree analysis was the Reactor Safety Study (U.S. Nuclear Regulatory Commission [1975b]) to determine the probability of operating accidents in a commercial nuclear power plant. An official appraisal of this work is found in Lewis *et al.* [1978]. A risk analysis, using event trees, of possible accidents at a proposed LNG terminal is presented in Section 6.8.

Sabotage or any deliberate disruption or destruction of an energy facility should at least preliminarily be considered in siting. In many cases, it is likely that such concerns are site independent and consequently not relevant to the evaluation of sites. If this is not the case, it still may be extremely difficult to model formally incidents of this kind in a meaningful way. If there appears to be a method for obtaining likelihoods for different acts of sabotage, it can be used. On the other hand, it may both be easier and more appropriate to consider this possibility qualitatively, and combine it informally with the results of the formal analysis.

6.5.3 CASES OF ASSESSED HEALTH AND SAFETY IMPACT

There are very few siting studies which have included the health and safety implications of each of the sites explicitly. Perhaps this is due to the difficulty of the task or to its political sensitivity. However, both of these reasons indicate the need to include such information in site evaluation.

---

† A recent report sponsored by the National Research Council focuses on the relationship between earthquake likelihoods and safer siting of critical facilities (see Panel on Earthquake Problems Related to the Siting of Critical Facilities [1980]).

For modeling purposes, it may be reasonable to categorize health and safety impacts on the basis of whether they result from normal operations or accidents. Except for considering the probability of an accident (e.g., nuclear core meltdown or a dam failure), most of the work concerning the health and safety impacts of proposed energy facilities is deterministic. That is, one estimate—more or less an average—of the impact is given. Furthermore, the impact is usually calculated using data averaged over fuel composition, meteorology, demography, and so on. Such impact information is included in Table 6.5 and papers by Hub and Schlenker [1974] and Black, Niehaus, and Simpson [1978], for example. This information may be useful in national planning models and examination of national or regional policy, but it is not very helpful in evaluating the differences in health and safety impacts between proposed sites for a particular facility. Site-specific models must be built for this comparison.

In some cases, it is possible to include the health and safety impacts of both accidents and normal operations into one index without explicitly building a model for each. This is the approach used in the Washington Public Power Supply System nuclear siting study discussed in Chapter 3. The index was the site population factor which indicated the relative size of the population exposed at the candidate sites. However, this index did not account for the likelihood of exposure, the meteorological effects, or the consequences of exposure. In that case, however, a sensitivity analysis indicated that such considerations would be unlikely to affect the site rankings.

Accidents

For certain types of accidents, the generic models are appropriate as a component of a site-specific model. For example, in comparing proposed nuclear power sites of similar technical design, the probability of a core melt may reasonably be assumed to be equal in all cases. An estimate of such a likelihood may come from a study such as the Reactor Safety Study (U.S. Nuclear Regulatory Commission [1975b]). Of course, one should then consider the probabilities of various levels of radioactive releases, the uncertainties about meteorological conditions at the time of the release, the population characteristics of the area, and imprecision in the dose–response relationship.

Two site-specific studies of potential accidents are discussed in detail in Sections 6.7 and 6.8. The first concerns the probability of a landslide at a proposed geothermal power plant. For this particular case, the area which could be affected by a landslide is uninhabited. However, the approach used to estimate the probability is relevant to accidents initiated by natural events.

In Section 6.8, the probabilities of various accidents are included in calculating the public risks of a proposed LNG import terminal. The basic initiating accidents in this case are due to human or equipment error. They involve airplane crashes and ship collisions. Then the meteorological and demographic characteristics of the area were characterized, based on data, and the human impacts of various conditions (e.g., being near an LNG fire) were conservatively estimated to give estimates of possible fa-

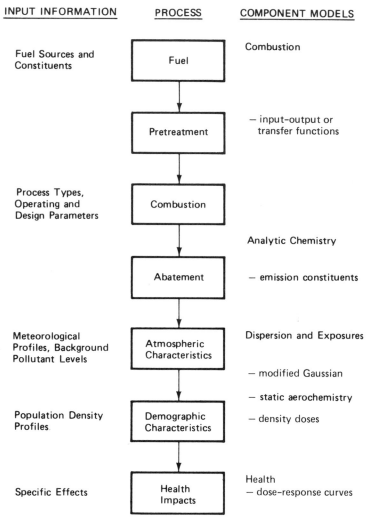

Fig. 6.8. Schematic description of a model for specifying the health impacts of any proposed power plant. (Adapted from Gruhl [1978].)

talities. If various sites were proposed, similar studies could and should be conducted for each of the candidate sites.

Normal Operations

Three similar site-specific models of human health impacts are illustrated in North and Merkhofer [1976], Morgan *et al.* [1978], and Gruhl [1978]. Since aspects of Gruhl's model were introduced in Section 6.2.3 concerning pollutant emissions, we will use it for discussion. Gruhl's model includes all the uncertainties about the inputs (e.g., the percentage of various pollutants in coal) in a particular case and calculates the implied uncertainties in the outputs (e.g., fatalities). A schematic description of the model indicating the components of the process resulting in health impacts about which there are uncertainties is shown in Fig. 6.8. Each of the components is modeled using site-specific data and professional judgment.

As an example, Gruhl calculates annual public health fatalities from the routine operation of a 1900 MWe coal-fired power plant. The illustration specifies a particular coal (southern West Virginia bituminous) which reduces uncertainty about fuel composition, heat characteristics, and so on. The height of the stack is specified as 235 m, no precleaning of the coal or scrubbers are used, and the capacity factor is 70%. The population density in the vicinity of the plant is added (90% of the Indian Point region in New York) with standard meteorological characteristics. A dose–

TABLE 6.6

ANNUAL PUBLIC HEALTH FATALITIES FROM A PARTICULAR 1900 MWE
COAL-FIRED FACILITY[a]

		Distribution level			
	Minimum	1 Standard deviation low	Median	1 Standard deviation high	Maximum
(Fractile)	(0.0)	(0.16)	(0.5)	(0.84)	(1.0)
Fatalities	0.10	0.97	2.94	6.26	9.70

Percent of uncertainty attributible to various pollutants

Pollutant	Percent	Pollutant	Percent
Nickel	64.0	SO_x	2.9
Beryllium	22.1	Arsenic	1.1
Particulates	5.1	Uranium/radium	0.6
NO_x	3.2		

[a] From Gruhl [1978].

response model of fatalities which is linear and additive over pollutants is used.

In addition to a probability distribution of possible fatalities, the model provides a list of the pollutants contributing most heavily to the uncertainty. For the specific case above, the minimum, maximum, and median possible impacts and the impacts one standard deviation above and below the median (0.84 and 0.16 fractiles) are given in Table 6.6. The main causes of these uncertainties are also given in the table. As Gruhl states, the uncertainties of the impacts of nickel and beryllium are partly due to uncertainties about the content of these elements in the coal, but are primarily a result of the current uncertainties about removal of the elements during combustion.

6.6   Quantifying Public Attitudes

There are two fundamental motivations for involving the public and concerned groups in the siting processes for energy facilities. The first is to articulate the public's values for use in evaluating candidate sites. Since a principal reason for the facilities being sited is to provide goods and services to the public, it seems reasonable to account for public values in the siting process. It is hoped that this will lead to improved decisions. The second motivation is to insure that the public and concerned groups feel that their attitudes and values do matter and that they are fairly and reasonably considered in siting. Such involvement results in improved psychological well-being for the public through increasing understanding of and trust in the institutions and processes by which siting is conducted.†

Recently there has been increased concern for public participation in major decisions which affect them. A recent example is contained in The Earthquake Hazards Reduction Act of 1977 (PL 95-124). In Section 5(h). it states

> the President shall provide an opportunity for participation by appropriate representatives of state and local governments, and by the public, including representatives of business and industry, the design profession, and the research community, in the formulation and implementation of the program.
>
> Such non-Federal participation shall include periodic review of the

† Rosenbaum [1977] presents a recent historical perspective on citizen involvement in U.S. governmental decision processes.

program plan, considered in its entirety, by an assembled and adequately staffed group of such representatives. Any comments on the program upon which such group agrees shall be reported to the Congress.

The major difficulty concerning public involvement in the siting of energy facilities is not whether such involvement should occur, but how it should occur. As clearly stated by Nagel [1978],† a senior executive of the American Electric Power Service Corporation,

> Public concern and involvement in the siting process is essential in a free society. In a complex, industrial (but orderly) society, however, complex issues require the application of specialized knowledge by those trained and experienced. In other words, a specialized technical activity such as power system planning cannot be carried out in an open forum or in the atmosphere of a town hall. This means that the entire intervention process needs to be circumscribed by certain rules, so that its duration and scope are limited and the issues raised are relevant to the matter at hand. The alternative can be nothing less than confusion and chaos.

The procedures outlined in this section are meant to provide for an orderly and responsible inclusion of public attitudes in the site evaluation process. This seems to address both of the motivations discussed above. However, it should be reiterated that this chapter concerns the description of site impacts with regard to several concerns including public attitudes. It may also be appropriate to include public involvement in aspects of the candidate site selection process and the setting of objectives. Such involvement may affect the candidate sites being evaluated and the attributes. It would not affect the evaluation process itself as opposed to the evaluation. The procedures discussed below are appropriate regardless of the degree of public involvement in other aspects of the siting process.

6.6.1 MEASURING PUBLIC ATTITUDES

The measurement of the environmental, economic, socioeconomic, and health and safety concerns is in terms of possible levels of physical impact. These impacts are, for instance, indicated by costs measured in dollars, forest destroyed measured in acres, or fatalities measured in numbers. The measurement of public attitudes is in terms of possible levels of psychological impacts. These impacts must be measured by indicators of public values. The data on public values are treated as informa-

† Copyright 1978 by the American Association for the Advancement of Science.

tion by the client exactly as the data on physical impacts are. The client must place values on the various possible levels of these impacts, both physical and psychological.

The relationship between the physical and psychological impacts can be clearly illustrated with the help of Fig. 6.9. This figure illustrates, using abbreviated identification phrases rather than objectives, the overall objectives hierarchy for an energy facility siting problem. The Xs in the figure represent specific attributes with which to measure impacts.

As stated in Section 1.2, the client is the decision maker for our siting problem. Hence, in all cases, the ultimate value judgments for evaluating sites are those of the client or clients. Some of these value judgments will involve "weighting" the five major concerns associated with the top row of the hierarchy in Fig. 6.9. There are two extremes worth noting for discussion. The first occurs when all the weight is on the four physical impact concerns—environment, economics, socioeconomics, and health and safety—and none on public attitudes. The second occurs when the situation is reversed with all of the weight on public attitudes. These two cases lead to the objectives hierarchies of Fig. 6.10. The two extremes correspond to conceptually and procedurally different approaches for incorporating public attitudes into siting problems. However, as we will see, many of the same considerations are required in both cases.

When no weight is explicitly placed on public attitudes, as in Fig. 6.10a, many of the value judgments made in evaluating possible impacts implicitly consider members and groups of the public. For instance, value judgments related to economic impacts should consider the shareholders of the client company and prices paid by consumers. The value judgments about socioeconomics will consider the attitudes of individuals in local communities near proposed sites. The "weight" given to environmental impacts will consider environmental groups. However, each of these considerations will necessarily be implicitly accounted for in the client's values. Furthermore, the relative weights of the different considerations implicitly address the concern of the client for the various concerned groups as well as their relative size and other factors.

With the hierarchy in Fig. 6.10b, all the implicit considerations mentioned above are explicit. It is the public itself which places the values on the possible impacts. If the values for different impacts are those of different groups, the client will necessarily need to integrate these—using explicit value judgments—to obtain the overall value structure for evaluating the candidate sites. The general problem of integrating the values of different groups will not be considered here since it is discussed in detail in Section 7.4. Procedures for quantifying public attitudes, for the public as a whole or for subgroups, are considered in Section 6.6.2.

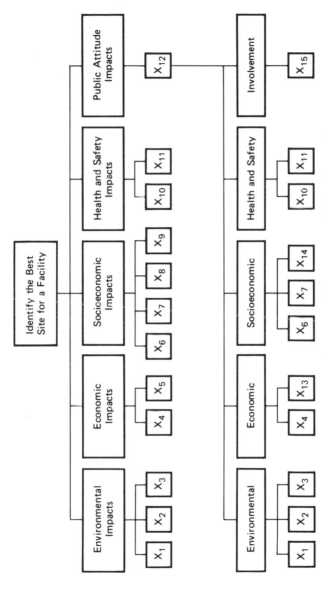

Fig. 6.9. General objectives hierarchy for an energy facilities siting problem.

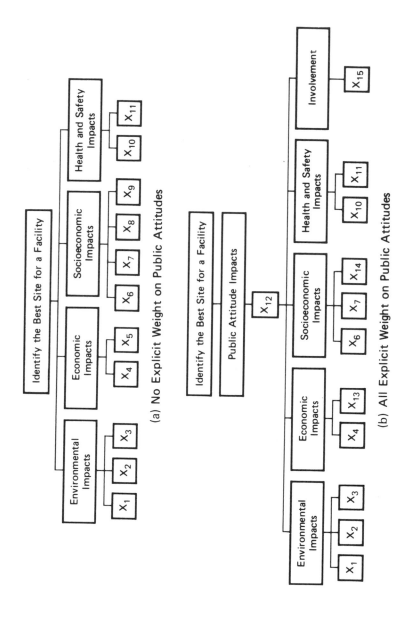

(a) No Explicit Weight on Public Attitudes

(b) All Explicit Weight on Public Attitudes

Fig. 6.10. Two special objectives hierarchies for an energy facilities siting problem.

In most important siting problems, it would seem that neither of the extreme objectives hierarchies of Fig. 6.10 would be as appropriate as an intermediate case corresponding to Fig. 6.9 with only part of the weight directly on public attitudes. This stems from the fact that there are two important advantages to each of the extreme hierarchies, and all four are associated with the intermediate case.

One reason for not using public attitudes exclusively as an indicator of desirability is that the complexity of certain aspects of the problem requires professional knowledge beyond that accessible or perhaps even communicable to the public. A classic example would involve proxy attributes such as air pollution levels. To place value judgments on such levels requires a knowledge of the undesirable biological processes which may result from different pollution levels. Even more understandable attributes, such as the number of workers temporarily relocating in a town during plant construction, require knowledge of the social and economic impacts resulting from various levels for proper evaluation. A second reason for using direct value assessments by the client is the difficulty of obtaining reliable and comprehensive value information from the public. Relying exclusively on public attitudes would require the assessment of complete utility functions over all the attributes. By making some of the value assessments required, the client can effectively use less sophisticated indicators of public attitudes, as represented by attribute X_{12} in Fig. 6.9. As indicated in Section 6.2.2, one such attribute might simply be the percentage of the public finding a particular site acceptable.

The two major reasons for explicitly including value judgments of the public can be clearly seen from Fig. 6.10. The first is that some members of the public may have some value judgments which are entirely different from those which would be chosen by the client. The attributes used to indicate desirability may even be different. In Fig. 6.10, note, for example, that attribute X_{13} replaces attribute X_5 in measuring economic impact of the plants. Attribute X_5 may be the return on invested capital to the energy company, whereas X_{13} may be the cost per unit to the consumer of the end product. With regard to socioeconomic impacts, the client may use (because of regulatory considerations) attributes X_8 and X_9 of negative impacts on a local community, whereas the community may be more interested in the positive effects such as jobs provided as indicated by attribute X_{14}. A second reason for including public attitudes explicitly is to promote involvement of the public in the siting process. This is indicated by attribute X_{15} in Fig. 6.9. If all the environmental, economic, socioeconomic, and health and safety concerns are valued directly by the client, the public attitude impact becomes essentially one of public involvement. In this case, attribute X_{12} is equivalent to X_{15}.

If no weight is put on public attitudes, it is worthwhile to note that the public may still be involved in the siting process in the selection of sites and objectives. In this case, no explicit value is placed on involvement. This is particularly appropriate where the degree of public involvement at all sites is the same, since, in this case, the degree of public involvement could not be helpful in differentiating between candidate sites.

6.6.2 MODELS OF PUBLIC ATTITUDES

There are four major issues which must be addressed in developing models of public attitudes. These are

1. whose values should be quantified,
2. what attributes should be used to quantify them,
3. how should the value information be gathered, and
4. how should individual values be combined into public values.

The first three will be addressed here. The last falls under the category of evaluating site impacts and is addressed in detail in Section 7.4. However, we will give a perspective on that problem here.

Information about individual values can be combined into group values in one step or in several steps. With one step, all individual values are combined directly into an overall value index. The decision about the procedure to be used for combining individual values requires value judgments about the relative "importance" of the different individuals. These additional value judgments are made by the client.

Normally, if more than one step is used to aggregate individual values, two steps are utilized. The first is to integrate values of individuals identified as being in the same group of concerned individuals (e.g., conservationists, shareholders, community near the proposed site). The second step involves combining the values of the various groups into an overall public attitude index. Both steps, again, require value judgments from the client about the relative "importance" of the entities being integrated. Thus, regardless of the aggregation procedure used, the client will necessarily make the same types of value judgments and this is the topic of Chapter 7.

Whose Values to Quantify.   The values of the public affected by the siting decision are those which should be considered for quantification. This "public" could include a large number of individuals, many of whom can be identified by the groups to which they belong, either by definition or by voluntary membership. These concerned groups include consumers of the product of the proposed energy facility, shareholders of the company proposing the facility, management and employees of that company, conservationists, business interests, and groups living near proposed facilities.

Because many of the interests of these groups could be accounted for within the environmental, economic, socioeconomic, and health and safety concerns, it may not be worthwhile to quantify directly attitudes of each of these groups. For instance, it may be worthwhile to include directly attitudes about the community adjacent to each proposed facility site. These attitudes could vary greatly over the set of sites and the local communities are, in a way, impacted to a greater extent than other groups. On the other hand, attitudes of consumers may be relatively site independent and each consumer is often affected by siting in only a minor way, if at all.

The judgments about whose values to quantify and include directly in the description of site impacts is the responsibility of the client. In exercising its choice, the client must consider the advantages and disadvantages of increased public participation in the site selection process and the requirements promulgated by regulations governing siting.

What Attributes to Use. Basically, the types of attributes which can be used to quantify public attitudes can be categorized as follows:

1. binary yes/no (or site acceptance/rejection),
2. ordinal ranking of the proposed sites, and
3. cardinal ranking of the proposed sites.

With binary attributes, the attribute might be the percentage of the public (or a group) which accepts the particular site. In using such an attribute, care must be taken to insure that all individuals using it (i.e., responding to questions or evaluating decisions) understand its meaning. For instance, an attribute may be the percentage of the individuals who prefer a facility at the particular site to no facility at all. However, this might be mistaken for the percentage of individuals who prefer the particular site to all other sites, given a facility will be built on one of the candidate sites. Clearly, other things being equal, if 80% of one local community accepts a nearby site, and 55% of another accepts a site in its vicinity, the former site is preferred to the latter. It is interesting to point out that the manner in which individual attitudes are combined is decided *de facto* by the choice of the attribute in this case.

With both the ordinal and cardinal rankings, information would be gathered from individuals. With ordinal rankings, each candidate site is assigned a number indicating the order of preference. With cardinal ranking, not only is the order indicated, but the differences in relative desirability among the sites is also provided. It is much easier to obtain an ordinal ranking. Such information may be provided by direct judgmental questions about the sites or more formally with the use of a value function. Cardinal rankings, in addition to the ordinal ranking information, may directly use judgments about the differences in value between sites. Formally, this may be done using measurable value functions or utility functions. Definitions and formal procedures for assessing value functions, measurable value functions, and utility functions are found in Chapter 7. With both ranking attributes, regardless of the manner in which the information was obtained, the client's value judgments must be used to combine individual attitudes into public attitudes as indicated in Section 7.4.

Gathering the Value Information.   There would appear to be three distinct ways in which to gather the information on public attitudes. The most direct way is to ask members of the public to express their attitudes and values. Those members would then respond about their own attitudes, rather that about the attitudes which they felt were the prevailing views. Because of the magnitude of the task, it would probably be necessary to survey only an appropriately selected sample of the public.† These attitudes would be statistically processed to provide a picture of the overall public attitude. Because of possible errors in sampling, as well as the inherent uncertainty of extrapolating from sample results, the public attitude would be probabilistically described using the sample as a basis.

The second way of quantifying public attitudes uses members of the public to provide the public attitude directly. Such a procedure has the advantage of requiring smaller sample sizes and allows more time for a more thorough quantification of attitudes. It is particularly useful when the public is segmented into relatively homogeneous groups and when spokespersons for the groups can be identified. For example, the president or board members of an environmentalist group may be qualified to articulate that group's attitudes toward particular sites. Lee [1979] presents a case study on marine mining using this general approach.

A major disadvantage of both of these methods is that they may not be practical for selecting among several sites. To gather the data the company proposing the facility must specify all the candidate sites being examined. This will probably lead to large increases in the land price and upset citizens at each candidate site. This would occur even though all but one of the candidate sites will be rejected. To avoid this problem, it may be appropriate to use experts' knowledge quantifying the public's attitudes, at least for those which require candidate site identification. Some

† It is interesting to note, however, that the citizens of Austria voted on November 5, 1978, whether to authorize the operation of the nation's first completed nuclear power plant. The results were 50.47% against and 49.53% in favor (*Nuclear News* [1978]).

public attitudes, such as those relating to produce cost, are not site dependent even though the consumer costs may be influenced greatly by the sites. These value attitudes can be solicited independent of the candidate sites. When experts are utilized to articulate public viewpoints prior to site selection (and even perhaps when not), it may be appropriate and advisable to include the local public in the decision processes which remain (e.g., mitigation of detrimental local socioeconomic impacts) after a proposed site is selected.

6.6.3 A CASE OF ASSESSED PUBLIC ATTITUDES

To date, there have been no attempts to quantify public attitudes and include them as part of the information in a decision analysis of a proposed energy facility siting problem as outlined by Fig. 6.9. However, there have been a number of case studies on similar problems with aspects relevant to the quantification of public attitudes. Two such cases are briefly described in Chapter 7. Section 7.7.3 involves the site selection of terminal facilities on Kodiak Island, Alaska to support offshore oil operations. Using experts' judgments, attitudes of native Indians, fisherman environmentalists, and petroleum companies were quantified. Section 7.4.4 describes the elicitation of values of various interested groups in a major policy decision affecting them.

Recent work by Ford and Gardiner (see Ford [1979], Ford and Gardiner [1979]) illustrates some important aspects and provides valuable experience for the elicitation of value attitudes of communities in the vicinity of proposed major energy facilities. In particular, they have analyzed various planning options for communities which are located near major energy developments. Their first study was conducted with several individuals from Farmington, New Mexico which is located in a region where uranium mining and processing is rapidly expanding.

The first step in the overall process was to develop an impact model of the consequences, to Farmington, of various policies. These consequences were described in terms of several attributes measuring the availability of public facilities, retail and service facilities, permanent housing, and mobile homes and the impacts on the property tax rate and influx of workers. The second step involved the quantification of utility functions for several key individuals in Farmington to use in evaluating these possible impacts. The individuals included the mayor, a county commissioner, a county planner, an environmental researcher, and two energy company officials.

In Ford and Gardiner's work, there was no attempt to forge all of the individual's utility functions into an overall indicator of the public's values. Instead, all of the options were evaluated with each individual's utility function. An attempt was made to identify an alternative which would result in a consensus with respect to the action to be taken. Specifically, for the case studied, it was concluded that with effective planning of the work force size and with significant front-end financing to mitigate possible detrimental boomtown effects, all of the participants preferred the proposed development to no development.

A similar study by Ford and Gardiner has recently been used in Price, Utah. Both the Farmington and Price cases are based on earlier experiences of Gardiner and Edwards [1975] with the California Coastal Commission. With regard to siting of energy facilities using decision analysis, these studies help to demonstrate the willingness and ability of local officials to participate effectively in providing values to be used in evaluating the candidate sites. When considering that the alternative is to include local values informally, the contribution seems significant.

6.7 Constructing a Model of a Potential Accident: A Landslide Case†

As mentioned throughout this chapter, an accident can have a significant effect on the impacts related to all the siting concerns: environmental, economic, socioeconomic, and health and safety. However, because of their rare occurrence and site-specific nature, there is usually a lack of available data for determining directly the likelihoods of such accidents. This section illustrates how to utilize professional experience, judgment, and the data that is available to build a model to specify this likelihood. This likelihood would then be incorporated in a larger model to describe the overall siting impacts.

The case used to demonstrate the development of an accident model concerns the probability of an earthquake-induced landslide at a proposed geothermal power plant in California. The major uncertainties in the model stem from two sources. First, both the occurrence and magnitude of earthquakes are random phenomena. Second, the evaluation of slope

† This section is liberally adapted from Keeney and Lamont [1979]. Individuals making substantial contributions to this study include John Barneich (project manager), John Hobgood, Dennis Jensen, Alan Lamont, Ed Margason, Jug Mathur, Albert Ridley, Jan Rietman, Chuck Taylor, and Tom Turcotte, all of Woodward-Clyde Consultants, who performed the geologic, seismological, and engineering analyses for the study and provided the probability assessments discussed. This work was part of an investigation conducted for the Pacific Gas and Electric Company.

stability from borings and soil samples is still an uncertain science. The final result of any such investigation requires professional judgment from an engineer, which integrates the results of slope stability analyses and other pertinent information. Professional probability assessments provide a means of quantitatively expressing these judgments and explicitly incorporating them into the overall analysis.

6.7.1 DESCRIPTION OF THE GEYSERS PROJECT

The geysers project was part of a feasibility study for using steam from an uncontrolled steam well to run a 5 MWe geothermal power plant. The site is located on an old landslide on the side of a hill. Initial geotechnical studies and analyses indicated that the landslide last moved between 800 and 5000 years ago and that the site had very low potential for sliding under normal conditions. However, as illustrated in Fig. 6.11, three major

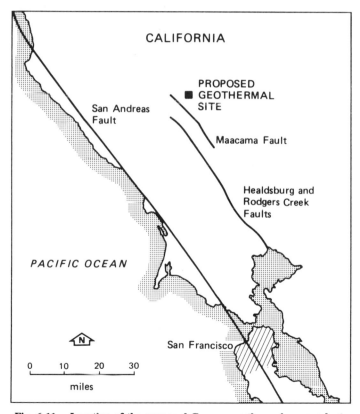

Fig. 6.11. Location of the proposed Geysers geothermal power plant.

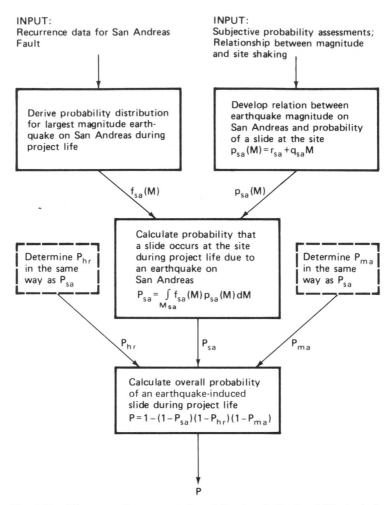

Fig. 6.12.   Diagrammatic representation of the steps in the Landslide Analysis.
(See text for definitions of variables.)

active faults are near the site. The San Andreas is approximately 30 miles away at its closest point. The Healdsburg and Rodgers Creek faults, which for the purposes of this study are treated as one fault due to their similarity in behavior, are approximately 10 miles away, and the Maacama fault is approximately 4.5 miles away. Each of these faults is capable of producing an earthquake severe enough to reactivate the landslide. This study was undertaken to determine the likelihood of such an occurrence.

It is a reasonable assumption that earthquake activity on one fault is probabilistically independent of earthquake activity on the other faults. Thus the analysis was decomposed as illustrated in Fig. 6.12. A statistical model of earthquake occurrence was used to determine the probability distributions $f_{sa}(M)$, $f_{hr}(M)$, and $f_{ma}(M)$ for the largest magnitude earthquake M to occur during the project life (30 years) on the San Andreas, Healdsburg-Rodgers Creek, and the Maacama faults, respectively. The probabilities, denoted P_{sa}, P_{hr}, and P_{ma}, that a slide would occur because of such earthquakes on each of the faults were calculated separately as follows.

Data about the shaking caused by various earthquakes, slope stability analyses, and professional judgments about the relationship between shaking and slides provided information for the probabilities $p_{sa}(M)$, $p_{ma}(M)$, and $p_{hr}(M)$ that a slide would occur at the site due to a magnitude M earthquake on the respective fault. Then the probability P_i that an earthquake on fault i would induce reactivation of the slide at some time during the project life is calculated from

$$P_i = \int_M f_i(M) p_i(M) \ dM, \qquad i = \text{sa,ma,hr}. \tag{6.17}$$

The overall probability P of a slide is one minus the probability of no slide, which is

$$P = 1 - (1 - P_{sa})(1 - P_{ma})(1 - P_{hr}). \tag{6.18}$$

6.7.2   THE LANDSLIDE MODEL

The functions f_i and p_i were developed by assuming that the most significant parameters affecting sliding are the earthquake magnitudes, the peak site acceleration, and the duration of site ground motion. This required the quantification of three relationships:

1. the likelihood of various sized earthquakes on each fault during the proposed plant life,

2. the site acceleration and shaking duration caused by earthquakes of various magnitudes on each fault, and

3. the probability of sliding as a function of the site acceleration and shaking duration.

We will discuss these in order.

Earthquake Occurrence.   Using seismological data and the Gutenberg and Richter [1954] model of earthquake occurrence, the relation between recurrence interval and magnitude for any fault is given by

$$R_i(M) = \exp(b_i M - a_i), \tag{6.19}$$

where $R_i(M)$ is the expected time until an earthquake with magnitude greater than M occurs on fault i. For the three faults in question, the values for a and b are given in Table 6.7.

From (6.19), one can derive the cumulative probability function for the largest earthquake to occur on a particular fault in T years,

$$P(\text{largest earthquake} \leq M \text{ in } T \text{ years})$$
$$= \exp\{-T[\exp(a_i - b_i M)]\}. \tag{6.20}$$

Differentiating this provides the probability density function for the largest earthquake to occur on the fault in question during T years,

$$f_i(M) = b_i T \exp[a_i - b_i M - T \exp(a_i - b_i M)]. \tag{6.21}$$

Earthquake Magnitude and Site Acceleration and Shaking Duration. Determining $p_{sa}(M)$, $p_{ma}(M)$, and $p_{hr}(M)$ involved relating various magnitude earthquakes to durations and site accelerations. This was accomplished using both published data relating peak acceleration to the distance from faults and historical data on maximum earthquake intensities experienced in the region. These results and the sources of the data are given in Table 6.8.

Eliciting the Professional Judgments about Sliding. Geotechnical engineers developed information relevant to the relationship between the probability of sliding and the acceleration and duration by making extensive field and laboratory investigations of the site. This included a geological survey of this and other slides in the area and three types of slope stability analyses (modified Bishop, sliding wedge, and infinite slope). However, such studies are inherently unable to predict exactly what will happen during an earthquake. A format is required for the geotechnical engineer to quantify his or her interpretation of the analytical results of

TABLE 6.7

PARAMETERS FOR EARTHQUAKE
OCCURRENCE

Fault	a	b
San Andreas[a]	3.93	1.22
Maacama[b]	3.27	1.66
Healdsburg-Rodgers Creek[b]	5.68	1.66

[a] Source: Greensfelder [1977].
[b] Source: Ryall *et al.* [1966] and Greensfelder [1977].

TABLE 6.8

EARTHQUAKE ACCELERATION AND DURATION DATA

Fault	Earthquake magnitude	Peak site acceleration[a] (g s)	Shaking duration[b] (sec)	Probability of slide (from Fig. 6.13)
San Andreas	7.5	0.25	≤40	0.0
	8.5	0.33	40	0.16
Maacama	6.0	0.40	7	0.02
	7.0	0.45	15	0.20
Healdsburg-Rodgers Creek	6.5	0.30	≤20	0.0
	7.0	0.35	20	0.03

[a] Source: Schnabel and Seed [1972].
[b] Source: Dobry *et al.* [1970].

the stability analyses in terms of a probability of sliding. This is the essential thrust of the analysis presented here.

The professional judgments about sliding were assessed in interviews with three engineers using techniques of eliciting probabilities such as those discussed in Section 6.1. Three types of questions were asked. First, the probabilities of sliding given earthquake duration and peak accelerations were elicited. Second, the engineers were asked whether the general shapes of the relations implied by their answers to the first questions were consistent with their general knowledge of dynamic slope behavior. Finally, a series of questions was asked to check the internal consistency of the responses to the first series of questions. The specific questions asked are discussed below.

The engineers were asked what the probability of a slide would be assuming that an earthquake induced a peak acceleration of $0.5g$ and had a duration of 40 sec at the site. They chose 0.6 as their response based on their analyses and experience. This implies that the stability is marginal, but it is somewhat more likely to be unstable than stable. When the duration was 40 sec and the acceleration was $0.4g$, the response was 0.3. Next, they were asked what is the level of acceleration such that they would not expect any chance of a slide after 40 sec. The response was $0.25g$. These points are plotted in Fig. 6.13 and the 40 sec curve was drawn through them.

To qualitatively check its shape, the engineers were asked if each additional unit increase in acceleration, with duration fixed, was more likely to induce a slide than the last unit increase. A positive response indicated the convex shape of the curve shown. Other points on the 40 sec curve were then checked to make sure the curve represented the judgment of the geotechnical engineers. Next, the entire process was repeated for

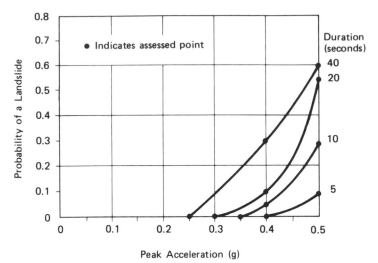

Peak Acceleration (g)

Fig. 6.13.   Relationship between peak acceleration and shaking duration and the probability of a landslide.

earthquake durations of 20, 10, and 5 sec, yielding the results indicated in Fig. 6.13.

These results were cross-checked by questioning the engineers involved about their judgments of the probability of a slide when the peak accelerations were held constant but the total durations were gradually increased. It was discovered during the questioning that this was a more natural way for the engineers to think about the problem since it allowed a better separation of the effects of peak acceleration and duration. There were a few inconsistencies between the results of this line of questioning and the results of the first series of questions. These inconsistencies were discussed and adjustments were made by the engineers until they agreed that the final results reflected their interpretation of the site exploration and analytical results. The curves in Fig. 6.13 represent the final adjusted results.

6.7.3   THE PROBABILITY OF LANDSLIDES

To illustrate the procedure for developing the probability of a slide as a function of the earthquake magnitude, consider the San Andreas fault. From Table 6.8, a magnitude 8.5 earthquake is expected to produce a peak acceleration of 0.33g and a duration of shaking of 40 sec at the site. From Fig. 6.13, it is seen that there is a probability of 0.16 that this earth-

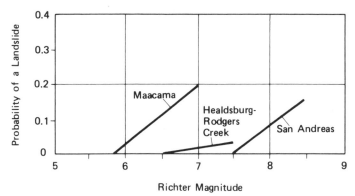

Fig. 6.14.   Relationship between earthquake magnitude and the probability of a landslide.

quake will cause a slide. This point is then plotted in Fig. 6.14. Additional points were determined in a similar way. Then, a linear equation of the form

$$p_i(M) = r_i + q_i M, \qquad M_{\min}^i \leq M \leq M_{\max}^i, \qquad (6.22)$$

was fitted to the points and the constants r_i and q_i were determined. Note that the probability of a slide for magnitudes below $M_{\min}$ is zero. For magnitudes above $M_{\max}$, the largest credible magnitude on the fault in question, the probability of the earthquake itself is zero. The derived linear relationships between earthquake magnitude and the probability of a landslide are shown in Fig. 6.14. The values for r, q, $M_{\min}$, and $M_{\max}$ are given in Table 6.9.

To calculate the probability P_i that a slide will be caused by an earthquake on a fault i in T years, (6.21) and (6.22) are substituted into (6.17) to give

$$P_i = \int_{M_i} (r_i + q_i M) b_i T \exp[a_i - b_i M - T \exp(a_i - b_i M)] \, dM. \quad (6.23)$$

TABLE 6.9

PARAMETERS FOR THE PROBABILITY OF SLIDING
GIVEN EARTHQUAKE MAGNITUDE

Fault	r	q	$M_{\min}$	$M_{\max}$
San Andreas	−1.20	0.16	7.5	8.5
Maacama	−1.05	0.18	5.9	7.0
Healdsburg-Rodgers Creek	−0.19	0.03	6.5	7.5

Using the appropriate values of a, b, q, r, M_{max}, and M_{min}, the resulting probabilities of sliding due to earthquakes in the 30-year project life are

$P_{sa} = 0.007$ for the San Andreas,
$P_{ma} = 0.003$ for the Maacama, and
$P_{hr} = 0.002$ for the Healdsburg-Rodgers Creek.

Finally, substituting the above probabilities into (6.18), we find the overall probability of a slide in the 30-year period is

$$P = 1 - [(1 - 0.007)(1 - 0.003)(1 - 0.002)] = 0.012. \quad (6.24)$$

This probability incorporates the uncertainties about earthquake occurrence and the interpretation of slope stability analyses. It provides an input estimate that helps calculate impacts of siting a geothermal plant at the proposed site.

6.8   Constructing a Model of Public Risks: An LNG Case†

Even though the domestic supply of natural gas is declining, there are strong arguments in favor of its continued use: it burns cleanly; an efficient distribution system exists; and consumers prefer gas as the fuel for heating homes and other buildings. One solution to the declining domestic supply problem is to import natural gas—the liquification and transportation of natural gas is a well-developed technology. The major concerns about the large-scale importation of liquefied natural gas (LNG) are related to safety. This section presents an analysis of the public risks associated with operating a proposed facility for importing liquefied natural gas. In Section 6.8.1 the specific problem is described. The risk analysis approach for indicating public safety impacts is outlined in Section 6.8.2. The details of the case follow.‡

† This section is liberally adapted from Keeney, Kulkarni, and Nair [1978, 1979]. The study described, conducted by Woodward-Clyde Consultants, was part of a safety analysis of the El Paso LNG Company. Mr. Ivan Schmitt, Vice President, was responsible for the Safety Analysis report. Mr. Randall I. Cole assembled the team of consultants and was a constant source of advice and contributions. Mr. Joe Porricelli of Engineering Computer Opteconomics provided all the ship accident and spill information, and Dr. Hal Wesson of Wesson and Associates provided input about LNG operations and the vapor cloud and ignition models. Their experience and knowledge were critical to the study. Other members of Woodward-Clyde Consultants' staff who made significant contributions to the study were Steven James, Craig Kirkwood, Ram Kulkarni, Kesh Nair, John Pelka, and John Mc-Morran.

‡ The astute reader will have noted the use of a nonword in this paragraph.

6.8.1 THE PROPOSED LA SALLE TERMINAL

The El Paso LNG Company and its subsidiaries plan to build and operate a marine terminal and LNG vaporization facility, named La Salle Terminal, in Matagorda Bay, Texas. Matagorda Bay, illustrated in Fig. 6.15, is located approximately 120 miles southwest of Houston on the Texas Gulf Coast. Although there are no cities or major towns in the vicinity of the proposed terminal, there is the small town of Port O'Connor and the village of Indianola.

The La Salle Terminal is designed to process 30 million m³ of natural gas per day. This LNG, which will be obtained from Algeria, will be delivered to the terminal by a fleet of LNG carriers. There will be approximately 143 LNG carrier arrivals in the facility per year, and each carrier will deliver approximately 125,000 m³ of LNG per arrival. After an LNG carrier berths, the LNG will be transferred via cryogenic piping to three LNG storage tanks, each of 100,000 m³ capacity. The LNG will be indepen-

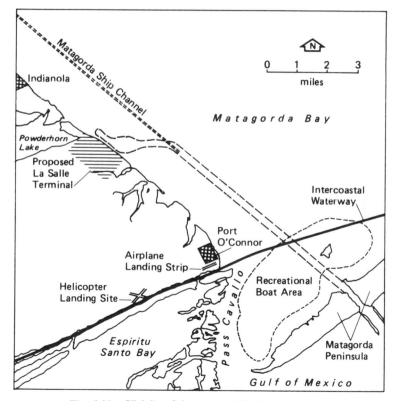

Fig. 6.15. Vicinity of the proposed La Salle Terminal.

dently withdrawn from these tanks, revaporized, and then piped to consumers via a high-pressure intrastate pipeline.

All phases of the design, construction, and operation of the La Salle Terminal and LNG carrier fleet will employ special safety procedures and techniques. The La Salle Terminal LNG storage tanks will be located in diked areas and will be designed to minimize any possible spillage. The LNG carriers have special double-hull and double-bottom construction, and will use sophisticated anticollision and navigational systems. Special U.S. Coast Guard operating procedures which apply to LNG carriers in U.S. waters and Matagorda Bay will be followed. Fire-protection and control equipment designed to deal with LNG fires will be constructed and maintained at the facility.

All of these safety considerations are designed to reduce to an extremely low level the likelihood of an LNG accident that could result in consequences to the public. However, such an event is possible. The estimation and evaluation of the risks to the public from the operation of such a facility is an important element in determining the feasibility of the project and in obtaining approval from the appropriate regulatory agencies.

An initiating event such as a ship collision or a terminal-equipment malfunction resulting in a spill of LNG must take place to create a potential for public fatalities. Because of the nature of the circumstances required to cause such a spill, it is very likely that the spill would be immediately ignited, resulting in a spill pool fire. In the unlikely event that the spill does not immediately ignite, the very cold ($-260°F$) LNG rapidly evaporates, forming a negatively buoyant LNG vapor cloud. Under certain weather conditions, this cloud may be carried to an area occupied by the public. The mixture of vaporized LNG and air is such that the vapor cloud may ignite as a result of contact with an ignition source such as a gas pilot light, the flame of a cigarette lighter, or an electric spark. If the vapor cloud does not ignite within a certain length of time, the ratio of vaporized LNG to air will be too low to allow ignition and the cloud will disperse. Individuals caught in a spill pool or vapor cloud fire or exposed to high thermal radiation levels causing severe burns may become fatalities.

6.8.2 THE RISK ANALYSIS MODEL

A schematic representation of the La Salle risk analysis model is given in Fig. 6.16. The model can be divided into three components:

1. development of accident scenarios and their associated probabilities,
2. quantification of public risks, and
3. evaluation of public risks.

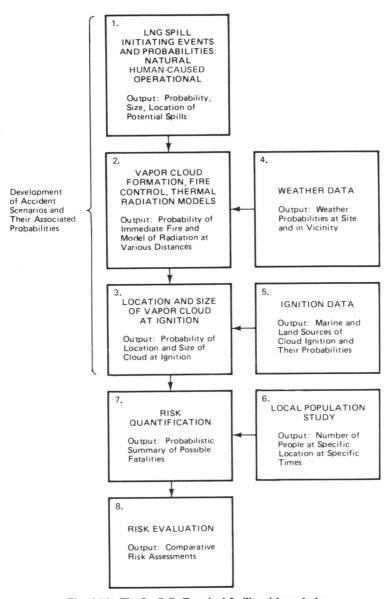

Fig. 6.16   The La Salle Terminal facility risk analysis.

Because of the complexities of any risk analysis, simplifying assumptions must be made. The spirit of this analysis required that all assumptions be stated explicitly and conservatively. That is, these assumptions will cause the analysis to overestimate the public risks.

*Development of Accident Scenarios and Their
Associated Probabilities*

An accident scenario is a sequence of events that might result in public fatalities. One representative accident scenario could be described as follows: an LNG carrier collision occurs in the harbor, releasing an LNG spill of a certain size. There is no immediate ignition, so a vapor cloud forms. The wind is from the east at 10 miles/hr with Pasquill stability† class D. The eighth ignition source ignites the vapor cloud. Elements 1–5 of Fig. 6.16 provide the information necessary for formulating the accident scenarios. Figure 6.17 represents accident scenarios as specific sequences of branches in an event tree.

Each element of the accident scenario must take place either simultaneously or in a specific order. In the risk analysis model, the annual probability of a particular scenario involving a vapor cloud is the product of the following factors:

1. annual probability of the initiating accident,
2. probability of no immediate ignition for that accident,
3. probability of the wind direction,
4. probability of the wind speed and stability class given that wind direction, and
5. probability that the nth ignition source ignites the vapor cloud.

Similarly, the probability of a particular accident scenario which results in a pool fire is equal to the annual probability of the initiating accident multiplied by the probability of immediate ignition.

Quantification of Public Risks

Four categories of public risk were examined in this analysis: societal risk, individual risk, group risk, and the risk of multiple fatalities.

Once the probabilities of all accident scenarios were determined, the probability $P(x)$ of x fatalities was calculated from

$$P(x) = \sum_i P(x/A_i)P(A_i), \tag{6.25}$$

where $P(x/A_i)$ is the probability of exactly x fatalities resulting from accident scenario A_i and $P(A_i)$ is the annual probability of scenario A_i. The expected number of fatalities $F(A_i)$ due to accident scenario A_i was then

$$F(A_i) = \sum_x xP(x/A_i). \tag{6.26}$$

† The Pasquill stability class is an indicator of atmospheric mixing potential. The mixing potential decreases from class A to class G. In this study, conditions are such that only classes A–F are necessary.

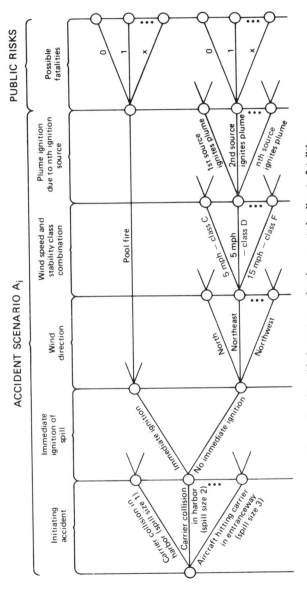

Fig. 6.17. Simplified event tree showing events leading to fatalities.
Note: At each node, probabilities are assigned to the various events.

Equations (6.25) and (6.26) provided the basis for quantifying all of the public risks.

Societal risk, indicated by the expected number of annual fatalities F, is

$$F = \sum_i P(A_i)F(A_i). \tag{6.27}$$

Individual risk R is measured by the annual probability that an exposed individual will be a fatality due to possible risks from LNG. This is found by dividing the expected number of annual fatalities by the total number N of people exposed, yielding

$$R = F/N. \tag{6.28}$$

A similar calculation was used to indicate group risk. This was measured by the probability R_G that a member in the particular group G would be a fatality in a specific year. This is found from

$$R_G = F_G/N_G, \tag{6.29}$$

where F_G is the expected fatalities per year to group G and N_G is the number of people in group G.

The risk of multiple fatalities is indicated by the probability that the number of fatalities in a given year is equal to or greater than a specific level y. This can be calculated directly from (6.25). It is simply the sum of the probability of y fatalities, $y + 1$ fatalities, and so on. Hence

$$P(\text{fatalities} \geq y) = \sum_{x \geq y} P(x). \tag{6.30}$$

Risk Evaluation

There are no generally accepted criteria for evaluating public risks. The approach used in this analysis was to compare risks generated from this project with existing risks to the public, including risks from alternate energy sources, and with acceptable levels of public risks suggested in the literature.

6.8.3 DEVELOPMENT OF ACCIDENT SCENARIOS AND THEIR ASSOCIATED PROBABILITIES

The elements of this step of the risk analysis consisted of determining: (1) the probability, size, and location of potential spills; (2) the probability of vapor cloud formation and downwind travel and the probability of an immediate fire; and (3) the probability of the location and size of cloud at ignition.

Probability, Size, and Location
of Potential Spills

The events that can cause spills were categorized as follows:

1. natural hazards (e.g., hurricanes, earthquakes) which affect the facilities,
2. external human-caused hazards (aircraft crashes) which affect the facilities,
3. LNG carrier fleet accidents, and
4. La Salle Terminal accidents.

Several analyses were conducted to determine the probabilities that each of the events would result in LNG spills. Particular attention was focused on those spills that would have a potential of resulting in public fatalities. The initiating accidents found to result in significant spills are summarized in Table 6.10 and discussed below.

Natural Hazards. The potential risks due to various natural hazards, including earthquakes, severe winds, storm waves and tsunamis, and

TABLE 6.10

INITIATING ACCIDENTS CONTRIBUTING TO PUBLIC RISK

Location of accident	Cause and size of spill	Annual probability of accident event	Spill size
Entrance	Most credible spill due to collision	3.46×10^{-5}	19,400 m³ instantaneously
	Maximum credible spill due to collision	9.28×10^{-6}	38,800 m³ in 9 min
	Spill due to ramming	7.59×10^{-8}	10,000 m³ in 12 min
	Spill due to aircraft crash	1.21×10^{-8}	10,000 m³ in 12 min
Harbor	Most credible spill due to collision	6.49×10^{-7}	19,400 m³ instantaneously
	Maximum credible spill due to collision	1.76×10^{-7}	38,800 m³ in 9 min
	Spill due to ramming	2.10×10^{-7}	10,000 m³ in 12 min
	Spill due to aircraft crash	9.23×10^{-8}	10,000 m³ in 12 min
Pier	Most credible spill due to collision	1.01×10^{-7}	19,400 m³ instantaneously
	Maximum credible spill due to collision	2.73×10^{-8}	38,800 m³ in 9 min
	Spill due to ramming	1.10×10^{-5}	10,000 m³ in 12 min
	Spill due to aircraft crash	5.93×10^{-7}	10,000 m³ in 12 min

meteorites were investigated. Essentially, none of these natural events contributes significantly to the public risk in comparison to the risks associated with other types of accidents. A summary of the risks from natural hazards follows.†

Using conservative estimates, the likelihood that an earthquake would produce a ground acceleration at the La Salle Terminal site large enough to exceed the design specification of the storage tanks is approximately 10^{-11} per year. Even if the tank did rupture with such an earthquake, the analysis for the onshore facilities indicated that public risk was essentially nil. There was a higher probability of pipe breaks than of storage tank failure due to ground motions, but the analysis indicated that the results were inconsequential to public risk.

The main source of severe winds in the vicinity of Matagorda Bay is hurricanes. The primary concern in the case of wind is its effect on the storage tanks, since all LNG carriers will leave and remain outside of Matagorda Bay if winds greater than 60 mph are forecast or observed. The tanks would be designed to withstand an instantaneous gust of 217 mph and a 1 min wind of 166 mph. Meteorological data (Glenn and Associates [1976]) indicate that a 1 min wind of 166 mph occurs once every 100 years, which is twice as frequent as an instantaneous gust of 217 mph. Hence, it was assumed that a storage tank may begin to fail with probability 1.5×10^{-2} in any given year. However, even if the tank failed completely, either a pool fire would burn on the site or the wind turbulence would disperse the vapor plume before it passed terminal boundaries.

There are no known faults capable of generating significant tsunamis in Matagorda Bay. Extrapolating data from the storm-wave analysis, the likelihood of damage that could result in public risk from terminal accidents was found negligible. Operating policy requires that in storm conditions all LNG ships will leave Matagorda Bay. Since sufficient warning of impending large waves would be available, the possibility that they would cause ship accidents and contribute to public risk is assumed to be negligible.

The probability of a meteorite penetrating a ship's tanks were calculated to be over two orders of magnitude smaller than the probability of ship collisions and, hence, negligible.

External Human-Caused Hazards. The likelihood of an airplane crashing into the storage tanks, the LNG pipelines at the terminal, or a tanker was calculated. Because there are no major airports in the vicinity of Matagorda Bay, aircraft-accident statistics for general aviation were

† Detail of these analyses are provided in Appendix A of El Paso Atlantic Company *et al.* [1977].

used in these calculations. The main contributing factor to aircraft–ship accidents in the ship channel is the operation of small planes at the local Port O'Connor airstrip. There is also a helicopter landing site in Port O'Connor, but it appeared that flight patterns could be arranged to avoid operations in the area where the LNG carriers will operate.

If an airplane crashed into the storage tanks or LNG pipelines, there is no risk to the public, for reasons discussed earlier. For crashes into ships, it is assumed that the spill would be 10,000 m³ in 12 min.

The potential risks due to sabotage were qualitatively examined. It would appear essentially impossible to eliminate the likelihood of sabotage by determined saboteurs. However, it is very likely that immediate ignition would occur due to the violence required to release the LNG in the first place. In the broader sense, the decision to allow LNG import terminals anywhere in the United States implies that such risks must be taken.

Onshore and Marine Terminal Accidents. Wesson and Associates, Inc. conducted an analysis of the potential consequences resulting from accidental LNG spill events in the onshore and marine terminal facilities. Specifically, the downwind vapor dispersion characteristics and thermal radiation levels were calculated for postulated spill events including maximum credible spills on the pier, spills in the unloading and sendout transfer piping systems, and spills from the LNG storage tanks and associated piping.

The postulated LNG storage tank spill is impounded in a diked area surrounding each LNG tank. LNG spill collection troughs, runoff trenches, and remote impounding basins are provided for other spill events. All spill impounding areas, including the LNG storage tank diked areas, are provided with automatic vapor dispersion and fire control systems. These systems insure that no thermal radiation or vapor dispersion hazards exist at any plant boundary line under any set of weather conditions.

Marine Accidents. Engineering Computer Opteconomics, Inc. conducted an analysis of ship accidents leading to an LNG cargo release. Specifically, the probabilities of collisions with other vessels, rammings of nonvessel stationary or floating objects, and groundings were calculated for three segments of Matagorda Bay:

1. the entranceway, which is the immediate ship approach route to and through the cut in Matagorda Peninsula;
2. the harbor, which is the ship channel within Matagorda Bay; and
3. the piers, which are the waters between the ship channel and the berth.

The analysis of spills due to collisions indicated that either one or two cargo tanks may be involved. A one-tank cargo involvement was considered the most credible spill event due to collision, and would involve 19,400 m³ of LNG. The maximum credible spill event due to collision was the simultaneous rupture of two adjacent cargo tanks, and could involve 38,800 m³ of LNG. The analysis of spills due to ramming indicated that the most credible spill was 10,000 m³. The analysis of grounding indicated that the expected spills due to cargo tank rupture did not pose any significant hazard.

Vapor Cloud Formation and Downwind Travel

If a spill of sufficient size occurs without immediate ignition, an LNG vapor cloud may form and travel downwind. The characteristics of the cloud depend for the most part on the size and rate of the spill and the surface area on which the LNG is spilled. Since land spills have been ruled out as significant to public risk, the focus here is on water spills. The travel of the vapor cloud depends on the wind speed and direction and the stability of the weather.

For the most credible spill of one tank (19,400 m³ spilled instantaneously), the vapor could travel up to 5.88 miles with a 10 mph wind of stability class D if it is not ignited. This would require roughly 35 min. Then the cloud would disperse as the LNG–air mixture was reduced to below the lower flammable limit of 5% average concentration.

Likelihood of Immediate Ignition. The potential for a fire associated with spills depends on the presence of both flammable vapors and the ignition source. Vaporization would commence immediately following a cargo tank penetration. Ignition of vapor can be caused by any heat source of sufficient temperature and duration. The primary ignition sources would be the friction and sparking generated by the immense forces needed to attain penetration. Temperatures could be expected to range from 1600 to 2700°F, far in excess of the 1000°F ignition temperature of methane and air. Secondary ignition sources such as boilers, galley fires, electrical cables, and light fixtures will also be present and would be exposed as a result of collisions, rammings, or aircraft–ship crashes. Therefore it is extremely likely that immediate ignition will occur under such circumstances. In the analysis, based on experts' judgments, it is assumed that the probability of ignition following such a penetration is 0.99.

Wind Direction, Speed, and Weather Stability. Because wind direction, wind speed, and stability class are not independent of each other, the weather conditions in the vicinity of the La Salle Terminal were tabulated using data from Glenn and Associates [1976] and the National Oceanic and Atmospheric Administration [1971]. First, the probabilities of eight wind directions (north, northwest, west, etc.) were determined. Then, conditioned on each of these directions, the joint probability of the wind speed (in 5 mph increments) and stability class were ascertained.

Size and Location of the Vapor Cloud
at Ignition and Its Effects

A vapor cloud can be ignited from numerous sources such as spark plugs, open flames, pilot lights, and electrical sparks due to short circuits. Because of the difficulty of tabulating all possible ignition sources, it was conservatively assumed that each building and each recreational boat or commercial fishing ship had only one source. Assuming fewer sources is conservative because it implies increased likelihood that a vapor cloud would cover a larger area before ignition.

Vapor Cloud Ignition Model. The structure of the ignition model can be illustrated as follows. The probability that any individual source ignites a passing vapor cloud was defined to be p, and this was identical for each ignition source. Hence, if a cloud begins to cover an area with ignition sources, the probability that it ignites from the second source is $(1 - p)(p)$, which is the probability $(1 - p)$ that it does not ignite at the first source times the probability p that it does ignite at the second source. In general, the probability that a vapor cloud is ignited by the nth source is $(1 - p)^{n-1}(p)$.

Radiation Model. In this study, a thermal radiation level of 5300 Btu/ft^2 hr was used as the lower limit resulting in fatalities. This is conservative because one would actually expect only blistering of skin exposed for 5 sec to such thermal radiation. Using such a level results in the distances calculated in Table 6.11 for LNG pool and cloud fires. Because these do not vary greatly with weather conditions, these distances are assumed for all weather conditions.

TABLE 6.11

DISTANCES ASSUMED FOR THERMAL RADIATION FATALITIES

Incident	Distance from 5300 Btu/ft^2 hr exposure for 5 sec
10,000 m^3 in 12 min pool fire	1100 ft from spill center
19,400 m^3 instantaneous spill pool fire	2500 ft from spill center
38,800 m^3 in 9 min pool fire	2500 ft from spill center
Vapor cloud fire	525 ft from plume fire

6.8.4 RISK QUANTIFICATION

The purpose of the risk quantification was to determine the expected number of public fatalities per year and annual risk levels due to the LNG terminal operation. This required, in addition to the information previously developed, data on the population distribution in the vicinity of the La Salle Terminal.

Population Study

A detailed on-site study of the population living in and visiting the vicinity of the La Salle Terminal was conducted. The main goal was to determine how many people are at what locations and at what times. Because of large seasonal and weekly fluctuations in the population, population data were tabulated separately for all combinations of days and nights, weekdays and weekends, tourist season (May–October) and nontourist season. After a preliminary examination of population data, it was decided that three combinations—nontourist season, tourist weekends, and tourist weekdays—adequately described population distributions. The risk quantification considered each of these cases separately.

Data used to estimate populations included electricity use, current and projected school enrollments, number of boats, estimated beach visitors, fishing season dates, special events in the area, etc. For fatality calculations, the population study indicated that it was reasonable to assume 3.62 people per permanent household, 4.02 people per occupied transient household, 4.5 people per recreational boat, and 2.5 people per commercial fishing boat.

Illustrative Calculation

To indicate the structure of the risk analysis model, the contribution to the average number of fatalities made by one accident scenario is illustrated. Define accident scenario A_1 as a collision between an LNG carrier and another ship in the harbor at the intersection of the ship channel and intercoastal waterway. This collision produces an instantaneous one-tank spill of 19,400 m^3 of LNG. Furthermore, assume the spill does not immediately ignite, so that a vapor cloud forms. The wind is assumed to be from the east at 10 mph with a Pasquill stability class D. The eighth ignition source ignites the vapor cloud. The illustration is for a weekday during the tourist season.

The probability of this accident scenario is the product of the following factors:

1. the annual probability of a collision releasing one tank of LNG (most credible spill) in the harbor during a weekday in tourist season (Table 6.10),†

2. the probability $p = 0.01$ of no ignition given this collision,

3. the probability that the wind is from the east,

4. the probability that, given an east wind, it is 10 mph and of stability class D, and

5. the probability that the eighth ignition source ignites the vapor cloud.

The calculation, using the probabilities indicated, is as follows:

$$P(A_1) = (2.33 \times 10^{-7})(0.01)(0.131)(0.186)(0.0478)$$
$$= 2.71 \times 10^{-12}. \qquad (6.31)$$

To calculate the expected fatalities $F(A_1)$ from accident scenario A_1 on a tourist season weekday, we first determine the maximum extent of the flammable vapor cloud if it is not ignited. This cloud is superimposed on the map of Matagorda Bay as shown in Fig. 6.18.

Since 100 recreational boats are assumed to be in the designated region of Fig. 6.18, and since the projected cloud covers 2% of the region, the expected number of recreational boats in the plume's path is 2, assuming a uniform distribution of boats. No commercial boats are assumed to be in this area on tourist season weekdays.

On tourist weekdays, there is an average of 1000 daytime visitors in Port O'Connor. It is conservatively assumed that all of these visitors are on the beach on the east side of Port O'Connor. The vapor cloud in Fig. 6.18 covers 34% of this beach.

The number of permanently occupied houses covered by the maximum possible cloud is 37 and the number of houses occupied by transients is 103, for a total of 140 household ignition sources. The average weekday daytime occupancy of these houses is taken as 2.62 and 3.02, respectively. This assumes one person is away from each household. Averaging the expected weekday-daytime occupancy of the houses occupied permanently and by transients, an expected number of people per household source is determined to be 2.91.

In the event of a house-caused ignition, it is assumed that individuals in boats and daytime visitors within the vapor cloud, as well as all individuals in houses covered by the cloud at the time of ignition, are fatalities.

The contribution of accident scenario A_1 to expected fatalities is found

---

† The probabilities presented in Table 6.10 were multiplied by the ratio of each time period (nontourist season, tourist weekends, and tourist weekdays) in days to 365 to give the probability of an accident within a time period.

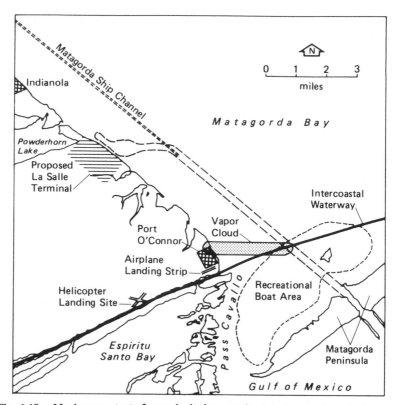

Fig. 6.18. Maximum extent of an unignited vapor cloud for illustrative calculations.

as follows. Since the vapor cloud is ignited by the eighth ignition source, it
is not ignited by the two boats but by the sixth house encountered. The
number of individuals within this cloud at the time of ignition is the sum of
the people in the two boats, on the beach, and in the six houses under the
cloud. This gives the expected number of fatalities in the cloud as

$$2(4.5) + 340 + 6(2.91) = 366.46. \qquad (6.32)$$

As indicated in Table 6.11, it is assumed that individuals within 525 ft of
a vapor cloud fire would be fatalities due to thermal radiation. Using the
average population density of Port O'Connor, it is assumed that 230 individuals could be within 525 ft of a plume fire. Twenty percent of these are
assumed to be out of doors at any given time. Hence, the total expected
fatalities from A_1 on a summer weekday is

$$F(A_1) = 366.46 + (0.2)(230) = 412.46. \qquad (6.33)$$

Multiplying (6.31) by (6.33) results in 1.118×10^{-9}, which is the contribution of accident scenario A_1 to the overall annual expected fatality rate F in (6.27).

Calculations of expected fatalities due to thermal radiation from an LNG pool fire are illustrated by the following example. The probability of an LNG carrier having a collision in the harbor during a tourist season weekend is 1.20×10^{-7}/yr. Based on data from the Matagorda Bay area, the average number of boats within 2500 ft of the center of that spill is 20. The average number of occupants in a recreational boat is assumed to be 4.5 and the average number of occupants in a small commercial fishing boat is assumed to be 2.5. Since slightly less than 75% of the total number of boats exposed to pool fires are recreational boats, an average occupancy per boat of 4.0 is used. The annual expected fatalities F_f due to a pool fire resulting from a collision in the harbor on a tourist season weekend is

$$F_f = (1.20 \times 10^{-7})(0.99)(20)(4) = 9.50 \times 10^{-6}. \tag{6.34}$$

Results

The annual risks for individuals in different groups are presented in Table 6.12. The probability of exceeding a specified number of fatalities in a given year is shown in Fig. 6.19.

The relative contributions to the annual expected fatalities of different accidents, seasons, locations, and spills are shown in Table 6.13. As seen in Table 6.13d, approximately 62% of the expected fatalities come from individuals in boats subjected to the pool fire thermal radiation resulting from immediate ignition of an LNG spill in the harbor, mostly during tourist weekends. The balance of the expected fatalities, 38%, is from the delayed ignition of vapor clouds. Approximately half of these are individuals in boats and half are individuals on shore.

The base case analysis assumed that the probability of ignition per

TABLE 6.12

SUMMARY OF PUBLIC RISKS

Group	Expected fatalities per year	Number of people in group sharing risk	Risk per person per year
Permanent population in Port O'Connor	2.0×10^{-8}	800	2.5×10^{-11}
Permanent population in Indianola	1.3×10^{-7}	80	1.7×10^{-9}
Daytime visitors	2.5×10^{-6}	2500	9.9×10^{-10}
Individuals in boats	1.35×10^{-5}	3000	4.5×10^{-9}
All individuals exposed to risk	1.7×10^{-5}	9000	1.9×10^{-9}

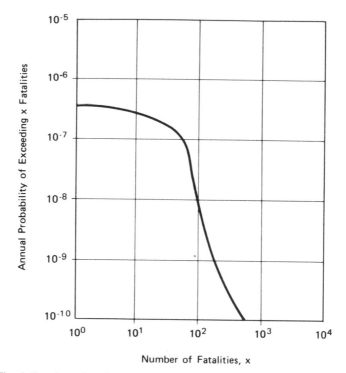

Fig. 6.19.   Annual probability of exceeding a specified number of fatalities.

source was 0.1 and that an LNG vapor cloud was ignitable until it diffused to 5% average methane concentration. Sensitivity analyses were conducted to examine the effect on public risk of more conservative levels of these two parameters. First, the probability of ignition per source was lowered to 0.01. Next, it was assumed that pockets of peak concentrations that are still ignitable exist in the vapor cloud until it reaches an average concentration of 2.5%.

Both cases showed a slight increase in the annual risk per person over that for the base case. For the two cases outlined above, the annual risks were 2.22×10^{-9} and 2.21×10^{-9} per person, respectively. Comparison with the risk of 1.9×10^{-9} per person per year for the base case shows that the public risks are not influenced significantly by the more conservative estimates of the two parameters.

6.8.5   RISK EVALUATION

The La Salle Terminal is designed to receive and vaporize approximately 30 million m³ of natural gas per day. For risk comparison, the energy

TABLE 6.13

ANALYSIS OF THE EXPECTED FATALITIES

	Expected fatalities per year	Percentage of total expected fatalities
a. Expected Fatalities for Different Seasons		
Season		
Nontourist (November–April)	8.21×10^{-7}	4.8%
Tourist Weekdays (May–October)	3.26×10^{-6}	19.2%
Tourist Weekends	1.29×10^{-5}	76.0%
b. Expected Fatalities for Different Accident Locations		
Accident Location		
Entranceway	2.45×10^{-6}	14.4%
Harbor	1.10×10^{-5}	64.8%
Pier	3.52×10^{-6}	20.8%
c. Expected Fatalities for Different LNG Spills		
Type of LNG Spill		
Most credible spill due to collision	9.70×10^{-6}	57.2%
Maximum credible spill due to collision	2.72×10^{-6}	16.0%
Spill due to ramming	3.63×10^{-6}	21.4%
Spill due to aircraft crash	8.99×10^{-7}	5.4%
d. Expected Fatalities for Different Accidents		
Type of Accident		
Spill resulting in pool fire	1.06×10^{-5}	62.4%
Spill resulting in vapor cloud	0.64×10^{-5}	37.6%

output of this natural gas is equivalent to about 18,1000 MWe electric power plants operating at 70% capacity (Van Horn and Wilson [1976]).

Buehring [1975] calculated that the expected number of deaths to the public due to transporting coal was 0.695 per year for a 1000 MWe coal facility. Multiplying by 18, this becomes 12.51 expected fatalities per year. For the equivalent energy output of La Salle Terminal, the expected fatality level is 0.000017. This is less than 0.001% of the expected fatalities from the equivalent consumer energy obtained from coal.

The group with the highest annual risk, 4.5×10^{-9} per year, is individuals using boats. For Port O'Connor residents, the annual risk per person is 2.5×10^{-11}. The National Safety Council [1975] provides data to calculate that the expected public risk per individual per year due to gas distribution systems in the U.S. is 5.15×10^{-7} and due to electric shock in electrically wired residences is 1.11×10^{-6}. Hence, for any individual exposed to a hazard because of the project, the probability that he or she will die due to a natural gas distribution accident is 271 times greater than the possibility of death due to the operations of the proposed La Salle Terminal. The probability that the individual will die next year due to electric shock at home is 584 times greater.

For the permanent population of Port O'Connor, the same model used to calculate the probability of airplane crashes into LNG carriers was used to calculate the probability that an airplane would crash directly into Port O'Connor and kill an individual. This probability was found to be 4.73×10^{-8} per year for a resident living 1 mile from the runway. This is over three orders of magnitude greater than the risk due to the operation of the proposed terminal. About 600 of the 840 residents of Port O'Connor live within 1 mile of the runway.

A criterion for risk acceptability has been proposed by Starr [1969]. It suggests that acceptability be determined by the annual probability of death per person exposed and the benefits of the activity, measured in arbitrary units. Because the evaluation of benefits is complex, the main use of Starr's criterion is in establishing a bound, of approximately 10^{-7}, for acceptable risk levels. The individual risk from the operations of the proposed La Salle Terminal is 1.9×10^{-9}, which is less than 2% of 10^{-7}. The societal risk due to operation of the La Salle Terminal is, consequently, much less than Starr's criterion for social acceptance.

6.8.6 CONCLUSIONS

The formal risk analysis described in the previous sections had several advantages:

1. data and professional judgments from experts in various fields could be integrated into a logical framework,
2. assumptions could be stated explicitly,
3. sensitivity analyses could be conducted to evaluate the significance of the assumptions,
4. the public risk could be systematically estimated and appraised, and
5. strategies to reduce risks could be developed.

The study was used in preparing the safety analysis report for submission to the Federal Power Commission (FPC). The Final Environmental Impact Statement issued by the FPC stated that the levels of public risk associated with the La Salle Terminal facility are acceptable.

Siting studies should include an analysis of the public risks of fatalities, injuries, and sickness associated with each candidate site. This would be very useful in the site comparison as well as in explaining and justifying the eventual site choice. The risk analysis presented here indicates the type of information which one can derive and the components in that derivation.

CHAPTER 7

EVALUATING SITE IMPACTS

One site is better than another for a particular energy facility only if it is judged to be so. Such a judgment requires values. In fact, the siting problem vanishes without values. In this chapter, we are concerned about:

1. What is it that is valued?
2. Whose values should be used?
3. How should the values be elicited?
4. How should they be incorporated into the analysis?

These questions are and must be addressed in every siting problem regardless of the evaluation procedure (including the informal nonquantified approach). There are no exceptions in this respect. However, with different procedures, the responses to the questions posed are different.

A major distinction between decision analysis and other approaches is the manner in which these questions are addressed. With decision analysis, they are addressed explicitly and formally. It is the impacts discussed in Chapter 6 that are valued in siting. These impacts are the possible consequences x of each site S_j which are quantified by the probability distribution $p_j(x)$. The values to be used are those of the client, who, in Section 1.2, was defined as the decision maker for our problem. As indicated, the client may wish to incorporate the values of others in his or her value structure. The values should be quantified using a utility function $u(x)$ which assigns a utility to each consequence. This utility function is a mathematical representation of the client's attitude toward risks, value tradeoffs among different impacts and groups of people, and preferences for impacts over time. Procedures for eliciting the utility function are discussed throughout this chapter. Then if the client feels that the axioms of decision analysis (for siting), discussed in the appendix, are responsible ones to follow in evaluating candidate sites, the expected utility Eu_j of each site S_j should be calculated from

$$Eu_j = \int u(x)p_j(x)\ dx \qquad (7.1)$$

and the sites ranked according to higher expected utilities.

With other approaches to siting, the impacts are also valued. However, rarely are the hard-to-quantify impacts, such as socioeconomics and some environmental concerns, formally included or the uncertainties made explicit. The values used are the client's, but they are seldom formalized to elucidate the value structure. The manner in which risks, value tradeoffs, and impacts over time are treated is not clear. Because the client's values are not formalized, elicitation procedures are not necessary. Then, of course, the values must be synthesized with the impacts in an informal manner which does not have the logical or theoretical basis that decision analysis has.

Issues Involved in Quantifying the Value Structure

The evaluation of impacts can be broken down into several components along the lines indicated by the value issues in Section 1.4. In particular, the later sections of this chapter address the following:

1. evaluation of impacts on multiple objectives (Section 7.2),
2. evaluation of impacts in different time periods (Section 7.3),
3. evaluation of impacts to different groups (Section 7.4), and
4. evaluation of risks and uncertainty (Section 7.5).

Each of these sections will include a presentation of formal models for quantifying preferences, assessment procedures for implementing the methodology, and illustrative examples. Section 7.6 illustrates in detail the art of quantifying preferences. Sections 7.2–7.6 concentrate on the manner in which the results should be implemented in the single-client formulation of the siting problem. For the multiple-client version, as characterized in Section 1.2, the assessment procedures need some adaptation, as discussed in Section 7.7.

The general problem addressed in this chapter is critical to the evaluation of alternatives in any decision problem. Several books, such as Fishburn [1964, 1970], Krantz *et al.* [1971], and Keeney and Raiffa [1976], have major parts which address evaluation in complex situations, and, in the past decade, numerous articles have appeared in the technical literature. Extensive recent reviews of this literature are Fishburn [1977] and Farquhar [1977]. There would be little use in repeating all these results here. Instead, we shall try to provide, in as concise a fashion as possible for each of the value issues above, the following:

1. a clear statement of the issue,
2. a presentation of concepts necessary to address the issue,
3. a summary of the main results useful in addressing it,
4. a procedure for assessing the value judgments necessary to use the results, and
5. an example illustrating the use of the results.

Appropriate references will be made to more detailed discussions on the topic.

7.1 Advantages of Formalizing the Value Structure

There are numerous significant advantages of formalization of the value structure. All of these stem from the two purposes of siting studies: to select the best site and to explain and justify the decision to regulatory authorities and others.

With regard to selecting the best site, in providing the utility function u, the client has an opportunity—and is practically forced—to develop a consistent set of values. Such a task is not easy because of the complexity of the problem. All the value issues in siting outlined in Section 1.4 are involved in this process. In the assessments, the problem is broken into parts. This allows experts to provide values for the appropriate parts. For instance, an ecologist could interpret and assign relative values to different impacts on the ecosystem and a socioeconomist could do the same for socioeconomic impacts. It is unlikely that one individual would adequately understand all the disciplines relevant to siting.

Once the utility function is first assessed, one has a clear basis from which to make improvements. Differences of opinion about values can be articulated unambiguously within this framework. This results in more focused discussion and, it is hoped, provides a means to resolution of the conflicts of value. At least, conflicts are clearly identified and the different viewpoints can be examined using (7.1). This expression provides a unique procedure, founded on basic axioms, for incorporating values into the site selection process. By its mathematical nature, it insures that the values are used consistently on all the sites so that no site-specific bias is introduced in the evaluation phase.

A tremendous advantage of a formalized value structure is the ability to do extensive and inexpensive sensitivity analysis. It is easy to investigate whether changing one value judgment would change the implications of the study and how they would be changed. It is easy to examine how much a value judgment can change before the implications (e.g., site rankings) change. And it is also easy to evaluate the alternative sites from the viewpoints of other concerned individuals with different value structures.

Because of all of these advantages, a decision analysis provides a more thorough explanation of the site selection process to a regulatory agency. When the value structure is explicit, it is less likely that hearings, or ultimately licensing, will be delayed because it was found that the values used were inappropriate from the agency's point of view. The sensitivity analysis provides a means of answering many "what if" questions of authorities quickly and accurately. Finally, because a complete decision analysis increases the likelihood that the best site will be identified and recommended, the probability that the site will eventually be approved by all necessary agencies should be increased.

7.1.1 NOTATION AND DEFINITIONS

To facilitate communication, it is appropriate to define precisely the problem being addressed in this chapter in terms of the notation to be used throughout. A set of attributes X_i, $i = 1, ..., n$ has been defined to measure achievement on each of the n lowest-level objectives. Thus a consequence $x \equiv (x_1, ..., x_n)$ can be used to describe the impact at any site, where x_i is a specific level of attribute X_i.

Because impacts with respect to a given objective may be valued differently if they occur at different times, it is often appropriate to treat impact x_i as a vector $(x_i^1, ..., x_i^t, ..., x_i^T)$, where t indicates the time period of concern. For example, if X_1 is costs in millions of dollars, then x_1^1 might be the cost in the first year of construction, x_1^2 the cost in the second year, ..., and x_1^{20} the cost in the 20th year, which may be the 11th year of operation of the facility.

Some of the attributes X_i may concern impacts to groups of people such as citizens in the vicinity of a proposed facility. Impacts to these groups may be measured by the preferences of the impacted groups. Other attributes may concern potential fatalities. Since the evaluation of such impacts is unique, Sections 7.4 and 7.5.5 are included.

7.1.2 QUANTIFICATION OF VALUE STRUCTURES

There are three different types of functions which will be used to represent preferences. These are the value function, the measurable value function, and the utility function. All may be useful on a particular siting problem. Each has slightly different properties and interpretations.

The simplest of the three is the value function, which will be denoted by v. A value function provides a preference order for the consequences x. A value function v assigns a real number $v(x)$ to each consequence, such that x is preferred to x' if and only if $v(x) > v(x')$ and x is indifferent to x' if and only if $v(x) = v(x')$. Thus the value $v(x)$ provides a ranking of the xs. Notationally

$$x > x' \leftrightarrow v(x) > v(x') \tag{7.2a}$$

and

$$x \sim x' \leftrightarrow v(x) = v(x'), \tag{7.2b}$$

where $>$ means is preferred to and $\sim$ means is indifferent to. With a value function, the difference in the v values does not necessarily have an interpretation. Also, when uncertainty is involved, the expectation of v—the vs weighted by the probabilities of their occurrence—has no meaning.

The measurable value function, denoted by w, is a special type of value function and does satisfy the condition (7.2) when w is substituted for v. In addition, the measurable value function provides a preference ordering of the differences in value (i.e., importance) between pairs of consequences. The measurable value function w assigns a real number $w(x)$ to each consequence such that (7.2) holds (with w replacing v) and such that the importance of changing from x to y is greater than that of changing from x' to y' if and only if

$$w(y) - w(x) > w(y') - w(x') \tag{7.3a}$$

and is the same if and only if

$$w(y) - w(x) = w(y') - w(x'), \tag{7.3b}$$

where y is preferred to x and y' is preferred to x'. With a measurable value function, the differences in w values do have an interpretation as defined in (7.3). The expectation of w has no meaning, as was the case with simple value functions.

The utility function, denoted by u, is also a special type of value function. It satisfies condition (7.2) when u is substituted for v, but it does not necessarily satisfy (7.3) when u is substituted for w. A utility function provides a means of obtaining a preference order for lotteries† over consequences. The utility function u assigns a real number $u(x)$ to each consequence such that (7.2) holds (with u replacing v) and such that lottery L_1 is preferred to lottery L_2 if and only if the expected utility Eu_1 of lottery L_1 is greater than the expected utility Eu_2 of L_2, and L_1 is indifferent to L_2 if their expected utilities are equal. Notationally,

$$L_1 > L_2 \leftrightarrow Eu_1 > Eu_2 \qquad (7.4a)$$

and

$$L_1 \sim L_2 \leftrightarrow Eu_1 = Eu_2. \qquad (7.4b)$$

None of the three functions for ordering preferences is unique. Any positive monotonic transformation of v results in another appropriate value function. Any positive linear transformation of w or u will result, in a different appropriate measurable value function or utility function, respectively.

Because of the uncertainty present in almost all major siting problems, we shall need to obtain a utility function to evaluate the candidate sites. However, an important step in this process can be the development of either a value function or a measurable value function. In addition, after the evaluation, it may be worthwhile to convert the results to a measurable value, provided by w, to interpret the preference difference between sites.

7.1.3   THE APPROACH

Now the problem of interest in this chapter is simple to state. We want to obtain the client's utility function. The approach used is to subdivide the assessment of u into parts, work on these parts, and then integrate them. This requires that the general qualitative value judgments of the client be stated and then quantified. The mathematical implications on the form of u are then derived.

As a simple example, we may find

$$u(x_1, x_2, x_3, x_4) = k_1 u_1(x_1) + k_2 u_2(x_2) + k_3 u_3(x_3)u_4(x_4), \qquad (7.5)$$

where u_i, $i = 1, \dots, 4$ are single-attribute utility functions, and k_1, k_2, and k_3 are scaling factors. It may be that X_1 represents costs, X_2 fatalities, and X_3 and X_4 the impacts to two separate groups of people. This clearly

---

† A lottery is defined by specifying a set of possible consequences and the probability that each will occur.

addresses four objectives, measured by x_1, ..., x_4, respectively. The k_i indicate the value tradeoffs. If $x_1 = (x_1^1, ..., x_1^T)$, then assessing u_1 includes impacts over time. The impacts are addressed with u_3 and u_4 and the evaluation of fatalities by u_2. The attitude toward uncertainty is embodied in each of the u_i by the nature of a utility function.

Experience has indicated that a few general value assumptions seem reasonable for a broad class of siting problems. Furthermore, these assumptions imply a robust utility function which can be used to formalize widely different value structures. The chapter introduces these general assumptions and results which follow from their use. Then, for any specific problem, the set of these assumptions which appropriately defines the client's values must be identified to indicate which particular form, such as (7.5), of the utility function is appropriate.

The forms which follow from such assumptions require many value judgments to make them specific. These value judgments are the degrees of freedom, so to speak, which provide for the aforementioned robustness. Each focuses on one value question important to the problem, such as the value tradeoff between dollar cost and environmental impact, the values of costs now versus those in the future, the equity to various impacted groups, and the attitude toward financial uncertainties. This in turn provides the means of properly including these crucial aspects of siting in a responsible, logical, and justifiable manner.

7.2 Evaluation of Multiple Objective Impacts†

Chapter 4 indicates that siting problems will usually involve at least five objectives, one for each general concern. There will also be cases involving more or less than these five objectives. These should be organized into an objectives hierarchy and attributes should be defined for each lowest-level objective in the hierarchy. If there are n lowest-level objectives, define them by X_i, $i = 1, ..., n$, and let x_i be a specific level of X_i. A possible consequence of any siting is $x = (x_1, ..., x_n)$. For now, each x_i will be treated as a scalar.‡ The x_i become vectors when impacts over time are introduced in Section 7.3. Our focus in this section is to determine the utility function $u(x)$.

---

† Sections 7.2–7.7 can be skipped without a loss of conceptual continuity. These sections provide detail on how to quantify values, as well as case studies.

‡ This assumption requires no loss of generality. If x_i were a vector, this could be treated by allowing n, the number of scalar levels in a consequence, to increase.

7.2.1 THE CONCEPTS FOR ADDRESSING
        MULTIPLE OBJECTIVES

The main concepts of multiattribute utility theory concern independence conditions. Subject to a variety of these conditions, the assessment of u can be divided into parts, each much easier to tackle than the whole. We would like to find simple functions f, f_1, ..., f_n such that

$$u(x_1, ..., x_n) = f[f_1(x_1), ..., f_n(x_n)]. \qquad (7.6)$$

Then assessment of u is reduced to the assessment of f and f_i, $i = 1, ..., n$. The f_i are single-attribute functions, whereas u and f are n-attribute functions. If f is simple, such as the case (7.5), then the assessment of u is simplified. The independence concepts discussed below imply the simple forms of f indicated in Section 7.2.2.

There are four main independence conditions relevant to the multiple objective issue. All four will be stated, briefly discussed, and then contrasted.

Preferential Independence. The pair of attributes $\{X_1, X_2\}$ is preferentially independent of the other attributes X_3, ..., X_n if the preference order for consequences involving only changes in the levels of X_1 and X_2 does not depend on the levels at which attributes X_3, ..., X_n are fixed.

Preferential independence implies that the indifference curves over X_1 and X_2 do not depend on the other attributes. This independence condition involves preferences for consequences differing in terms of two attributes, with no uncertainty involved.

The next assumption is also concerned with consequences when no uncertainty is involved. However, it also addresses strength of preferences (i.e., value differences) when changes occur in only one attribute level.

Weak-Difference Independence. Attribute X_1 is weak-difference independent of attributes X_2, ..., X_n if the order of preference differences between pairs of X_1 levels does not depend on the levels at which attributes X_2, ..., X_n are fixed.

There are two important assumptions relating to situations which do involve uncertainty. As such, the conditions use preferences for lotteries rather than consequences. A lottery is defined by specifying a mutually exclusive and collectively exhaustive set of possible consequences and the probabilities associated with the occurrence of each.

Utility Independence. Attribute X_1 is utility independent of attributes X_2, ..., X_n if the preference order for lotteries involving only changes in the level of X_1 does not depend on the levels at which attributes $X_2, ..., X_n$ are fixed.

The last independence condition concerns lotteries over more than one attribute.

Additive Independence. Attributes $X_1, ..., X_n$ are additive independent if the preference order for lotteries does not depend on the joint probability distributions of these lotteries, but depends only on their marginal probability distributions.

To get an intuitive feeling for these assumptions, let us illustrate them in simple cases. The substance of preferential independence can be indicated with a three-attribute consequence space as shown in Fig. 7.1.

To avoid subscripts, the attributes are denoted X, Y, and Z with corresponding levels x, y, and z. There are three X, Y planes shown in the figure. By definition, if $\{X, Y\}$ is preferentially independent of Z, then the preference order of consequences in each of these planes (and indeed in all possible X, Y planes) will not depend on the level of Z. For instance, suppose the consequences in the plane with Z set at z^0 may be ordered A, B, C, D, E, F, G, with H indifferent to G. Then, because of preferential

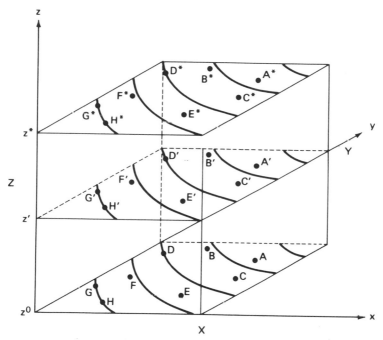

Fig. 7.1. Illustration of preferential independence.

independence, the consequences in the plane with Z set at z' must be A', B', C', D', E', F', G', with H' indifferent to G'. And also, with Z set at z^*, the order must be A*, B*, C*, D*, E*, F*, G*, with H* indifferent to G*.

An implication of preferential independence is that the indifference curves in all X, Y planes must be the same. Several indifference curves are illustrated in each of the three planes in Fig. 7.1 and it is easy to see that they are the same.

The usefulness of preferential independence is that it allows us to determine the preference order of consequences in only one X, Y plane and transfer this to all others. If $\{X, Y\}$ is preferentially independent of Z, it does not follow that any other pairs are preferentially independent. However, for any number of attributes, if two pairs of attributes overlap, and are each preferentially independent, then, as proved in Gorman [1968a,b], the pair of attributes involved in only one of the two given conditions (i.e., not in the overlap) must also be preferentially independent. This means, for our example, that if $\{X, Y\}$ is preferentially independent of Z, and $\{X, Z\}$ is preferentially independent of Y, then $\{Y, Z\}$ must be preferentially independent of X.

The next three independence assumptions can be illustrated most easily with two attributes as shown in Fig. 7.2. Here the attributes are X and Y with levels x and y. Weak-difference independence introduces the notion of difference in value between two consequences. The purpose is to develop theories which will allow us to make statements such as, "the difference between consequences A and B is more important than the difference between consequences C and D." Weak-difference independence is illustrated in Fig. 7.2, as follows. Suppose that, through a series of

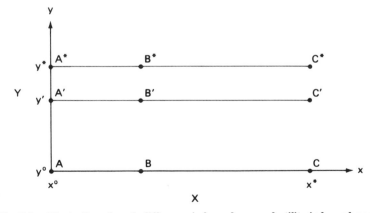

Fig. 7.2. Illustration of weak-difference independence and utility independence.

questions, it has been established that the preference difference between consequences A and B is equal to the preference difference between B and C. Because the level of Y is fixed at y^0 for all three of these consequences, the preference difference relationship can be translated to all other levels of Y if X is weak-difference independent of Y. In this case, the preference difference between A' and B' must equal that between B' and C', and the preference difference between A* and B* must equal that between B* and C*. With this condition, there is, however, no requirement that the preference difference between A and B be equal to that between A' and B', although this may be the case.

Weak-difference independence is not a symmetrical relationship. That is, the fact that X is weak-difference independent of Y does not imply anything about whether Y is weak-difference independent of X. In terms of the example, suppose y' had been chosen such that the preference difference between A and A' equaled that between A' and A*. Then, even if X is weak-difference independent of Y, it may or may not be that the preference differences between B and B' and between B' and B* are equal.

The last two independence conditions concern lotteries because we are interested in developing utility functions which can be combined with the probabilities to address the uncertainties present in siting problems. The utility independence notion is very similar to that of weak-difference independence. In Fig. 7.2, suppose the consequence B is indifferent to the lottery yielding either A or C, each with probability 0.5. Then if X is utility independent of Y, the same preference relationship can be translated to all levels of Y. This means, for instance, that B' must be indifferent to a lottery yielding either A' or C', each with probability 0.5, and that B* must be indifferent to the 50–50 lottery yielding either A* or C*.

The utility independence concept is also not symmetrical: X can be utility independent of Y, and Y need not be utility independent of X. However, suppose that Y is utility independent of X in Fig. 7.2 and that A' is indifferent to a lottery yielding either A* with probability 0.6 or A with probability 0.4. Then B' must be indifferent to a lottery yielding B* with probability 0.6 or B with probability 0.4. The corresponding relationship holds for the C's.

The additive independence condition is illustrated in Fig. 7.3. Consider the two lotteries L_1 and L_2 defined in the figure. Lottery L_1 yields equal 0.5 chances at the consequences (x^0, y^0) and (x', y') and lottery L_2 yields 0.5 chances at each of (x^0, y') and (x', y^0). Note that both lotteries have an equal (namely 0.5) chance at either x^0 or x' and also that both have an equal 0.5 chance at y^0 and y'. By definition then, the marginal probability distributions on each of the attributes X and Y are the same in both lot-

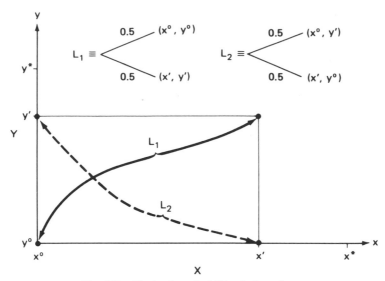

Fig. 7.3. Illustration of additive independence.

teries. Thus, if X and Y are additive independent, one must be indifferent between lotteries L_1 and L_2. This same indifference condition must hold if either or both of x' and y' are changed in Fig. 7.3, because L_1 and L_2 would still have the same marginal probability distributions on the two attributes.

There is no meaning attached to the statement that X is additive independent of Y. Either X and Y are additive independent or they are not.

More extensive discussions of all of these independence conditions are found in the technical literature. Some of the original sources are Debreu [1960], Luce and Tukey [1964], and Krantz [1964] for preferential independence; Krantz *et al.* [1971] and Dyer and Sarin [1977, 1979] for weak-difference independence; Keeney [1968], Raiffa [1969], and Meyer [1970] for utility independence; and Fishburn [1965, 1970] for additive independence. Keeney and Raiffa [1976] has detailed discussions of these conditions except for weak-difference independence.

7.2.2 THE MAIN RESULTS FOR ADDRESSING MULTIPLE OBJECTIVES

The main results for addressing multiple objectives are representation theorems stating conditions under which preferences can be represented in a convenient functional form. The following is essentially a list of the main representation theorems for structuring preferences. They will be presented without proof, although reference to sources with proofs will be made. The first four results examine the implications of each of the four independence assumptions taken singularly. Results combining independence assumptions will follow.

Result 1. Given attributes X_1, ..., X_n, $n \geqslant 3$, an additive value function

$$v(x_1, ..., x_n) = \sum_{i=1}^{n} k_i v_i(x_i) \tag{7.7}$$

exists if and only if $\{X_1, X_i\}$, $i = 2, ..., n$ is preferentially independent of the other attributes, where v_i is a value function over X_i and the k_i are scaling constants.

To use (7.7), one can determine v by assessing the v_i functions on a zero to one scale and the k_i such that

$$\sum_{i=1}^{n} k_i = 1. \tag{7.8}$$

Proofs of Result 1 are found in Debreu [1960], Fishburn [1970], and Krantz *et al.* [1971]. Examples of the assessment of v are found in Keeney and Raiffa [1976].

Result 2. Given attributes $X_1, ..., X_n$, $n \geqslant 2$, a multilinear measurable value function

$$w(x_1, ..., x_n) = \sum_{i=1}^{n} k_i w_i(x_i) + \sum_{i=1}^{n} \sum_{j>i}^{n} k_{ij} w_i(x_i) w_j(x_j)$$

$$+ \sum_{i=j}^{n} \sum_{j>i}^{n} \sum_{h>j}^{n} k_{ijh} w_i(x_i) w_j(x_j) w_h(x_h)$$

$$+ \cdots + k_{1...n} w_1(x_1) \cdots w_n(x_n) \tag{7.9}$$

exists if and only if X_i, $i = 1, ..., n$ is weak-difference independent of the other attributes, where w_i is a measurable value function over X_i and the ks are scaling constants.

To use (7.9), one specifies w by assessing the w_i on a zero to one scale and the ks such that they sum to one. A discussion of (7.9) is found in Dyer and Sarin [1979].

Result 3. Given attributes X_1, ..., X_n, $n \geqslant 3$, a multilinear utility function

$$u(x_1, \ldots, x_n) = \sum_{i=1}^{n} k_i u_i(x_i) + \sum_{i=1}^{n} \sum_{j>i}^{n} k_{ij} u_i(x_i) u_j(x_j)$$

$$+ \sum_{i=1}^{n} \sum_{j>i}^{n} \sum_{h>j}^{n} k_{ijh} u_i(x_i) u_j(x_j) u_h(x_h)$$

$$+ \cdots + k_{1 \ldots n} u_1(x_1) \cdots u_n(x_n) \qquad (7.10)$$

exists if and only if X_i, $i = 1, \ldots, n$ is utility independent of the other attributes, where u_i is a utility function over X_i and the ks are scaling constants.

Notice that the form of (7.10) is identical to that of (7.9). To determine u, one can assess the individual utility functions u_i on a zero to one scale and the scaling constants such that they sum to one. The Result 3 is proven in Keeney [1972].

Result 4.   Given attributes $X_1, \ldots, X_n$, $n \geq 2$, an additive utility function

$$u(x_1, \ldots, x_n) = \sum_{i=1}^{n} k_i u_i(x_i) \qquad (7.11)$$

exists if and only if the attributes are additive independent, where u_i is a utility function over X_i and the k_i are scaling constants.

Notice that (7.11) is a special case of (7.10) and u can be assessed accordingly. The original proof of (7.11) is found in Fishburn [1965].

As might be inferred from (7.9) and (7.10), the mathematical restrictions placed on the forms of w by weak-difference independence and u by utility independence are the same. Thus, the following important result, which was the basis for the WPPSS preference structure illustrated in Chapter 3, will be stated only for the utility function.

Result 5.   Given attributes $X_1, \ldots, X_n$, $n \geq 3$, the utility function

$$u(x_1, \ldots, x_n) = \sum_{i=1}^{n} k_i u_i(x_i) + k \sum_{i=1}^{n} \sum_{j>i}^{n} k_i k_j u_i(x_i) u_j(x_j)$$

$$+ k^2 \sum_{i=1}^{n} \sum_{j>i}^{n} \sum_{h>j}^{n} k_i k_j k_h u_i(x_i) u_h(x_h)$$

$$+ \cdots + k^{n-1} k_1 \cdots k_n u_1(x_1) \cdots u_n(x_n) \qquad (7.12)$$

exists if and only if $\{X_1, X_i\}$, $i = 2, \ldots, n$ is preferentially independent of the other attributes and if X_1 is utility independent of the other attributes.

As with the other utility functions, one can assess the u_i on a zero to one scale and determine the scaling constants k_i to specify u. The additional constant k is calculated from the k_i, $i = 1, \ldots, n$.

If $\Sigma\, k_i = 1$, then $k = 0$, and if $\Sigma\, k_i \neq 1$, then $k \neq 0$. If $k = 0$, then clearly (7.12) reduces to the additive utility function

$$u(x_1, \ldots, x_n) = \sum_{i=1}^{n} k_i u_i(x_i). \tag{7.13}$$

If $k \neq 0$, multiplying each side of (7.12) by k, adding 1, and factoring yields

$$ku(x_1, \ldots, x_n) + 1 = \prod_{i=1}^{n} [kk_i u_i(x_i) + 1], \tag{7.14}$$

which is referred to as the multiplicative utility function. The proof of Result 5 is found in Keeney [1974]. Pollak [1967] and Meyer [1970] each used a more complex set of assumptions to derive the form (7.12). It is easy to see from the form of (7.12) that it is a special case of the multilinear utility function (7.10). For the n-attribute case, the number of scaling factors required to specify (7.10) is 2^n-1, and to specify (7.12) is $n + 1$.

As indicated, if the condition that X_1 is weak-difference independent of the other attributes replaces the condition that X_1 is utility independent in Result 5, then the measurable value function will necessarily be additive or multiplicative. That is, the us in (7.13) and (7.14) can be replaced by ws. This is proved in Dyer and Sarin [1979].

The next result relates the different types of functions for ordering preferences.

Result 6.   Given attributes $X_1, \ldots, X_n$, $n \geq 3$, if $\{X_1, X_i\}$ is preferentially independent of the other attributes and X_1 is utilityindependent of the other attributes, then u must have one of the following three forms:

$$u(x) = -e^{-cv(x)}, \qquad c > 0, \tag{7.15a}$$

$$u(x) = v(x), \tag{7.15b}$$

$$u(x) = e^{cv(x)}, \qquad c > 0, \tag{7.15c}$$

where v is the additive value function (7.7).

In general, neither of the utility functions (7.15a) or (7.15c) are scaled from zero to one. A positive linear transformation in each case can lead to this scaling. For instance, adding 1 to the right-hand side of (7.15a) and dividing by $1 - e^{-c}$ yields the scaling zero to one.

A proof of Result 6 is found in Keeney and Raiffa [1976]. If the premise that X_1 is utility independent is replaced by the premise that X_1 is weak-difference independent in Result 6, then a result similar to (7.15) follows with $u(x)$ replaced by $w(x)$. This is proved in Dyer and Sarin [1979]. Bell and Raiffa [1979] argue that (7.15) should hold for any utility function and measurable value function which are consistently assessed.

7.2.3 Assessing the Value Judgments

To specify any of the functions v, w, or u using the results above, assessments are required to determine three types of information:

1. appropriateness of the assumptions,
2. the individual functions v_i, w_i, or u_i, and
3. the scaling factors.

Obtaining this information is as much an art as it is a science. Keeney and Raiffa [1976] devote a great deal of attention to how one should conduct these assessments and illustrate them for many real cases. Rather than repeat that here, we will follow a different tack. The approach for determining the necessary information will be described compactly in this section. This will cover the theory of assessment. Then in Section 7.6 we include part of a dialogue used to elicit a multiattribute utility function in an energy policy context. This should provide some insights into the art of assessment.

Verifying Independence Conditions.   All of the independence conditions are examined by looking for specific cases of the client's preferences which contradict the assumption in question. If none are found, the assumption is assumed to be appropriate for the problem.

As an example, consider investigating whether $\{X_1, X_2\}$ is preferentially independent of the other attributes X_3, ..., X_n. First X_3, ..., X_n are set at relatively undesirable levels (say x_3^0, ..., x_n^0) and the preferences in the X_1, X_2 plane are examined. One questions the client to find pairs of consequences in this plane which are indifferent. Suppose $(x_1, x_2, x_3^0, ..., x_n^0)$ is indifferent to $(x_1', x_2', x_3^0, ..., x_n^0)$. Then X_3, ..., X_n are changed to different levels (say x_3^*, ..., x_n^*) and the client is asked if $(x_1, x_2, x_3^*, ..., x_n^*)$ is indifferent to $(x_1', x_2', x_3^*, ..., x_n^*)$. A yes answer is consistent with preferential independence; a no answer is not. If such responses are consistent with preferential independence for several pairs of X_1 and X_2 and for several different levels of X_3, ..., X_n, then it is reasonable to assume that $\{X_1, X_2\}$ is preferentially independent of X_3, ..., X_n.

The verification of weak-difference independence or utility independence is identical in style so we shall discuss only the former case here. The utility independent case is illustrated in Section 7.6. Suppose we wish to ascertain if X_1 is weak-difference independent of X_2, ..., X_n. Let us define the range of X_1 to go from x_1^0 to x_1^*. We ask the client for a level x_1' such that the preference difference from x_1^0 to x_1' is equal to that from x_1' to x_1^*, given always that the other attributes are fixed at, say x_2^0, ..., x_n^0. Then we can change the levels of X_2, ..., X_n and repeat the process. If x_1' is still the level of X_1 such that the preference differences from x_1^0 and x_1' and from x_1' to x_1^* are equal, then it may be that X_1 is weak-difference independent of X_2, ..., X_n. If x_1' is not the level, then the condition cannot hold. If x_1' is found to be the level which splits the preference difference from x_1^0 to x_1^* for several levels of the other attributes, then it is reasonable to assume that X_1 is weak-difference independent of X_2, ..., X_n.

To examine the appropriateness of the additive independence condition, several pairs of lotteries with identical marginal probability distributions, such as those illustrated in Fig. 7.3, are presented to the client. To make this simpler, all attributes but two can be fixed for all the consequences in both lotteries of a given pair. If the levels of the attributes which differ in the consequences do cover the ranges of those attributes, and if each of the given pairs of lotteries are indifferent to the client, then it is probably appropriate to assume that X_1, ..., X_n are additive independent.

Assessing the Individual Functions.   The individual functions which we want to assess are the single-attribute value functions, denoted by v_i; the single-attribute measurable value functions, denoted by w_i; and the single-attribute utility functions, denoted by u_i. In general, each of these is determined by assessing a few points on the function for some x_is and then fitting a curve. However, since v_i is unique up to any positive monotonic transformation and w_i and u_i are unique up to positive linear transformations, the exact procedures differ for these two cases.

Suppose v_1 in (7.7) is to be assessed over the range from x_1^0 to x_1^*, where x_1^0 is least desirable and preferences increase to x_1^*. We can arbitrarily set the origin and scale of v_1 by

$$v_1(x_1^0) = 0, \qquad v_1(x_1^*) = 1. \tag{7.16}$$

The assessment requires that another attribute, say X_2, be used to help calibrate v_1. We want to assess the midvalue point, denoted by x_1', such that the client is willing to give up the same amount of X_2 to move from x_1^0 to x_1' as to move from x_1' to x_1^*. The same situation must hold regardless of what level is initially used for X_2. Different amounts of X_2 may be given up starting from different levels, but the amount must be the same in each case for both x_1^0 to x_1' and for x_1' to x_1^*. It follows then that

$$v_1(x_1') = 0.5. \tag{7.17}$$

The next step is to find the midvalue point for both of the ranges x_1^0 to x_1' and x_1' to x_1^*, using the same procedure. Let these be designated by x_1^q and x_1^b, respectively. Then

$$v_1(x_1^q) = 0.25 \quad \text{and} \quad v_1(x_1^b) = 0.75. \tag{7.18}$$

The process could continue but it is probably sufficient to construct a curve through the points in (7.16)–(7.18) to yield v_1. This is done in Fig. 7.4.

An example illustrating the simultaneous assessment of two v_i functions is provided in Keeney and Raiffa [1976]. This procedure, referred to as conjoint scaling, is an alternative to the midvalue technique discussed here.

With value functions, three basic steps are needed to provide v_i: scaling, assessing a few points, and selecting (i.e., drawing) a curve. With measurable value functions and utility functions, the order is reversed to scaling, selecting the shape of the curve, and assessing a few points to identify the specific curve. With both w_i and u_i, the shape of the curve has

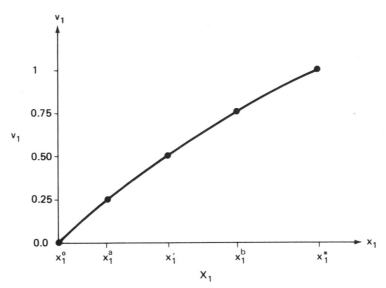

Fig. 7.4.   Assessing a single-attribute value function.
To assess v_1: (1) scale v_1, $v_1(x_1^0) = 0$, $v_1(x_1^*) = 1$; (2) assess midvalue points x_1', x_1^q, x_1^b. Then, $v_1(x_1') = 0.5$, $v_1(x_1^q) = 0.25$, $v_1(x_1^b) = 0.75$; (3) draw in the curve v_1.

an interpretation in terms of preferences which should be used in assessment. This meaning is the reason w_i and u_i are unique only up to positive *linear* transformations. Since the assessment ideas are similar, and since the assessment of u_i is discussed in detail in Section 7.5, which addresses risk attitudes, the assessment of w_i will be focused upon here.

Let us determine w_1 which has preferences increasing in the range x_1^0 to x_1^*. Then we can scale w_1 by

$$w_1(x_1^0) = 0, \qquad w_1(x_1^*) = 1. \tag{7.19}$$

To specify the shape of w_1, we investigate the qualitative character of the client's preferences. For instance, we can take the point $x_1' = (x_1^0 + x_1^*)/2$ halfway between x_1^0 and x_1^*, and ask for the midvalue point between x_1^0 and x_1'. Suppose it is one-third of the distance from x_1^0 to x_1'. Then we ask for the midvalue value point between x_1' and x_1^*. If it is also one-third of the distance from x_1' to x_1^*, a certain structure is implied since the ranges x_1^0 to x_1' and x_1' to x_1^* are the same. Suppose for any pair of points with this same range, the midvalue point is one-third of the distance from the less desired point to the more desired point. This would have very strong implications† for the shape of w_1. In this case, it follows that

$$w_1(x_1) = d + b\,(-e^{-cx_1}), \tag{7.20}$$

where d and b are scaling terms to obtain consistency with (7.19) and the measurable value function has an exponential form with one parameter c.

The parameter c is determined from knowing the midvalue point for one pair of x_1 levels. We could use the already determined point one-third of the distance from x_1^0 to x_1', for example. However, let us suppose we assess $\hat{x}_1$ to be the midvalue point for the range x_1^0 to x_1^*. Then, it follows from the definition of a measurable value function that

$$w_1(x_1^*) - w_1(\hat{x}_1) = w_1(\hat{x}_1) - w_1(x_1^0). \tag{7.21}$$

Combining this with (7.19) yields

$$w_1(\hat{x}_1) = 0.5, \tag{7.22}$$

which can be substituted into (7.20) to determine the parameter c. Remember the scaling parameters d and b are determined first from evaluating (7.19) with (7.20).

Assessing the Scaling Constants.   The scaling constants, designated by the ks in (7.7)–(7.14), indicate the value tradeoffs between the various

---

† If the midvalue value point does not remain one-third of the distance up the range, it may increase or decrease as one swings across the range. This also has strong implications for the shape of w_1. These implications are analogous to those for the single-attribute utility case discussed in Section 7.5.

pairs of attributes. Given attributes X_1, ..., X_n, there will be n scaling factors for the additive functions, $n + 1$ for the multiplicative functions, and $2^n - 1$ for the multilinear functions. For now, let us designate the number of scaling constants as r. To determine these, we need to develop r independent equations with the r scaling constants as unknowns and then solve them.

To do this, we have, in general, a function F over X_1, ..., X_n broken down into another function f with $f_1(x_1)$, ..., $f_n(x_n)$ and k_1, ..., k_r as arguments. Notationally,

$$F(x_1, ..., x_n) = f[f_1(x_1), ..., f_n(x_n), k_1, ..., k_r], \qquad (7.23)$$

where the form of f is determined from the independence conditions and the f_i are assessed as mentioned above. The easiest way to generate equations is to find two consequences x and y which are equally preferred by the client. Then, clearly, $F(x) = F(y)$ so

$$\begin{aligned} f[f_1(x_1), ..., f_n(x_n), k_1, ..., k_r] \\ = f[f_1(y_1), ..., f_n(y_n), k_1, ..., k_r], \end{aligned} \qquad (7.24)$$

which is one equation with the unknowns k_1, ..., k_r.

In practice, it is usually best to fix $n - 2$ of the attributes and vary just two to obtain a pair of indifference consequences. If these two attributes are X_1 and X_2, then the question posed to the client directly concerns the value tradeoffs between X_1 and X_2. The dialogue of an actual assessment in Section 7.6 illustrates the art involved in generating equations such as (7.24) using value tradeoffs. Operationally, if it turns out that some equations are redundant (i.e., not independent), additional equations can be generated as necessary using (7.24).

7.2.4 A MULTIATTRIBUTE ASSESSMENT
FOR A PROPOSED COAL-FIRED POWER PLANT

Throughout this book, there are numerous examples illustrating the assessment of values for the siting of proposed energy facilities. The two complete cases concern the nuclear power plant study of Chapter 3 and the pumped storage unit siting of Chapter 9. Other studies illustrate the applicability of and approach to structure and quantify values for dealing with public attitudes (Section 6.6.3), impacts over time (Section 7.3.4), and impacts to several groups (Section 7.4.4). Because of the uncertainties often involved, all of these cases involve the development of a multiattribute utility function.

As indicated earlier in this section, there are two other functions useful for structuring preferences. These are the value function and the measurable value function. Because of the major uncertainties inherent in most siting decisions, it will be necessary to construct a utility function for most siting studies. However, even when this is the case, the value functions can be very important for three main reasons:

1. The value attitudes of the client toward uncertainty can sometimes be addressed on the individual attributes and the overall site ranking be done using a value function.

2. For some attributes, rather than construct a utility function, the client may find it easier to construct a measurable value function from which the corresponding single-attribute utility function can be calculated.

3. The value functions can serve to provide several consistency checks on the appropriateness of the utility function if both are separately assessed.

The case study described here illustrates these points.

A study was conducted (see Sarin [1979]) to locate suitable sites for 1500 MWe coal-fired power plants in the northwestern United States. Thirteen candidate sites in the states of Washington, Oregon, Idaho, Wyoming, and Montana were identified in the screening process. Sites were evaluated using the six attributes described in Table 7.1. The consequences, in terms of levels of these attributes, were quantified using on-site data collection, existing literature, and experts' professional judgments.

Value assessments were conducted with members of the study team and with executives in the client firm. The numerical assessments indicated below are those of the client executives.

Structure of the Utility Function and Measurable Value Function. By questioning the client using the style discussed in Section 7.6, it was found that an appropriate utility function u was the multiplicative form

$$1 + ku(x_1, \ldots, x_6) = \prod_{i=1}^{6} [1 + kk_i u_i(x_i)] \qquad (7.25)$$

and that an appropriate measurable value function w was additive, so

$$w(x_1, \ldots, x_6) = \sum_{i=1}^{6} \lambda_i w_i(x_i), \qquad (7.26)$$

where u, u_i, w, and w_i are all scaled zero to one and the k_i and λ_i are positive scaling constants. Because w and $\log(1 + ku)$ are both value functions (i.e., they order the consequences) and additive in functions of the

TABLE 7.1

ATTRIBUTES FOR RANKING COAL POWER SITES

	Attribute	Measurement unit	Range Best	Range Worst
X_1	Air quality concentration ratio	Calculated concentrations as a percentage of allowable standard	25	100
X_2	Impact on salmonids and other fish	Average annual spawning escapement of adult salmonids and other fish lost	0	10,000
X_3	Biological impact[a]	Constructed scale with 0 representing the least impact and 10 the most severe impact	0	10
X_4	Socioeconomic impact[b]	Constructed scale of potential short- and long-term impacts caused by the construction and the operation of the plant	0	7
X_5	Environmental impact of transmission interconnection, water line, and railroad	Length of lines measured in miles passing through environmentally sensitive areas	0	50
X_6	Annual differential site cost	Annualized differential capital and yearly operating and maintenance costs measured in millions of 1978 dollars	0	60

[a] For example, level 6 on the constructed scale is described as "loss of 2.0 square miles of habitat, less than 85% of which is mature second growth and 15–50% is wetland or endangered species habitat."

[b] For example, level 4 on the constructed scale is described as "remote site, near a city with a population of 25,000. Assumptions: no company town is built; 75–90% of work force seeks to locate in community."

X_is, they must be positive linear transformations of each other as argued in Dyer and Sarin [1979]. Thus, to be consistent with the choice of scaling,

$$w(x_1, \ldots, x_6)[\log(1 + k)] = \log[1 + ku(x_1, \ldots, x_6)], \qquad (7.27)$$

so

$$\lambda_i w_i(x_i)[\log(1 + k)] = \log[1 + kk_i u_i(x_i)]. \qquad (7.28)$$

Substituting the level x_i^* of X_i into (7.28) such that $w_i(x_i^*) = u_i(x_i^*) = 1$ and solving for λ_i yields

$$\lambda_i = \log(1 + kk_i)/\log(1 + k). \qquad (7.29)$$

Substituting (7.29) into (7.28) and solving yields

$$u_i(x_i) = \frac{\exp\{w_i(x_i)[\log(1 + kk_i)]\} - 1}{kk_i}. \qquad (7.30)$$

Expression (7.30) gives us the mechanism for determining u_i from an assessed w_i for these attributes for which it is easier to obtain a measurable value function than a utility function.

Value Tradeoffs. The value tradeoffs necessary to quantify the scaling factors in both (7.25) and (7.26) are of the form indicated in Section 7.2.3. With all other attributes fixed at constant levels, two are varied until an indifference pair of consequences is identified. Thus, for instance, it was found that the consequences ($x_1 = 25$, $x_6 = 45$) and ($x_1 = 95$, $x_6 = 23$) were indifferent when other attributes were fixed.

Because more than one executive from the client firm expressed value tradeoffs and no consensus was identified, a range of possible values of the scaling factors was determined. This is illustrated in Table 7.2.

Individual Attribute Assessments. Single-dimensional utility functions were assessed for attributes X_1, X_2, and X_4. The impacts were probabilistically assessed for these attributes. Single-dimensional measurable value functions were assessed for attributes X_3, X_5, and X_6. The impact data did not include uncertainties with regard to these attributes. A utility function was also assessed for the cost attribute X_6.

TABLE 7.2

RANGE OF
SCALING CONSTANTS[a]

	Lowest	Highest
λ_1	0.13	0.25
k_1	0.18	0.35
λ_2	0.01	0.01
k_2	0.01	0.02
λ_3	0.01	0.01
k_3	0.01	0.02
λ_4	0.04	0.06
k_4	0.06	0.09
λ_5	0.01	0.01
k_5	0.01	0.02
λ_6	0.80	0.66
k_6	0.86	0.77
k	-0.65	-0.58

[a] All scaling constants are rounded off to two decimal places.

Using (7.30), utility functions u_3 and u_5 were calculated from w_3 and w_5 and the middle value of the k_i scaling constants in Table 7.2. By assessing w_3 and w_5, it was not necessary to ask the clients questions concerning their preferences for lotteries over attributes for which no uncertainty existed in the consequences. Furthermore, u_6 was also derived from w_6 to provide a check on the directly assessed u_6.

Ranking the Sites.   Using any possible set of the client's value tradeoffs in the ranges indicated in Table 7.2, the same six sites were always preferred to the other seven sites. It also happened that the site ranked first remained first for all client value tradeoffs.

However, because there are always factors important to siting not included in the formal analysis, a broad sensitivity analysis of the value structure was conducted. This showed that for a wide array of values, the top three sites were always among the same five of the candidate sites. As a result, it was recommended that more careful data be collected on those five sites. Particular attention was to be concentrated on quantifying the uncertainties about costs both because they may be large and because the scaling factors indicate the significance of costs to this coal-fired power plant siting.

7.3   Evaluation of Impacts over Time

With many objectives measured by an attribute X_i, there can be impacts over a relatively long time. This time may be the construction time for the facility, the operating lifetime of the facility, or longer, as is the case with the effects of some pollutants. It is often convenient to categorize the total impact time into periods, typically of a year. Then an impact x_i on X_i is actually a vector $(x_i^1, \ldots, x_i^t, \ldots, x_i^T)$ referred to as a time stream, where X_i^t is the impact on the ith objective in period t measured by x_i^t. The total time of concern is T periods. The problem addressed in this section is to evaluate the vector $x_i = (x_i^1, \ldots, x_i^T)$ to obtain either a value function $v_i(x_i)$, a measurable value function $w_i(x_i)$, or a utility function $u_i(x_i)$ over X_i. Given this situation, throughout this section, we drop the subscript i to reduce clutter on the page.

7.3.1   THE CONCEPTS FOR ADDRESSING IMPACTS OVER TIME

It should be obvious that the problem being addressed is structurally the same as the one considered in the last section. We want to find simple functions $f, f^1, ..., f^T$, such that

$$u(x^1, ..., x^T) = f[f^1(x^1), ..., f^T(x^T)]. \tag{7.31}$$

As a result, all of the concepts introduced in Section 7.2.1 are relevant to the problem of preferences for impacts over time. However, because of the special nature of the time problem—that each of $x^1, ..., x^T$ are measured with the same unit and contribute to the same objective—some additional assumptions may be appropriate in some cases. We shall discuss some important ones here.

Standard Discounting. There exists a discount rate r, where $0 < r < 1$, such that an impact x^t in any period t is equivalent to an impact rx^t in period $(t + 1)$ for all $t = 1, 2, ..., T - 1$.

Standard discounting is commonly used in evaluating time streams of money. The constant r is referred to as the standard discount rate. In Section 7.3.2 we shall investigate component assumptions necessary for standard discounting to be valid.

Stationarity. If two time streams have identical first period impacts, then the modified streams obtained by deleting the first period impacts and advancing all other impacts by one period must be preferentially ordered in the same way as the original streams.

This assumption says roughly that the relationship between periods 1 and 2 must be the same as that between 2 and 3, 3 and 4, and so on.

The previous two assumptions concerned preferences for time streams, as opposed to lotteries for time streams. Let us divide the attributes $X^1, X^2, ..., X^T$ into three groups $\{X^1, ..., X^j\}, \{X^{j+i}, ..., X^h\}, \{X^{h+1}, ..., X^T\}$ such that none of these are empty. And for simplicity, we shall define these groups as Y^1, Y^2, and Y^3.

Conditional Utility Independence. Attribute Y^1 is conditionally utility independent of Y^2, given Y^3, if the preference order involving lotteries over changes on the level of Y^1 does not depend on the levels of Y^2, given the levels of Y^3 are fixed.

The implication of this assumption is that preferences for lotteries over Y^1 levels can depend on Y^3 levels but not on Y^2 levels.

To introduce the fourth assumption, we need to define the notion of a state descriptor. Let us break the time stream into two sections $(x^1, ..., x^t)$ and $(x^{t+1}, ..., x^T)$. The most general case, when no assumptions are made, has preferences for lotteries over $(x^{t+1}, ..., x^T)$ depending on $(x^1, ..., x^t)$. However, it may be that these preferences depend on $(x^1, ..., x^t)$ only through some summary index, such as the impact in the most recent period (x^t), impact in the last two periods (x^{t-1}, x^t), or the average impact $(1/t) \Sigma x^j$. Such a summary, which usually has fewer dimensions than $(x^1, ..., x^t)$, is referred to as a state descriptor.

State Dependent Utility. Given $(x^1, ..., x^T)$, if preferences for lotteries over $(x^{t+1}, ..., x^T)$ depend on $(x^1, ..., x^t)$ only through a state descriptor, then utilities over $(x^{t+1}, ..., x^T)$ are state dependent on $(x^1, ..., x^t)$.

In a sense, if the set of attributes $X^1, ..., X^T$ is adequate for the siting problem being addressed, then it will always be true that any subset of the attributes is state dependent on the other attributes. The existence of the utility function $u(x^1, ..., x^T)$ implies this since, for instance, it can be thought of as a conditional utility function over $(x^{t+1}, ..., x^T)$, given $(x^1, ..., x^t)$, so that $(x^{t+1}, ..., x^T)$ is state dependent on $(x^1, ..., x^t)$. However, suppose that the preference order over lotteries on $(x^{t+1}, ..., x^T)$ is the same for two different streams $(x^1, ..., x^t)$ and $(\hat{x}^1, ..., \hat{x}^t)$. If this situation implies that the state descriptors for $(x^1, ..., x^t)$ and $(\hat{x}^1, ..., \hat{x}^t)$ are identical, then the state descriptor is efficient. Efficient state descriptors are desirable because they reduce the amount of assessment necessary to specify the utility function over $(x^1, ..., x^T)$.

Let us discuss these four assumptions.

Discounting is a very strong assumption commonly used to evaluate monetary streams. In this case, the standard discount rate r is often chosen to be (or at least be dependent upon) the interest rate for borrowing money. If one could borrow or lend any amount of money at the standard discount rate, then discounting may be appropriate if there are no transaction costs and, more importantly, if no uncertainty is present. However, at least with major energy facilities, there are always major uncertainties present. In such cases, because of the dynamic circumstances of costs, income, and other investment possibilities, standard discounting may not be an appropriate way to address the valve problem.

Another reason why discounting may not be appropriate for evaluating income is that time streams of money are often indicators of the future health of the company. For example, would a major company prefer income stream (100, 110, 125, 140, 160) or stream (160, 140, 125, 110, 100) if the numbers represented inflation adjusted net income in millions of dollars each year from a particular facility. The latter is preferred using standard discounting regardless of the discount rate, but the former may be preferred because it indicates growth and an improving future. The concepts of conditional utility and state dependent utilities can account for such preferences.

Stationarity, although weaker than standard discounting, is also a strong assumption. It in no way incorporates any learning of preferences or dependency of preferences on what has occurred in the past. For instance, suppose that x^1, x^2, ..., x^T represents pollution levels in successive time periods and that value tradeoffs between x^1 and x^2 have been assessed. Now if we examine value tradeoffs between x^2 and x^3, given that x^1 is fixed at various levels, we may find dependencies. If x^1 is high, it may be appropriate to give up more with respect to x^3 to reduce x^2, than would be the case if x^1 is low. It may be tolerable to have one year of high pollution, but two years in a row are unacceptable. Of course, whether such a preference is appropriate depends on the problem context. If this situation does prevail, then the value tradeoffs between x^2 and x^3 cannot be the same as between x^1 and x^2 in all cases, and, hence, stationarity is not a valid assumption.

Conditional utility independence and state dependent utilities do provide for dependencies such as that described. Conditional utility independence as stated can be viewed as a special but common and important case of state dependence. If utilities over $(x^{t+1}, ..., x^T)$ are not utility independent of $(x^1, ..., x^t)$ but are conditionally utility independent of $(x^1, ..., x^j)$ given $(x^{j+1}, ..., x^t)$, then we can conclude that $(x^{t+1}, ..., x^T)$ is state dependent on $(x^1, ..., x^t)$, where $(x^{j+1}, ..., x^t)$ is a state descriptor. However $(x^{j+1}, ..., x^t)$ is not necessarily an efficient state descriptor. For instance, it could also be the case that $(x^{t+1}, ..., x^T)$ is conditionally utility independent of $(x^1, ..., x^{t-1})$, given x^t, in which case x^t would be another state descriptor. It may not even be the most efficient state descriptor in spite of the fact that it is one dimensional. The utilities over $(x^{t+1}, ..., x^T)$ may depend only on whether x^t is greater or less than some critical level. If any state descriptors of $(x^1, ..., x^t)$ are equivalent, then by definition $\{X^{t+1}, ..., X^T\}$ is utility independent of $\{X^1, ..., X^t\}$.

These four assumptions are discussed in more detail in the technical literature. The stationarity condition has been investigated extensively by economists in considering impacts over time. Koopmans [1960, 1972] uses this condition to structure a value function for impacts over time, which is discussed in the next section. Conditional utility independence is introduced and discussed in Keeney and Raiffa [1976]. Bell [1977a, b] has investigated the concept both theoretically and practically. The concept of state dependence has been developed mainly by Meyer [1977].

7.3.2   THE MAIN RESULTS FOR ADDRESSING IMPACTS
        OVER TIME

All of the multiattribute results of Section 7.2.2 can be used for structuring preferences over time if the associated assumptions are applicable. The four additional results stated below are specifically relevant to preferences over time, but do have more general applicability, subject, of course, to the appropriateness of the assumptions.

Result 7.   Given attributes X^1, ..., X^T, $T \geq 2$, the value function

$$v(x^1, ..., x^T) = \sum_{t=1}^{T} r^{t-1}x^t, \tag{7.32}$$

where r is a constant, exists if and only if standard discounting holds.

In this case v is referred to as the net present value of the time stream $(x^1, ..., x^T)$ using standard discounting and r is the discount rate. Clearly, this value function is additive and, as such, is a special case of the additive value function (7.7). An alternative set of assumptions leading to (7.32) is discussed in Keeney and Raiffa [1976].

Result 8.   Given attributes X^1, ..., X^T, $T \geq 3$, the value function

$$v(x^1, ..., x^T) = \sum_{t=1}^{T} \lambda^{t-1}v'(x^t), \tag{7.33}$$

where v' is a single-attribute value function and λ is a constant, exists if and only if $\{X^t, X^{t+1}\}$, $t = 1, ..., T - 1$, is preferentially independent of the other attributes and stationarity holds.

The value function (7.33) is also a special case of the additive value function (7.7) but it is more general than (7.32). That is to say, the standard discounting assumption of Result 7 is a stronger assumption than the preferential independence and stationarity assumptions of Result 8. To assess (7.32), one needs only to determine the parameter r, whereas with (7.33), both the parameter λ and the single-attribute value function v' need to be assessed. The original formulation of Result 8 and a formal proof are found in Koopmans [1972].

Result 9.   Given attributes X^1, ..., X^T, $T \geq 4$, if $\{X^1, ..., X^{t-1}\}$ is mutually conditionally utility independent with $\{X^{t+1}, ..., X^T\}$ for all $t = 2, ..., T - 1$, then either

$$u(x^1, ..., x^T) = \sum_{t=1}^{T-1} u(x^t, x^{t+1}) - \sum_{t=2}^{T-1} u(x^t) \tag{7.34a}$$

or

$$u(x^1, ..., x^T) = \left[\prod_{t=2}^{T-1} (\lambda + u(x^t))\right]^{-1} \left[\prod_{t=1}^{T-1} (\lambda + u(x^t, x^{t+1}))\right] - \lambda, \quad (7.34b)$$

where λ is a

Result 10.   Given attributes $X^1, ..., X^T$, $T \geqslant 3$, if $(x^1, ..., x^t)$ is state dependent on $(x^{t+1}, ..., x^T)$ and $(x^{t+1}, ..., x^T)$ is state dependent on $(x^1, ..., x^t)$, where s^t and s^{t+1} are efficient state descriptions of $(x^1, ..., x^t)$ and $(x^{t+1}, ..., x^T)$, respectively, then one of the following must hold:

$$u(x^1, ..., x^T) = a(x^1, ..., x^t)$$
$$+ a'(x^{t+1}, ..., x^T) + f(s^t, s^{t+1}), \quad (7.35a)$$

$$u(x^1, ..., x^T) = b(x^1, ..., x^t)$$
$$\times b'(x^{t+1}, ..., x^T)f(s^t, s^{t+1}), \quad (7.35b)$$

$$u(x^1, ..., x^T) = a(x^1, ..., x^t)$$
$$+ b(x^1, ..., x^t)f(s^t, s^{t+1}), \quad (7.35c)$$

$$u(x^1, ..., x^T) = a'(x^{t+1}, ..., x^T)$$
$$+ b'(x^{t+1}, ..., x^T)f(s^t, s^{t+1}). \quad (7.35d)$$

Notice that in Result 10 state dependence was used only for the time periods up to and after period t. If this condition is to hold for all $t = 1, . . . , T$, then a, a', b, b', and f can be more thoroughly specified.

Results 9 and 10 are more general than the results providing the utility functions discussed in Section 7.2.2. They are only representative of the results which follow from conditional utility independence and state dependence assumptions. There are many results less general than (7.34) and (7.35) and yet more general than results relying on full sets of preferential independence or utility independence assumptions. These can be constructed using some independence assumptions and some conditional utility independence or state dependence assumptions, for example. Specific cases of these are found in Richard [1972], Oksman [1974], and Sections 6.10–6.12 and 9.3–9.6 in Keeney and Raiffa [1976].

The Practice of Evaluating Siting Impacts over Time.   Essentially all of the existing siting studies of energy facilities have given only a small amount of attention to the evaluation of impacts over time. This is the situation in the nuclear power study of Chapter 3 and the pumped storage study of Chapter 9. Furthermore, it appears that the treatment of the issue of values over time is almost always dealt with in the same oversimplified manner. After briefly describing the standard approach, we shall comment on the possible reasons for its prevalence.

Most siting studies treat the economic impact in different time periods differently from other impacts. The financial impacts over time are aggregated into one index using standard discounting as illustrated by Result 7. The only necessary value judgment concerns the proper discount rate for money and it is assumed that this includes all the value judgments necessary to appropriately value economic impacts over long time horizons.

Noneconomic impacts over time are usually not aggregated into a single index explicitly. The reported impact usually is the annual impact, such as the annual fatalities or the annual discharge of waste heat into a river. This is simply calculated as the average annual impact which is equivalent to standard discounting at an $r = 1$ discount rate (i.e., no discounting) in Result 7. Sometimes, when the annual impact is expected to change in a deterministic manner over time, the impact is reported separately for each time period. A good example of this concerns the annual size of the construction work force expected at the proposed Montague, Massachusetts nuclear power plant (see the draft environmental statement of the U.S. Nuclear Regulatory Commission [1975a]).

The major inadequacy of the standard approach to values over time is that it is inappropriate for addressing circumstances with large uncertainties. And yet recently, an electric utility industry executive stated "Uncertainty is the very essence of the problem facing utility management today" (Nagel [1978]). The same uncertainties face decision makers in all energy industries. There would appear to be three main reasons to explain why the evaluation of impacts over time is only superficially addressed in most siting studies:

1. Historically, the uncertainties were much smaller and economics was the main concern, so standard discounting provided an adequate method of addressing the problem.

2. The problem is very complex and, other than discounting or "annualizing," it is not clear what could or should be done.

3. Better approaches to what can be done have only recently been developed and knowledge of and experience with the approaches are not widely available.

The Section 7.3.3 summarizes procedures for implementing a more reasonable and more sophisticated approach to the time value problem. Section 7.3.4 describes a case study which utilizes such ideas.

7.3.3 ASSESSING THE VALUE JUDGMENTS

The specification of the value functions and utility functions given in (7.32)–(7.35) requires procedures to assess the appropriateness of the assumptions, the component utility functions, and the scaling parameters.

Verifying Assumptions. Standard discounting, when it is appropriate at all, is usually applied to monetary streams such as costs or profits. The appropriateness of standard discounting can be examined in this context as follows. Query the client as to what level of dollar costs x^1 (with $x^2 = 0$) is equal in value to $x^2 = 100$ dollar cost with $x^1 = 0$. Suppose the response is $x^1 = 90$. This implies that the standard discount rate must be 0.9 if it exists, since $90 = (0.9)(100)$. In this case, ask if $x^1 = 900$ dollars is equivalent in value to $x^2 = 1000$ dollars, and if $x^1 = 900,000$ dollars is equivalent in value to $x^2 = 1,000,000$ dollars, if indeed the cost range goes that high. All these responses must be yes if standard discounting is to hold.

The second requirement for standard discounting is the stationarity assumption. If $x^1 = 90$ is equivalent to $x^2 = 100$, then $x^2 = 90$ must be equivalent to $x^3 = 100$, and so on. All the same indifference pairs found between x^1 and x^2 levels must hold between x^2 and x^3 levels and, in general, between x^t and x^{t+1} levels. By trying several such pairs which cover the possible ranges of the attributes in question, we can assume stationarity if the same indifference relationship does hold for those pairs in each adjacent pair of time periods.

Conditional utility independence is verified in the same manner as utility independence. Suppose one wishes to ascertain whether Y^1 is conditionally utility independent of Y^2 given Y^3. One simply checks whether Y^1 is utility independent of Y^2, given a level of Y^3 which is held fixed. This is done as discussed in Section 7.2.3. Then one changes the level of Y^3 and repeats the procedure. If it is found that Y^1 is utility independent of Y^2 for several levels of Y^3 which cover the range of such levels, then it is appropriate to assume Y^1 is conditionally utility independent of Y^2, given Y^3. If Y^1 is not utility independent of Y^2 for any level of Y^3, then the conditional utility independence cannot hold.

The goal with state dependence is to find efficient state descriptors. If $\{X^{t+1}, \ldots, X^T\}$ is conditionally utility independent of $\{X^1, \ldots, X^j\}$, given $\{X^{j+1}, \ldots, X^t\}$, then an efficient description for preferences for lotteries over $(x^{t+1}, \ldots, x^T)$ does not contain $(x^1, \ldots, x^j)$. Thus conditional utility independence can be used to identify the attributes needed in an efficient state descriptor. For instance, if, whenever any of the attributes in the set $\{X^{j+1}, \ldots, X^t\}$ above is moved to the set $\{X^1, \ldots, X^j\}$, the conditional utility independence assumption does not hold, then the state descriptor for preferences over $(x^{t+1}, ..., x^T)$ will contain levels of $x^{j+1}, ..., x^t$. It may be that the state descriptor is the full vector $(x^{j+1}, ..., x^t)$ or a function of that vector with a smaller dimension. For example, if $\{X^{j+1}, ..., X^t\}$ also happened to be utility independent of the other attributes, such an efficient state descriptor might be $u'(x^{j+1}, ..., x^t)$, where u' is the utility function over attributes $X^{j+1}, ..., X^t$. The process of searching for an efficient state descriptor at this point is probably best carried out by trial and error, using the knowledge of the client and siting team to identify possible efficient state descriptors. To the extent that this fails, one will not go wrong in using the whole vector $(x^{j+1}, ..., x^t)$. This simply may result in more work than necessary in assessing the utility function u.

Assessing the Component Utility Functions. The value functions (7.32) and (7.33) are easily assessed. The single parameter in each case is assessed using one indifference pair, and v' in (7.33) is a single-attribute value function assessed as discussed in Section 7.2.3. In each case we obtain a one-dimensional index v over which we want to assess a utility function $u(v)$. It may be that Result 6 relating u and v (presented in Section 7.2.2) holds, in which case the problem is solved. If not, we need to directly assess utility for $5-10$ v values which cover the range. Then we fit a curve to these points as illustrated in Fig. 7.5. If we designate the best and worst v values as v^* and v^0, respectively, corresponding to x^0 and $x*$, we can arbitrarily set

$$u(v^0) = 0, \qquad u(v^*) = 1$$

to scale u.

The first part of u may be determined by assessing a utility function u^1 over the range of X^1, given that the other attributes are at their least desirable levels. Let us define x^+ as the consequence where X^1 is at its best level and the other attributes are at their worst levels and let us designate the value $v(x^+)$ as v^+. If a lottery yielding $x*$ with probability p and x^0 with probability $(1 - p)$ is indifferent to x^+ for sure, it follows that the utility of x^+ must be p. Therefore, $u(v^+) = p$, as shown in Fig. 7.5.

For any $(x^1, ..., x^T)$ such that its value falls between v^0 and v^+, $u(x^1, ..., x^T)$ can be determined as follows. Find the consequence, where X^1 is at level x^1 and the other attributes are at their worst levels, which is indifferent to $(x^1, ..., x^T)$. Then $u(x^1, ..., x^T)$ must equal $pu^1(\hat{x}^1)$. By improving the levels of the attributes $X^2, ..., X^T$ and repeating the procedure described above, we can get the utility of a different range of v. This should be chosen to overlap with the v^0-v^+ range to provide for consistent scaling. Then the

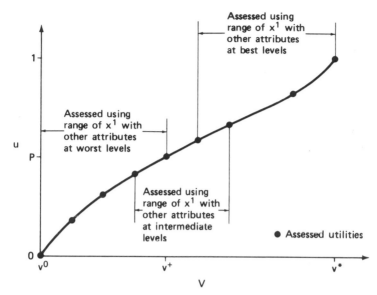

Fig. 7.5.   Assessing a utility function over value.

levels are again improved and the procedure repeated until X^2, ..., X^T are at their best levels. The result is as illustrated in Fig. 7.5. A more detailed discussion of this procedure is found in Keeney and Raiffa [1976].

If some independence conditions, such as those discussed in Section 7.2, are appropriate when applied to the component utility functions of Results 9 and 10, then it may be possible to assess these multiattribute utility functions by assessing a few single-attribute utility functions and some scaling constants. Otherwise, it will be necessary to assess directly some utility functions of more than one attribute. The procedure for this is simple to describe, but not so easy to utilize in practice.

Let us illustrate the concept with a simple case $u(x^1, x^2)$. The idea is to obtain the utilities of a few points in (X^1, X^2) space and then fit a function to these points. The function is the utility function used in evaluating the alternatives. The utilities of the points can be determined exactly as discussed for the one-attribute utility case in Section 7.5. See Bell [1979b] for a systematic approach to this problem.

Because of (1) the variety of independence and conditional independence assumptions, (2) the possibility of using different assumptions over parts of the overall space, and (3) the fact that there is some flexibility in choosing a reasonable utility function for a client, it is unlikely that one will need to resort to holistic assessment of multiattribute utility functions by curve fitting as described above. In other terms, collectively, the models of values consistent with some set of our possible assumptions is very robust. Part of the art of assessment is to find the appropriate set of assumptions to simultaneously imply a reasonable model and a reasonable assessment effort.

Assessing the Scaling Constants. The procedure for obtaining the scaling constants is exactly as described in Section 7.2.3. If there are r scaling constants, one develops a set of r independent equations with the r unknown scaling constants. Each equation is generated by equating the utility of two situations for which the client expressed indifference. For instance, after the component utility functions are assessed, if $(x^1, ..., x^T)$ is indifferent to $(y^1, ..., y^T)$, then setting $u(x^1, ..., x^T) = u(y^1, ..., y^T)$ results in one equation with some unknown scaling constants.

7.3.4 AN ASSESSMENT INVOLVING PREFERENCES
OVER TIME: THE SPRUCE BUDWORM

The forest pest case briefly summarized here demonstrates the ideas discussed earlier in this section. The methodology and procedures are relevant to energy facility siting and initial efforts are underway to utilize the ideas in energy siting problems.

The forests of New Brunswick, Canada have been frequently attacked by an insect pest known as the spruce budworm. Historical evidence taken from the trees themselves indicates that the forests have been periodically devastated with great regularity for more than 500 years. When a major outbreak of budworm began in the late 1940s, DDT and other insecticides were introduced to minimize disruption of the important lumber industry. Spraying was used continually through the 1970s. However, in the ensuing years, spraying costs spiraled, the lumbering industry became more important to the overall economy, and information and concern about insecticides increased greatly. In addition, the spraying grew less and less effective, and a major outbreak of budworm was underway in 1975. The Canadian government reviewed the problem and examined policy options. How best could the cutting and spraying of trees prevent a severe outbreak of the budworm, and, hence, both maintain a viable lumber industry and preserve the forest for recreational purposes?

As part of the attempt to answer this question, a detailed simulation model of the New Brunswick forests has been built at the Institute for Resource Ecology of the University of British Columbia and at the Inter-

national Institute for Applied Systems Analysis (see Holling [1978]).†
This model examines the impacts that different alternatives might have on
several critical variables, such as the environmental quality of the forest,
indicated by the percentage of trees in certain stages of growth and health;
the existence of budworm in varying stages of their life cycle; lumber
company profits; and employment by the lumbering industry. Given all
this information, the problem is to "make sense of it" to aid policy
makers in reaching a responsible decision.

Bell [1977b] reports, in detail, the steps taken to derive relevant attributes from the several output variables and to quantify the value structure, reflecting the preferences of one member of the simulation project
for the condition of the forest and the economy. The attributes were profit
(calculated from the dollar value of timber logged minus expenses including spraying costs), employment (the measure was percentage of mill
capacity used), and the recreational value of the forest (indicated by two
constructed indices representing the areas of "good" and "bad" recreational potential in the forest). Bell assessed a utility function for these four
attributes over time.

It was first established that the decision maker's preference for the time
stream of recreation and the two streams of profit and employment were
mutually utility independent. Thus, Result 3 of Section 7.2 allowed one to
assess separately utility functions for the recreational component and the
profit–employment component of the overall utility function. The utility
function was

$$u(p, e, r) = k_1 u_R(r) + k_2 u_S(p, e) + k_3 u_R(r) u_S(p, e), \qquad (7.36)$$

where p, e, and r are time streams of levels of profit, employment, and
recreation, respectively; u_R and u_S designate the recreational and social
utility functions; and the ks are scaling constants.

To assess the recreational utility function over the several periods, the
utility independence assumptions implying the appropriateness of the
multiplicative utility function from Result 5 of Section 7.2 were verified.
The recreational utility function could then be derived using only consistently scaled marginal utility functions for the areas of "good" and "bad"
forest in a single period. It is of interest to note that standard discounting
of the temporal stream of recreational indices (or transforms of these indices) was found to be an inadequate representation of the decision maker's
judgments.

The assessment of the social utility function u_S over profit and employ-

† The models were designed under the leadership of Professor Buzz Hölling. Team
members included Bill Clark, Ray Hilborn, Dixon Jones, Zafar Rashid, Viele Schnäpse,
Carl Walters, and Ralph Yorque.

ment streams was more complex. It was necessary to account for interdependencies of preferences in adjacent time periods, since achieved levels in period t set up aspiration levels for period $t + 1$. The decision maker proved to be risk prone in uncertain situations involving either levels of employment that were lower than the level in the previous year or employment levels that were much lower than the following period; otherwise he was risk averse. This complication was handled using assumptions based on conditional utility independence introduced in Section 7.3.2. Because of its relevance to impacts over time in siting decisions, we shall focus on this assessment.

Let P_t and E_t be the attributes of profit and employment in time period $t = 1, ..., T$. The utility function to be assessed could then be written as $u_S(p_1, e_1, p_2, e_2, ..., p_T, e_T)$. Bell first verified that the sets of attributes $\{P_1, E_1, ..., P_{t-1}, E_{t-1}\}$ and $\{P_{t+1}, E_{t+1}, ..., P_T, E_T\}$ were mutually conditionally utility independent, given $\{P_t, E_t\}$ for $t = 2, ..., T - 1$. Thus, Result 9 could be used and it was later verified that the correct form of u_S was

$$u_S(p, e) = \frac{\Pi_{t=1}^{T-1}[\lambda + u_t(p_t, e_t, p_{t+1}, e_{t+1})]}{\Pi_{t=2}^{T-1}[\lambda + u_t(p_t, e_t, p_{t+1}^0, e_{t+1}^0)]} - \lambda, \qquad (7.37)$$

where u_S is scaled so that $u_S(p_1^0, e_1^0, ..., p_T^0, e_T^0) = 0$ and $u_t = u_S$ with the attributes at the base level left out of the argument; for instance, $u_1(p_1, e_1, p_2, e_2) = u_S(p_1, e_1, p_2, e_2, p_3^0, e_3^0, ..., p_T^0, e_T^0)$.

Thus, completely specifying the form of u_S required determination of the functions u_t, $t = 1, ..., T - 1$. Bell's extensive questioning resulted in the assumptions that for the attributes $\{P_t, E_t, P_{t+1}, E_{t+1}\}$, the attributes P_t and $\{P_{t+1}, E_{t+1}\}$ were mutually conditionally utility independent, given E_t, and that P_{t+1} and $\{P_t, E_t\}$ were mutually conditionally utility independent, given E_{t+1}. From this, Result 9 could again be used to structure the u_t, $t = 1, ..., T - 1$ so that

$u_t(p_t, e_t, p_{t+1}, e_{t+1})$
$$= \frac{[\lambda_t + u_t(p_t, e_t)][\lambda_t + u_t(p_{t+1}, e_{t+1})][\lambda_t + u_t(e_t, e_{t+1})]}{[\lambda_t + u_t(e_t)][\lambda_t + u_t(e_{t+1})]} - \lambda_t. \qquad (7.38)$$

Furthermore, it was appropriate to assume stationarity of preferences, so that u_t and u_{t+1}, $t = 1, ..., T - 1$ were positive linear transformations of each other. It follows that the assessment of u_S using (7.37) and (7.38) requires the direct assessment of two two-attribute utility functions and three scaling factors.

The assessments of these utility functions were conducted as described in Section 7.2. An interesting feature was the dependency of the employment utility function in period $t + 1$ on the employment level in period t. If e_t is the employment level in period t, preferences for employment levels above e_t in period $t + 1$ indicated risk aversion.† Preferences for employment levels below e_t indicated risk proneness.

A Perspective. After reading this case, one might feel that the methodology is too complicated to be useful in siting problems. It is certainly complicated. On the other hand, the evaluation of sites is an important and complex problem—too complex to relegate to oversimplified methods. The difference in value of two time streams could be equivalent to tens or hundreds of millions of dollars and yet these differences may go undetected in informal or simple discounting methods. Particularly with the more sophisticated, less intuitive evaluation procedures, any implications of the evaluation should be appraised carefully using the professional judgment of experts, the client, and the analyst. The likelihood that this careful scrutiny may result in important new insights leading to better decisions and improved consequences is what justifies the level of effort required.

7.4 Evaluation of Impacts over Multiple Groups

The construction and operation of major energy facilities impacts and/or is of concern to several groups of people. The most obvious of the groups are people living near the proposed facility, environmentalists, consumers, recreational and sports organizations, regulatory interests, unions, and the business community. It is important to address the interests of these groups in the siting of major energy facilities. This poses a particularly complex problem because

1. any decision (or no decision) impacts several groups, and these groups have interests directly in conflict with each other,
2. some of these groups themselves have multiple conflicting objectives, and
3. the uncertainties about the consequences of any decision are large.

In addition, there may be major disagreements within groups about the desirability of any particular proposed facility. In this case, we would prefer to divide the group into smaller groups, each with a relatively homogeneous value structure.

† See Section 7.5.2 for definitions of these attitudes toward risk.

Because of the assumption of homogeneity within a group, a value structure can be quantified to represent each group. These values can be usefully integrated into the overall site evaluation in two ways:

1. by appraising each alternative site from the viewpoint of each group separately and

2. by amalgamating the implications of these results into an index indicating impact on the "involved" public.

The value structure should be quantified as a utility function in order to evaluate the alternatives involving uncertainty. Procedures for doing this are suggested in Section 7.4.3. Of course, a value function may be used as part of the means to reach this end. The procedures discussed in Section 7.2 will help in this regard.

Suppose we have a utility function u_g for each group $g = 1, ..., G$. Then it is straightforward to evaluate the alternative sites from gs point of view by calculating expected utilities using u_g and conducting a sensitivity analysis. Amalgamating the various viewpoints is a problem analogous to the multiple objectives problem. We want an overall utility function u_I which is some function of the group utility functions. That is, we want a simple function f such that

$$u_I(x) = f[u_1(x), u_2(x), ..., u_G(x)], \qquad (7.39)$$

where x represents a possible consequence of any siting alternative.

The formulation (7.39) is a restriction which we feel is appropriate for the energy siting problems being considered in this book. There are several other related analytical formulations for combining individual sets of preferences into a group preference. These involve the combination of rankings, value functions, expected utilities, and so on according to a wide array of proposed assumptions. See, for instance, Nash [1950, 1953], Arrow [1951], Luce and Raiffa [1957], Sen [1970], Pattanaik [1971], Kirkwood [1972], Fishburn [1973], and Keeney and Kirkwood [1975].

7.4.1 THE CONCEPTS FOR ADDRESSING IMPACTS TO MULTIPLE GROUPS

Throughout this section, we shall always make the assumption that u_I and u_g, $g = 1, ..., G$ in (7.39) are utility functions. In addition, four assumptions will be of interest.

Universal Indifference. If two alternatives are indifferent to each individual group, then they are indifferent to the collection of groups.

This assumption says, in technical terms, that if the expected utilities of two alternatives using each u_g are the same, then the expected utilities for both alternatives must be the same using u_I.

Pareto Optimality. If one alternative is better than a second alternative for at least one group and no worse for any other group, then the collection of groups must prefer the former alternative.

Pareto optimality is a cornerstone of much of welfare economics. It allows one to eliminate inferior alternatives and results in a set of "nondominated" alternatives from among which the best should be chosen.

The next assumptions require a few concepts and definitions. Suppose there are only two groups represented with utility functions u_1 and u_2 and that these utility functions are scaled zero to one. That is, the worst consequence x^0 (which may be different for the two groups) is assigned a zero utility and the best consequence x^* is assigned a utility of one. Now let (u_1, u_2) represent the situation where, as a result of x occurring, the respective utilities to the groups are u_1 and u_2.

Consider the following two lotteries referred to as a symmetrical pair of lotteries. Lottery L_1 results in either $(u_1 = 0, u_2 + 0)$ or $(u_1 = u_1', u_2 = u_2')$, each with probability 0.5. Lottery L_2 results in either $(u_1 = u_1', u_2 = 0)$ or $(u_1 = 0, u_2 = u_2')$, each with probability 0.5. Notice that alone, each group should be indifferent between lotteries L_1 and L_2, since they result in the same expected utilities. However, we shall define lottery L_1 to be more equitable than lottery L_2 since the utilities of each possible outcome are more balanced. Now the third assumption can be stated.

Equity. If all groups except two are indifferent between any two alternatives (i.e., lotteries), and if these two alternatives represent a symmetrical pair of lotteries for these two groups, the more equitable alternative is preferred.

One might have just the opposite preference for a symmetrical pair of lotteries for the following reason. With the more equitable lottery, there is a 50% chance that neither of the two groups will be pleased, whereas with the other lottery, it is certain that one of the groups will do as well as the choice permits. Thus, one might be conservative and choose the lottery which guarantees "some success." By using equity as a desired criterion, the client is in a sense "putting all the eggs in one basket." However, this attitude exists in degrees and can be greatly tempered by the scaling constants in the resulting utility functions. The opposite of the equity assumption is conservatism, defined as follows.

Conservatism. If all groups except two are indifferent between any two alternatives (i.e., lotteries), and if these two alternatives represent a symmetrical pair of lotteries for these groups, the more conservative (and hence less equitable) alternative is preferred.

The latter two assumptions could have been stated to allow for simultaneous changes of the preferences of each group, but this would not have simplified the understanding of the main concepts. As stated, they are easy to interpret and yet they have the same implications for u as more generally stated assumptions.

7.4.2 THE MAIN RESULTS FOR ADDRESSING IMPACTS OVER MULTIPLE GROUPS

To many individuals, all of the assumptions above would appear to be reasonable. However, they are inconsistent as a set. Both universal indifference and Pareto optimality imply the same form for the overall utility function, but this form does not promote equity or conservatism. And clearly, equity and conservatism are in direct conflict.

Result 11. Given either universal indifference or Pareto optimality, the overall utility function u_I is given by

$$u_I(x) = \sum_{g=1}^{G} \lambda_g u_g(x), \tag{7.40}$$

where $\lambda_g > 0$, $g = 1, ..., G$, are scaling factors.

If the u_g are represented on a zero to one scale, then $\Sigma\lambda_{g=1}$ in order to scale u_I from zero to one. The proof that universal indifference results in (7.40) is found in Harsanyi [1955]. More recently, Kirkwood [1979] has shown that the same result follows from Pareto optimality.

One can simply evaluate any symmetrical pair of lotteries with (7.40) and conclude that equity is not consistent with (7.40). There are, however, several utility functions which do promote equity. One result which seems to be quite useful uses the preferential independence and utility independence assumptions from Section 7.2. To do this, we shall need to think of U_g as being an attribute measured on the u_g scale.

Result 12. Given attributes $U_1, ..., U_G$, $G \geq 3$, if $\{U_1, U_g\}$, $g = 2$, ..., G is preferentially independent of the other attributes, U_1 is utility independent of the other attributes, and equity or conservatism is desired, then

$$\lambda u_I(x) + 1 = \prod_{g=1}^{G} [\lambda\lambda_g u_g(x) + 1], \tag{7.41}$$

where $\lambda > 0$ (for equity), $-1 < \lambda < 0$ (for conservatism), and $\lambda_g > 0$, $g = 1, ..., G$, are scaling factors.

The function (7.41) is referred to as the multiplicative utility function. It requires the same information as (7.40), namely the G utility functions u_g, $g = 1, ..., G$, and the scaling constants λ_g. The additional constant λ is calculated from the λ_g.

7.4.3 ASSESSING THE VALUE JUDGMENTS

In this section we shall discuss the client's inclusion of the viewpoints of different impacted groups into a siting problem. Their viewpoints are being represented by their preferences; there is still one client (in Section 7.7, the extension of these ideas to several clients will be discussed). Because of our perspective, the selection of the assumptions and the assessing of the u_g and λ_g are the responsibility of the client. That is, the U_g attributes are analogous to the X_i attributes from the client's perspective. The u_I is an aggregate indicator of the feelings of the impacted groups.

Verifying Assumptions. The easiest way to verify appropriate assumptions is to explain them to the client and ascertain if they are appropriate. In particular, the universal indifference and Pareto optimality assumptions are readily understood. Because they both have the same implication, one should be used as a check on the other.

To examine universal indifference, all the u_g levels except two are fixed and a symmetrical pair of lotteries is formed from those two. It is best to make the differences in utility for each of these two as distinct as possible. Therefore the client is asked to choose between a lottery yielding a 50–50 chance at $(u_1 = 0, u_2 = 0)$ or $(u_1 = 1, u_2 = 1)$ and a lottery yielding a 50–50 chance at $(u_1 = 0, u_2 = 1)$ or $(u_1 = 1, u_2 = 0)$. If the client is indifferent between the two lotteries, and if indifference holds for other symmetrical pairs of lotteries, it is reasonable to assume that universal indifference is appropriate.

Pareto optimality can be examined by comparing a lottery to a certain consequence. The client may be asked to choose between $(u_1 = 0.4, u_2 = 0.4)$ and a lottery yielding a 50–50 chance at $(u_1 = 0, u_2 = 0.4)$ or $(u_1 = 0.9, u_2 = 0.4)$. Note that $u_2 = 0.4$ regardless. In the sure situation, $u_1 = 0.4$ and in the lottery, the expected utility using u_1 is 0.45. Therefore group 1 should prefer the lottery and group 2 should be indifferent. Stated another way, the lottery dominates the sure situation, which should therefore be eliminated from further consideration if Pareto optimality holds. However, from the client's viewpoint, it may be preferable to violate Pareto optimality and select the sure situation because of the inequity inherent in the lottery. Such a choice would suggest that the equity assumption might be appropriate. On the other hand, if the client preferred the lottery to the sure situation, and if such a preference was indicated in several other similar cases, then it would be reasonable to assume that Pareto optimality was appropriate.

If universal indifference and Pareto optimality are inappropriate assumptions, then either equity or conservatism must hold or the client's basic attitude for combining the utility functions changes over their domains. Comparing the extreme pair of symmetrical lotteries provides about the best indicator for equity or conservatism. With a choice between lottery L_1 yielding equal chances at $(u_1 = 1, u_2 = 1)$ and $(u_1 = 0, u_2 = 0)$ and lottery L_2 yielding equal chances at $(u_1 = 1, u_2 = 0)$ and $(u_1 = 0, u_2 = 1)$, a preference for lottery L_1 indicates that equity is probably appropriate. A preference for lottery L_2 indicates conservatism. A few more such comparisons should be sufficient to justify either case.

Assessing the Utility Functions.   The utility functions u_g, $g = 1$, ..., G represent the preferences of the groups. However, the client's preferences for the impacts to these groups are what is important in this section. The client, therefore, is responsible for obtaining the u_g. In this sense, this is analogous to assessing utility functions over the X_i attributes in Section 7.2. The u_g may be assessed in several ways, which differ mainly in the degree to which all the members of the various groups are involved. At one extreme, one might try to assess utility functions for several members of a group and then combine these into a group utility function. This would be time consuming, expensive, and probably unreliable. At the other extreme, the client could postulate a utility function for group g and use it. This too would be unreliable and would probably be of no use in justifying the approach. Options in between these two seem better.

For instance, the siting team and client may use their collective knowledge of the various groups to structure a functional form of a utility function meant to represent the values of each group. Different forms may, of course, be necessary for different groups. The parameters of these utility functions would be determined with a certain amount of direct or indirect input from the groups being considered. They should be determined by individuals familiar with the values and viewpoints of the groups. These individuals may or may not be group members and may or may not work for the client or be on the siting team. For instance, the viewpoint of a local community might be represented by the mayor and a few council members. The selection of these "knowledgeable individuals," as they are referred to by Gros [1975], would necessarily depend on their availability and willingness as well as their expertise. However, with a thorough sensitivity analysis, it is quite likely that the viewpoints of most important groups can be adequately accounted for in the analysis. The standard to be met is whether such groups are explicitly considered in any manner.

Assessing the Scaling Constants.   The λ_g terms in Results 11 and 12 are scaling constants representing the relative weights given to the various groups. These constants result from the value judgments of the client. As Kirkwood [1972] pointed out, they should depend on two considerations:

1. the implications of the range of possible impacts to each group and
2. the client's concern for that group.

For instance, if the range of impacts is small for group 1, the scaling factor λ_1 may be relatively small compared to the other λ_g, even if the group is felt to be important by the client. The fact that the range of impacts is large for group 2 may not lead to a relatively large λ_2, because the client may not consider group 2 as important as other groups. Importance can result from different factors such as the size of the group, the group origins (i.e., local versus outsiders), and the political power of the group.

The process for determining the λ_gs, is the same as that described in Section 7.2.3 for obtaining the k_i scaling factors. One wishes to generate G equations with G unknowns, using Result 11 or 12, and then solve for the λ_g. It is simplest to fix the utilities of $G - 2$ of the groups and vary the remaining two until an indifferent pair is found. For instance, if $(u_1', u_2', u_3, ..., u_G)$ is indifferent to $(u_1'', u_2'', u_3, ..., u_G)$, equating their respective utilities yields one equation with unknowns λ_1 and λ_2. In determining this indifference, the client may need to examine particular sets of consequences x which will yield u_1', u_1'', u_2', and u_2'' in order to get an intuitive feeling for their meaning.

In some cases, the λ_g, or at least the relative values of some of them, may be easily determined by a simple rule. For instance, it may be appropriate to rank the relative values of the λ_g by the sizes of the groups they represent. Then a group of twice the size would have a λ_g twice as large. To implement such a scheme, it is sufficient to estimate the size of the various groups.

7.4.4   AN ASSESSMENT INVOLVING PREFERENCES OVER GROUPS: SKEENA RIVER SALMON MANAGEMENT†

The Skeena River and its tributaries in British Columbia, Canada, is an important salmon fishing area. Salmon fishing, which is the basis of the area's economy, currently provides about 5000 jobs. This includes the fish-

---

† This section is liberally adapted from Keeney [1977a].

ermen themselves, people working in canneries, and individuals earning a living from tourism as a result of recreational fishing. Policy decisions indicating, for example, (1) who can fish, (2) what they can fish (types or size of salmon), (3) where they can fish, (4) which methods they can use, and (5) when they can fish, impact directly or indirectly everyone living in the Skeena area. Other options such as the development of artificial spawning grounds are also possible and need to be evaluated. Such possibilities have many parameters (size, construction type, cost). If one should decide to construct spawning grounds, how should they be designed?

This section concentrates on the development of the multiattribute utility function for examining policy affecting salmon fishing in the Skeena River. The basic ideas could be applied to energy siting problems exactly as summarized here. As indicated, the multiattribute utility model allows one to focus explicitly on value tradeoffs and equity considerations among groups. The case also illustrates the potential of the theory for conflict illumination and resolution.

The Interest Groups and Their Objectives

The decision maker for such policy problems is the Canadian Department of the Environment (DOE). There are five main groups whose preferences are important to DOE. Four of these—lure fishermen, net fishermen, sport fishermen, and Indians—are directly involved in fishing. The fifth "regional" group includes all those individuals whose welfare is linked to fishing, such as cannery employees and motel operators.

The lure and net fishermen fish for a livelihood, using lures and nets, respectively. The lure fishermen operate near the mouth of the Skeena and the net fishermen a littler farther upstream in a controlled area. Upstream from them are the sport fishermen and still farther upstream are the Indians. The latter two groups fish mainly for pleasure and food.

The objectives hierarchy for this problem is illustrated in Fig. 7.6. As decision maker, DOE has two major objectives: to satisfy each of the five interest groups as much as possible and to minimize its own (government) expenses. The degree to which the net fishermen are satisfied depends, of course, on how well their own objectives are satisfied. As indicated in the figure, their main interests are to maximize the income per net fisherman, optimize their fishing time (i.e., neither too much nor too little work), and maximize the diversity of the catch. The last objective is a proxy indicator for their psychological well-being. Knowing that the river is healthy (i.e., supporting many species) provides both future flexibility and future security. The lure fishermen have analogous objectives. The Indians and sport

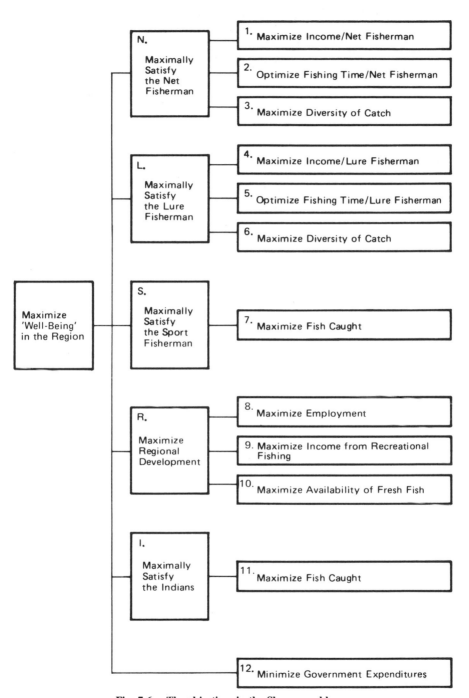

Fig. 7.6. The objectives in the Skeena problem.

fishermen are interested in maximizing their fish catch. The region wants to maximize economic benefits from employment and recreational sources as well as to have an abundance of fresh fish to eat.

Let us elaborate on the complexity inherent in such a problem. Suppose DOE is considering implementing an innovative licensing policy. One scenario that may result from that policy is that only a few additional fish will return from the ocean, whereas another scenario may lead to large increases in adult salmon in the Skeena. The uncertainties here are large. If DOE changes licensing strategies, this may increase administrative (government) costs and simultaneously lead to better harvests for lure and net fishermen. However, there may be less fish for the sport fishermen and the Indians. These groups would then be displeased. The overall impact on the region might be more employment in canneries, etc., but less recreational income. What should DOE do? Somehow they must measure each of the possible impacts, balance these in some fair way, and decide "with all pros and cons considered, whether to go ahead with the new licensing strategy or not."

For each of the lowest level objectives of the DOE's hierarchy, Table 7.3 lists an attribute and ranges of the possible impacts on each attribute. The consequence of any decision taken by DOE can be represented by the 12-tuple $(x_1, x_2, \ldots, x_{12})$, where x_i indicates a specific level of attribute X_i. For decision making, we want DOEs governmental utility function $u_G(x_1, x_2, \ldots, x_{12})$.

Structuring the Utility Function

One should interpret the results in this section as preliminary. All the assessments are based on discussions with Dr. Ray Hilborn of the Univer-

TABLE 7.3

ATTRIBUTES FOR THE SKEENA PROBLEM

Attribute	Worst level	Best level
X_1 = annual income/net fisherman	0	$25,000
X_2 = annual days fishing/net fisherman	100	0
X_3 = species of salmon in the Skeena	1	10
X_4 = annual income/lure fisherman	0	$25,000
X_5 = annual days fishing/lure fisherman	100	0
X_6 = species of salmon in the Skeena	1	10
X_7 = annual sport fisherman catch of salmon	0	1,000,000
X_8 = employment	0	5,000
X_9 = annual revenue due to recreation	0	$10,000,000
X_{10} = cost of fresh salmon/lb	$10.00	$0.20
X_{11} = annual Indian catch of salmon	0	100,000
X_{12} = annual expenditures (millions of dollars)	$10	0

sity of British Columbia. Similar assessments, conducted with Dr. Carl Walters of the same institute, are described in Keeney [1977a]. Both Hilborn and Walters were working on a model of salmon in the Skeena River. In making the utility assessments, each used his knowledge of the "Skeena Problem" to respond in the way he expected the groups and DOE to respond.

The utility functions u_N, u_L, u_S, u_R, and u_I for the net fishermen, lure fishermen, sport fishermen, region, and Indians, respectively, were assessed quickly and roughly using the procedures discussed in Section 7.2. Those procedures provided a basis on which to improve the component utility functions necessary for constructing a first cut of DOE's preferences. Next, the preferences of the five groups were integrated into a single indicator of preference $u(u_N, u_L, u_S, u_R, u_I)$. Then we combined this with attribute X_{12}, the cost to the Canadian government, to obtain an overall DOE utility function $u_G(u, x_{12})$.

To integrate the preferences of the interest groups, we first investigated whether the assumptions necessary for Result 12 seemed reasonable. Notationally, this required that, for each u_j, $j = N, L, S, R, I$, we create an associated attribute U_j. In this context, the preferential independence conditions imply that tradeoffs between any two groups (measured on the U_j scales) do not depend on how well satisfied the other three groups are. The utility independence condition means that if the consequences to only one group are uncertain, we shall always choose among alternatives using only the outputs to that group.

We decided that the utility independence of each U_j seemed reasonable. However, the preferential independence assumptions were not completely justified, since value tradeoffs between the net fishermen and the region, for example, depended on the lure fishermen. If the lure fishermen were doing poorly, there was a willingness to give up more in regional development to move net fishermen from a bad to a good position (e.g., from a 0.2 to a 0.8 utility measured by u_N) than there was if lure fishermen were doing well. That is, conservatism seemed appropriate since there was a premium on having at least one group of commercial fishermen at a good level.

Next, the possibility of combining u_N and u_L into a commercial fishermen's utility function u_C with associated attribute U_C was investigated. Value tradeoffs between U_C and U_R were examined and they did seem to be preferentially independent of $\{U_S, U_I\}$. Thus, we could use Result 3 of Section 7.2 to combine u_N and u_L into u_C and then use Result 12 to combine u_C, u_S, u_I, and u_R into u.

In notational form, from these assumptions, we concluded that

$$1 + \lambda u(u_C, u_S, u_R, u_I) \\ = (1 + \lambda \lambda_C u_C)(1 + \lambda \lambda_S u_S)(1 + \lambda \lambda_R u_R)(1 + \lambda \lambda_I u_I) \quad (7.42)$$

and

$$u_C(u_N, u_L) = \lambda_N u_N + \lambda_L u_L + (1 - \lambda_N - \lambda_L) u_N u_L, \qquad (7.43)$$

where the us are all scaled zero to one and the λ_js are scaling factors. Given (7.42) and (7.43), we need to assess the u_js and λ_js.

Assessing the Utility Function

Assessments for the utility functions of the various groups were made using procedures and formulations described in Section 7.2. The unique aspect of the assessment concerned obtaining the λ_js. Because the procedure is similar, we will simply state that $\lambda_L = \lambda_N = 0.6$ and illustrate the assessment with the other $j = $ S, R, I, C.

First, we wanted to rank the λ_js. This was done with the aid of Table 7.3. We asked, "Given that all the attributes $X_1, \ldots, X_{11}$ are at their worst level, which one of the four attribute sets $U_C \equiv \{X_1, X_2, X_3, X_4, X_5, X_6\}$, $U_S \equiv X_7$, $U_R \equiv \{X_8, X_9, X_{10}\}$, or $U_I \equiv X_{11}$ would you rather move up to its best level if you could move only one?" Hilborn chose the attributes $\{X_8, X_9, X_{10}\}$ associated with regional development, so λ_R, the regional development scaling factor, must be the largest λ_j. We then asked, "Which of the remaining three attribute sets would you prefer to move to its best level?" The answer was the commercial fisherman's attributes, implying λ_C is second largest. Continuing in the same manner led to

$$\lambda_R > \lambda_C > \lambda_I > \lambda_S. \qquad (7.44)$$

To determine values for those λ_js, we first examined value tradeoffs between U_R and U_C. However, one cannot answer questions directly in terms of u_R and u_C so we returned to the basic attributes of each group. To identify the DOE's tradeoffs, it is often best to use the attribute with the largest scaling factor for each group. The region weighted employment highest in U_R and the commercial fishermen weighted annual income highest in U_C. Thus we looked at the value tradeoffs between X_1 and X_8. Specifically, we found ($x_1 = 0$; $x_8 = 800$) indifferent to ($x_1 = \$25,000$; $x_8 = 0$). Equating the utilities of these two points using (7.42) and (7.43), while assuming the other attributes are fixed at their worst levels (because of preferential independence), yields

$$\lambda_R u_R(x_8 = 800; x_9 = 0; x_{10} = \$10)$$
$$= \lambda_C \lambda_N u_N(x_1 = \$25,000; x_2 = 100; x_3 = 1) \qquad (7.45)$$

Evaluating (7.45) with the utilities of u_R and u_N results in one equation relating λ_R and λ_C. Similar relationships between other λ_js were developed using value tradeoffs between the specific attributes. These equations were solved to yield the λ_js necessary to specify u.

To aggregate u and x_{12} into an overall utility function u_G for DOE, it seemed reasonable to assume that U and X_{12} were utility independent of each other. Thus from Result 3 in Section 7.2,

$$u_G(u, x_{12}) = hu + h_{12}u_{12}(x_{12}) + (1 - h - h_{12})\, uu_{12}(x_{12}), \qquad (7.46)$$

where h and h_{12} are constants. Using assessment procedures similar to those described above, which address the value tradeoffs between DOE costs and group benefits, the constants were evaluated as $h = 0.9$ and $h_{12} = 0.4$.

More details on the assessments described here are found in Keeney [1977a]. Subsequent to the work outlined here, Hilborn and Walters [1977] conducted workshops to attempt to assess directly the utility functions of members of the various groups concerned about salmon policy on the Skeena River. The overall study of the Skeena salmon management options is discussed in Holling [1978].

7.5 Quantifying Attitudes toward Risks

In this section, risk is to be considered synonymous with uncertainty. The attitude toward risk (or uncertainty) is quantified in the utility function u. Of course, other considerations involving value judgments are also quantified in u. These include the value tradeoffs between attributes, the relative values of impacts in different time periods, and the values assigned to different groups, all of which were discussed in Sections 7.2–7.4. For all of these cases, we have managed to decompose the overall utility function into a form involving (usually) single-attribute utility functions and scaling factors. The value tradeoffs for attributes, time periods, and groups are quantified by the scaling factors, whereas the attitude toward risk is quantified by the form of the single-attribute utility functions.

For this discussion, we can characterize attributes as those with natural and with constructed scales. The natural scales include both continuous cases, such as costs in dollars, and discrete cases, such as number of fatalities (see Section 7.5.5). The constructed scales usually define clearly only a few points on the scale, such as the site environmental impact index in the nuclear power siting study of Chapter 3. The utility functions for natural attributes are assessed using the assumptions and results described in this section. The utility functions for constructed indices are assessed point by point as indicated for the site environmental attribute in Section 3.4. Procedures for both cases will be given at the end of this section.

7.5.1 THE CONCEPTS FOR ADDRESSING RISK ATTITUDES

There are attributes, such as profits, for which preferences increase as the attribute level increases. That is, the more profits, the more preferred the situation is, assuming that all other factors are held fixed. In an analogous manner, preferences may decrease as attribute levels increase for attributes such as acres of forest destroyed. And sometimes, preferences may both increase and decrease depending on the domain of an attribute level. For example, up to a point of, say 15% growth, it may be desirable for the number of jobs in an area near the construction of an energy facility to increase. However, preferences for additional growth would actually decrease because of the disruption that such growth would cause to the community. The point is that simple assumptions about the direction of increasing preferences can almost always be easily ascertained.

The important concepts about risk attitudes concern risk aversion, risk neutrality, and risk proneness. To discuss these concepts, we need to define a nondegenerate lottery, one where no single consequence has a probability equal to one of occurring. There must be at least two consequences with finite probabilities of occurring. The following assumptions are mutually exclusive and collectively exhaustive when applied to any particular lottery.

Risk Aversion.   One risk is averse if and only if the expected consequence of any nondegenerate lottery is preferred to that lottery.

For example, consider the lottery yielding either a 1 or 2 billion dollar cost, each with a one-half chance. The expected consequence of the lottery is clearly 1.5 billion dollars. If the client is risk averse, then a consequence of 1.5 million must be preferred to the lottery.

Risk Neutrality.   One is risk neutral if and only if the expected consequence of any nondegenerate lottery is indifferent to that lottery.

And as no surprise, the remaining assumption is risk proneness.

Risk Proneness.   One is risk prone if and only if the expected consequence of any nondegenerate lottery is less preferred than that lottery.

Given any single-attribute utility function, a measure developed by Pratt [1964] can be used to indicate its degree of risk aversion. The measure may be positive, zero, or negative, indicating risk aversion, risk neutrality, and risk proneness, respectively. Pratt also introduced more sophisticated concepts of decreasing risk aversion, etc., which will not be discussed here. A summary of Pratt's original results, as well as several examples illustrating their use, is found in Keeney and Raiffa [1976].

7.5.2   The Main Results for Quantifying Risk Attitudes

The general shape of the utility function is completely determined by the attitude toward risk. This can all be stated in one concise result.

Result 13.   Risk aversion (neutrality, proneness) implies that the utility function is concave (linear, convex).

These three cases are illustrated for both increasing and decreasing utility functions in Fig. 7.7. We have assumed that the domain for attribute X ranged from a minimum x^0 to a maximum x^* and that u was scaled from zero to one.

In theory, by using the more sophisticated risk attitudes, such as decreasing risk aversion, we can not only specify the general shape of the utility function, but also an exact functional form. However, experience has indicated that such fine tuning is rarely required for the individual utility functions when they are part of a multiattribute formulation. It will almost always suffice to use a single parameter utility function, where the single parameter quantifies the client's degree of risk aversion for the attribute in question.

As indicated in the nuclear power siting and pumped storage case studies in Chapters 3 and 9, respectively, the exponential or linear utility functions are a fairly robust set of single forms for characterizing single-attribute utility functions. We can formalize this by the following.

Result 14.   Classes of risk averse, risk neutral, and risk prone utility functions are

$$u(x) = a + b(-e^{-cx}), \qquad (7.47a)$$

$$u(x) = a + b(cx), \qquad (7.47b)$$

and

$$u(x) = a + b(e^{cx}), \qquad (7.47c)$$

respectively, where a and $b > 0$ are constants to insure that u is scaled from zero to one (or on any scale desired) and c is positive for increasing utility functions and negative for decreasing ones.

The parameter c in (7.47a) and (7.47c) indicates the degree of the client's risk aversion. For the linear case (7.47b), parameter c can be set at $+1$ or -1 for the increasing and decreasing cases, respectively. More details about the exponential utility functions and discussions of other single-attribute utility functions are found in Pratt [1964] and Keeney and Raiffa [1976].

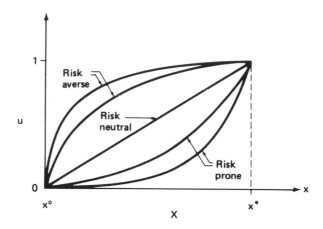

(a) Increasing Utility Functions

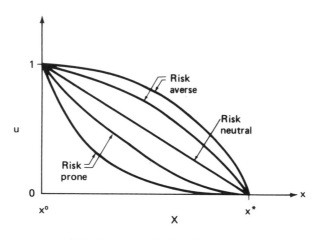

(b) Decreasing Utility Functions

Fig. 7.7. Risk attitudes and utility functions.

7.5.3 ASSESSING THE VALUE JUDGMENTS

There are two types of value judgments needed to determine the individual utility functions. The first specifies the risk attitude and therefore determines the general shape of the utility function. The second identifies the specific utility function of that general shape.

Examining Risk Attitudes. Suppose we want $u(x)$ for attribute X for $x^0 < x < x^*$. And since it is trivial to ascertain whether larger levels of X are preferred to smaller, let us assume larger levels are less preferred as in the case with costs. To begin examining risk attitudes, we take a 50–50 lottery at the extremes of X and compare it to the expected consequence. That is, the client is asked whether a 50–50 chance at each of x^0 and x^* is preferred to, indifferent to, or less preferred than the sure consequence $\bar{x} \equiv (x^0 + x^*)/2$. A preference for the sure consequence indicates that risk aversion may hold.

Next, the same line of questioning is repeated for the lower- and upper-half ranges of X. The lottery yielding equal chances at x^0 and $\bar{x}$ is compared to the expected consequence $(x^0 + \bar{x})/2$. Preference for the sure consequence again indicates risk aversion. Similarly, a preference for the sure consequence $(\bar{x} + x^*)/2$ to a 50–50 lottery yielding either $\bar{x}$ or x^* also indicates risk aversion. If assessments for the entire range plus the upper and lower halves are consistent in terms of their risk implications, risk aversion is probably a very good assumption to make. If different implications are found, and a reexamination indicates no errors in understanding, it is appropriate to divide the domain of X and search for sections exhibiting different risk attitudes. For instance, it may be that from x^0 to x' the client is risk averse, but from x' to x^*, risk neutrality is appropriate.

Selecting the Risk Parameter. We have now determined that the risk attitude which implies one form of (7.47) is probably reasonable. If the form is (7.47b), no additional assessments are necessary. Parameter c is set at $+1$ or -1 depending on whether the utility function is increasing or decreasing. Then the constants a and b are simply set to scale u from zero to one.

For the risk averse and risk prone cases, a little more effort is required. Suppose that the attribute is such that preferences increase for greater levels of the attribute and that the client is risk averse. Then from Result 14, it follows that a reasonable utility function is

$$u(x) = a + b\,(-e^{-cx}), \qquad b > 0, \quad c > 0. \tag{7.48}$$

If $u(x)$ is to be assessed for $x^0 \leqslant x \leqslant x^*$, we might set

$$u(x^0) = 0 \qquad \text{and} \qquad u(x^*) = 1 \tag{7.49}$$

to scale u. Next, we shall need to assess the client's certainty equivalent for one lottery. In other words, we need to know a certainty equivalent $\hat{x}$ which the client finds indifferent to the lottery yielding either x' or x'', each with an equal chance, where x' and x'' are arbitrarily chosen. Then the utility assigned to the certainty equivalent must equal the expected utility of the lottery, so

$$u(\hat{x}) = 0.5\,u(x') + 0.5\,u(x''). \tag{7.50}$$

Substituting (7.48) into (7.49) and (7.50) gives us three equations with the three unknown constants a, b, and c. Solving for the constants results in the desired utility function. As shown in the pumped storage case in Section 9.4.2, the actual determination of the constants is simplified if x' and x'' are chosen to be x^0 and x^*, respectively.

Now let us return to the case of a constructed index with clearly defined levels ordered x^0, x^1, ..., x^6, x^*, where x^0 is least preferred and x^* is most preferred. Then we can again set our scale by (7.49) and assess $u(x^j)$, $j = 1, ..., 6$ accordingly. For each x^j, we want to find a probability p_j such that x^j for sure is indifferent to a lottery yielding either x^* with probability p_j or x^0 with probability $(1 - p_j)$. Then equating utilities,

$$u(x^j) = p_j u(x^*) + (1 - p_j)u(x^0) = p_j, j = 1, ..., 6. \tag{7.51}$$

For both the natural and constructed scales, once a utility function is assessed, there are many possible consistency checks to verify the appropriateness of the utility functions. One may compare two lotteries or a sure consequence and a lottery. The preferred situation should always correspond to the higher computed expected utility. If this is not the case, adjustments are necessary in the utility function. Such checking should continue until a consistent set of preferences is found.

7.5.4   CASES QUANTIFYING ATTITUDE TOWARD RISK

There are numerous examples in this book illustrating the applicability of the techniques discussed in this section. For example, the utility functions for each of the attributes in the nuclear power plant siting study of Chapter 3, the pumped storage siting study of Chapter 9, and all of the examples in this chapter were assessed using these techniques. The utility assessment for the cost attribute in the pumped storage study (see Section 9.4) is a good illustration of a case involving a natural attribute. Similarly, the utility assessment for the site environmental impact of the nuclear siting study (see Section 3.4) best illustrates the technique for constructed attributes.

7.5.5   THE RISK OF POSSIBLE FATALITIES

One of the important risks inherent in any siting study is the potential loss of human lives. These fatalities could result from accidental releases of radioactivity at nuclear power plants, from sulfur dioxide pollution at coal or oil power plants, from dam breaks at hydroelectric power plants, or from lack of power due to unreliability of solar power plants. One point is that absolutely no alternatives are free from the risk of public fatalities. A second point is that it is often the case that the potential for human fatalities is not exactly the same for all of the potential sites in each study. If this is the case, it is certainly appropriate to include an attribute in the analysis to capture this aspect of the problem (see Starr [1979]).

As discussed in Section 6.5, there are basically two ways to include the risk of public fatalities in an analysis. One is to use a proxy attribute meant to measure one aspect of the mechanism by which the fatalities could occur. This was done using the site population factor in the WPPSS nuclear siting study. The site population factor essentially added up the population in the vicinity of a proposed power plant, using a greater weighting for individuals nearer to the facility. The other approach is to measure directly the potential fatalities with "number of fatalities" as the attribute. A model is then built, as discussed in Chapter 6, to provide a probability distribution over the number of fatalities. The following three assumptions, each of which seems appealing, may be useful for structuring the utility function over fatalities.

Risk Equity.   If the individual risk, defined as the probability that the individual will be a fatality, is fixed for all but two individuals, and if these two individuals do not have identical risks, it is desirable to transfer some risk from the higher risk individual to the lower risk individual.

Risk equity roughly states that it is preferable to spread a fixed total risk (i.e., the sum of the individual risks) to as many people as possible.

Individual Indifference.   If each individual has the same marginal risk in two situations, the situations should be indifferent.

The individual indifference condition was discussed in the related context involving impacts over groups in Section 7.4. In this context, it is equivalent to a desire to minimize total risk.

Catastrophe Avoidance.   A p chance at n fatalities is preferred to a $0.1p$ chance at $10n$ fatalities for all p and n.

Catastrophe avoidance tries to minimize the likelihood of major disasters at the expense of several smaller ones, possibly resulting in more expected fatalities.

Result 15.   With regard to a utility function over fatalities, risk equity implies risk proneness, individual indifference implies risk neutrality, and catastrophe avoidance implies risk aversion.

This result is proved in Keeney [1980]. The interesting aspect of this result is that the three assumptions are mutually contradictory. For any particular problem, only one may hold for any particular domain of the attribute. This result merely serves to reinforce the fact that the difficulty in
dealing with potential fatalities is significant. However, by better understanding the assumptions consistent with any representation of the utility
function, we are better able to appraise which form may be appropriate in
a specific case. At least, we can perform a sensitivity analysis to see if the
form of the utility function really matters for the problem being addressed.

7.6   The Art of Assessing Preferences†

Sections 7.2–7.5 compactly describe the "science" of assessing the information necessary to quantify preferences. Numerous examples
throughout this chapter and the cases in Chapters 3 and 9 also summarize
such assessments in actual problems. In this section, we examine the
"art" of assessing preferences. This is done by providing parts of a dialogue of an actual assessment, with comments on why various questions
were asked. The case chosen was not specifically concerned with energy
siting, but with energy policy. It is closely related to the siting topic of this
book, and as such, it seems appropriate to illustrate the assessment procedure desirable for evaluating proposed energy sites.

Outline of This Section.   This section describes in detail the assessment of a multiattribute utility function, over 11 measures of effectiveness, used to indicate the environmental impact of alternate energy development scenarios in the state of Wisconsin. The Wisconsin effort,
directed by Professor Wes Foell, has specified possible consequences of
several energy alternatives. This section describes the first attempt to
quantify the preference structure for that problem.

Section 7.6.1 summarizes aspects of the problem which concern us. In
Section 7.6.2 the assessment of Dr. William Buehring's utility function is
illustrated in detail. The discussion essentially presents the dialogue
between Buehring and myself the first time the assessment was done. Section 7.6.3 gives some follow-up assessments conducted a week later. In
the interim, Dr. Buehring had assessed Wes Foell's utility function over
the same 11 measures and had several discussions about preferences over
the attributes. In Section 7.6.4 we calibrate Dr. Buehring's utility func-

† This section is liberally adapted from Keeney [1977b].

tion. The final section discusses several aspects of the assessment procedure.

Related Results and a Perspective. The assessment protocol described in this section is not the only available procedure. There are both simpler procedures and good reasons for using them. For example, Professor Ward Edwards has utilized rather simple procedures to quantify preferences involving multiple objectives and applied the results successfully to several problems (see Edwards [1977]). In addition, there are several studies, including those of Fischer [1976, 1979], von Winterfeldt and Edwards [1973a, b], Dawes and Corrigan [1974], and Otway and Edwards [1977] supporting the fact that these simpler assessments are often just as appropriate for evaluating alternatives as are utility functions assessed by more complicated procedures. Why, then, should one use the more complicated procedure described here?

First, I do not think one should always use the procedure described here. I do think it should be used sometimes. The choice of an appropriate procedure should depend on the problem, the client, the time available, and several other factors.

I feel there are two basic reasons for quantifying one's preferences. One is to develop a preference model (i.e., objective function) for examining alternatives. The other is to force the decision maker to think hard about the problem: What are the objectives, why does the client care about that, how much does he or she care, and so on. It is sometimes important to initiate a deep, soul-searching process on the part of the client. The results of this process effectively define a major aspect of the problem: What it is that the client wishes to accomplish.

All assessment procedures result in a preference model. However, the more involved procedure described here does force deeper thinking. It also requires that one verify the qualitative assumptions made about the preference structure. Sometimes the additional time and effort is worth it. On complex, million dollar siting decisions, I personally do not find it unreasonable that the key decision maker and his or her aides collectively spend, say, 40 hours (or maybe even a few person-months) thinking hard about and quantifying their preferences. They are often spending months and hundreds of thousands of dollars trying to quantify impacts, and their preferences would indicate which impacts are important.

7.6.1 IMPACTS OF ALTERNATIVES FOR ELECTRICAL ENERGY PRODUCTION IN WISCONSIN

Over the past several years, Buehring and Foell and others (see Buehring and Foell, [1974], Buehring, Foell, and Dennis [1974], Foell

[1979]) have tried to assess the impact of various alternatives for producing electrical energy in Wisconsin from now until the year 2000. Rather than go into any detail, let me just briefly mention aspects of their work relevant to the discussion here. The primary policies which Buehring and Foell are examining differ in terms of two main characteristics: the total electrical power generated and the percentages generated from nuclear and fossil sources.

At the beginning of their work, a set of desired objectives of an energy policy was generated. The process was essentially creative, as discussed in Chapter 5. Alterations were made after discussions, etc., to arrive at a reasonably comprehensive set of objectives. The next step involved specifying attributes (i.e., measures of effectiveness) to measure the degree to which these objectives were met. These attributes, indicated by X_1, ..., X_{11}, are listed in Table 7.4. Also in the table are the units used to measure the attributes as well as a range for the possible impacts of any of the alternatives. It was simple to check that, in fact, for all attributes except electricity generated, less of an attribute was preferred to more. Hence, for later purposes, we list best and worst levels rather than maximum and minimum levels in Table 7.4.

Buehring [1975] has specified in great detail exactly what impacts each attribute is meant to capture. Let us simply try to clarify a few aspects here. Fatalities include deaths due, for example, to working in the coal mines, transporting the fuel, nuclear power plant disasters, and prolonged pollution. Permanent unusable land may result from storage of radioac-

TABLE 7.4

ATTRIBUTES FOR EVALUATING ENERGY POLICY

Attribute	Measure (units)	Level	
		Worst	Best
X_1, fatalities	Deaths	700	100
X_2, permanent land use	Acres	2000	0
X_3, temporary land use	10^3 Acres	200	10
X_4, water evaporated	10^{12} Gallons	1.5	0.5
X_5, SO_2 pollution	10^6 Tons	80	5
X_6, particulate pollution	10^6 Tons	10	0.2
X_7, thermal energy needed	10^{12} kWhr (thermal)	6	3
X_8, radioactive waste	Metric tons	200	0
X_9, nuclear safeguards	Tons of plutonium produced	50	0
X_{10}, chronic effects	Tons of lead	2000	0
X_{11}, electricity generated	10^{12} kWhr (electric)	0.5	3

tive waste at a location. Attribute X_2 measures the impact due to the loss of usable land, whereas X_8 is meant to indicate the implications (e.g., genetic impact) resulting from the waste itself. Attributes X_3 and X_4 measure the land and water resources unavailable because of electrical power generation. Attributes X_5 and X_6 are intended to capture all of the undesirable effects of air pollution other than health impacts (measured by X_{10}) and deaths from acute SO_2 exposure (measured by X_1). Attribute X_7 indicates the thermal power needed to generate the electrical power measured by X_{11}. Together, these provide an indication of the waste and efficiency of the system. In addition, X_{11} is a proxy indicator of the desirable impacts on quality of life due to more energy. Attribute X_9 is used to indicate the vulnerability of the system to theft of nuclear material.

The next and major part of Buehring and Foell's work was to estimate the impact of the various alternatives being investigated in terms of the 11 attributes of Table 7.4. This was done by trying to trace back from the generation of electrical power to all of the impacts produced on the way, including the steps of obtaining, transporting, using, and disposing of the fuel. After this was done for each option, a method was needed for aggregating all of the data in a responsible fashion to select a reasonable, if not the best, policy. This purpose is served by the utility function.

7.6.2 THE FIRST ASSESSMENT OF BUEHRING'S UTILITY FUNCTION

Assessing a utility function is a process where the assessor asks the client a series of questions about his or her preferences. From the responses, the assessor constructs the client's utility function. First, one asks some questions to determine the general shape of the utility function. Then one asks more specific questions to quantify a specific utility function. Finally, there should be consistency checks and modification. My experience has been that almost invariably, in multiattribute contexts, the client will make some modifications to the preferences as first articulated. This should not be disturbing, since a major purpose of the assessment is to force the client to understand the implications of his or her preferences in these very complex situations. Since the problem is complicated, it is unlikely that one can immediately articulate consistent preferences which correctly represent the individual's feelings.

The assessment process is dynamic. What the assessor does next depends on how the person whose preferences are being assessed responds to the current question. It does not depend solely on the answer itself, but on other factors such as the assessor's perception of the ease the client had in responding, on his or her understanding of the question, and on the desirability of going into more detail. At any point at which the assessor feels that a question was misunderstood, and hence wrongly answered, he or she can ask a similar question to verify this intuition. If there was a misunderstanding, questions should be repeated.

In spite of these dynamic aspects, one can more or less follow a pattern in assessing utility functions. Once the attributes are specified, the assessment process (as outlined in Section 2.3.4) might be broken into five parts:

1. familiarization with the terminology and motivation for the assessment,
2. verification of independence assumptions concerning preferences to specify the general value structure,
3. assessment of the value tradeoffs among attributes to determine the scaling constants,
4. assessment of the individual attribute utility functions, and
5. checking for consistency and modification.

In the dialogue which follows, we essentially cover parts 2–5. Before this assessment, Buehring read several sources to familiarize himself with the terminology and procedures used in such assessments. When we met for the utility function assessment, Buehring and I first went over the meanings of all of the attributes. Then we discussed the point of view that Buehring should take in the assessments. That is, should he articulate his own preferences, or those he feels the government has or should have, or others? Clearly, the different perspectives might lead to different responses. We concluded that, for this assessment, the preferences should be his. With this, we were ready.

Verification of Preferential Independence Assumptions

Keeney: So, let's begin. First we can examine the general preference structure. As a start, consider attribute X_1, fatalities, versus X_2, the permanent unusable land. We will use Fig. 7.8 to help in questioning. Note that the range of fatalities, 100–700, is plotted on the abscissa and that the amount of land permanently unusable, 0–2000 acres, is on the ordinate. So I guess the the best point is ($x_1 = 100$, $x_2 = 0$), is that right?

Buehring: Yes.

K: And for now, let us assume that all of the other attributes X_3, X_4, ..., X_{11} are at their worst levels as defined in Table 7.4. Now, suppose you are at the best point A (Fig. 7.8); would you rather lose 600 people and move to point B or lose 2000 acres and go to C?

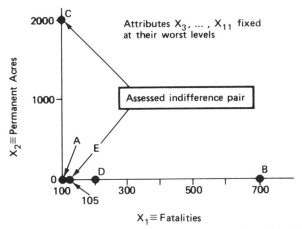

Fig. 7.8.   Value tradeoffs between fatalities and permanent land use with other attributes at their worst levels.

B: Lose the 2000 acres.

K: [That question seemed to be easy to answer since it came quickly. Thus, I lowered the fatalities greatly.†] Okay, would you rather lose 100 additional people (point D) or the 2000 acres?

B: I would still rather lose the acres, all of them.

K: And not that this number needs to be precise—you can certainly change any numbers anywhere in the process—how many people on a first guess would you be willing to give up to be indifferent to these 2000 acres?

B: That's pretty tough. That's permanent commitment for land. But, relative to fatalities, it just doesn't seem that important to me.

K: How about 110 people, point E?

B: It's probably in that neighborhood. It's very small. I'm trying to think whether it's bigger than 101. . . . I guess it is bigger than 101. Maybe 105, how does that sound?

K: That's fine for now. One thing that comes to mind is: what is included in attribute X_2? Is it only concerned with the loss of use of the land and not with the psychological worry that an individual who lives near a radioactive waste facility may feel, for example?

B: No, this isn't supposed to include that. That problem of waste is captured by attribute X_8, which includes both the high-level and low-level waste.

---

† The assessment procedure is dynamic. The next question often depends on the answer to the previous one. Some of my thoughts which influenced the choice of a question are given in brackets.

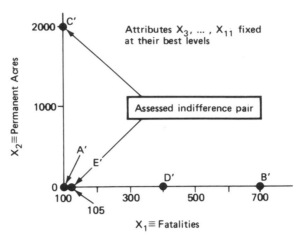

Fig. 7.9. **Value tradeoffs between fatalities and permanent land use with other attributes at their best levels.**

K: Fine, let's proceed. We want to move over to Fig. 7.9 now and ask essentially the same question with all of the other attributes, X_3–X_{11}, at their best levels. Which would you rather do, go from ($x_1 = 100$, $x_2 = 0$), point A', up to 700 on X_1 (point B') or up to 2000 on X_2 (point C')?

B: 2000.

K: And what if I made this 400 on X_1 (point D') or 2000 on X_2?

B: Still 2000.

K: [The responses seemed easy so I jumped ahead.] Is there any reason why the indifference point in this case should be different from that in Fig. 7.8, that is, the 105 for X_1?

B: It's essentially the same, I think.

K: This says one thing, that the consequences ($x_1 = 105$, $x_2 = 0$) and ($x_1 = 100$, $x_2 = 2000$) must be on the same indifference curve for the two levels of the other attributes specified. Of course, it says nothing about any of the rest of the preference structure. But let me save the general questions until after we try another specific case.

Next, we considered some preference tradeoffs between X_2 and X_3, temporary unusable land. The results are illustrated in Figs. 7.10 and 7.11. They indicated that indifference curves over X_2 and X_3 were the same regardless of where the other attributes are. This is the preferential independence condition.

K: Do you think, as a general rule, that the tradeoffs between X_2 and X_3 don't depend on the other attributes?

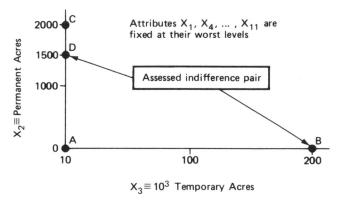

Fig. 7.10.   **Value tradeoffs between temporary and permanent land use with other attributes at their worst levels.**

B: Yes, that's true ... as long as the other attributes are fixed.

K: This indicates that the pair $\{X_2, X_3\}$ is preferentially independent of the other attributes. So let's go on. Which one of these first three attributes is easiest for you to think about? I shall then use that to examine additional value tradeoffs.

B: Oh, I see. Frankly, with X_1, I feel that the indifference levels would be very low; I think one of the other two would be better.

K: Then I'll take X_3, and we'll examine the value tradeoffs between X_3 and X_4. Let me begin with a naïve question. Is it better to have less water evaporated than more, always?

B: Indeed.

K: So the best consequence in Fig. 7.12 is point A. Here we have fixed

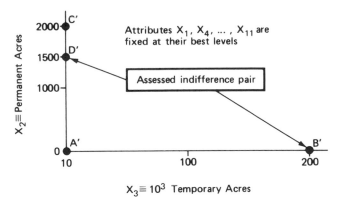

Fig. 7.11.   **Value tradeoffs between temporary and permanent land use with other attributes at their best levels.**

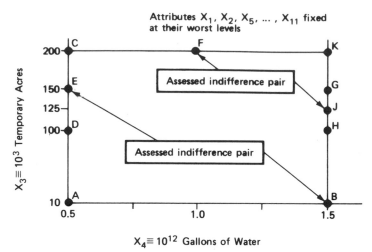

Fig. 7.12.   **Value tradeoffs between water evaporated and temporary land use with other attributes at their worst levels.**

all attributes but X_3 and X_4 at the worst levels. Which would you prefer, point B or point C?

B: I would say that the water loss is preferred; I'd rather lose the additional trillion gallons than the 200,000 acres.

K: So how about if you compare point B to point D where X_3 is at 100,000 acres? This is actually a change of 90,000 acres … from 10,000 at point A.

B: Well, 90,000 acres versus a trillion gallons … that's a lot of water. I guess I'd go with the land loss. I'd rather lose the land than a trillion gallons.

K: Since D is preferred to B which is preferred to C, there must be a point between C and D where you are indifferent. What I usually do then is take half of the difference. I say usually because, for instance, if you easily answered that B was preferred to C, but had a hard time deciding D was preferred to B, this would imply that the indifference level was near 100,000. Then, I might give you 170,000 to make the answer a little easier and help you to converge. [Discussion like this is meant to take the mystery out of the assessment process, to help develop rapport and to give one a break now and then.]

B: I see, that's a good policy.

K: So how about x_3 = 150,000 (i.e., point E) or the 1.5 trillion gallons?

B: I'd say that's about as close as I could define it. I'm indifferent.

K: Moving to Fig. 7.13, if we change all of the attributes other than X_3 and X_4 to their best levels, do you see any difference?

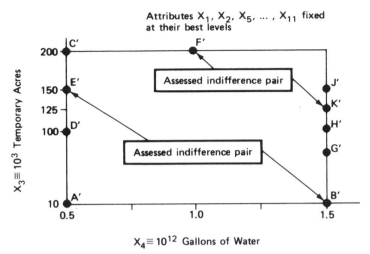

Fig. 7.13.   Value tradeoffs between water evaporated and temporary land use with other attributes at their best levels.

B: No, I don't.

K: Fine. Now let's back up to Fig. 7.12 and consider point F versus point G. [Strictly speaking, one should ask questions in different ranges of the $\{X_3, X_4\}$ consequence space to verify preferential independence conditions. Responses indicating that such assumptions are reasonable in part of the $\{X_3, X_4\}$ space cannot necessarily be extrapolated to all of the $\{X_3, X_4\}$ space. The analyst's judgment must be used to decide exactly how much can be implied by specific responses.]

B: Okay, I have a choice of 200,000 acres and one trillion gallons versus 150,000 acres and 1.5 trillion gallons. In this case … I'm not sure if I'm confused, but I think I'd take F. Maybe I should sit down with my pencil and think about this a little bit.

K: One way to look at this is as follows. You are at F. Are you willing to give up an additional 0.5 trillion gallons in addition to the one trillion already in order to reduce land temporarily unusable from 200,000 to 150,000 acres?

B: Would I go that way? Ummm, I don't think I would.

K: You'd stay at F.

B: I think I would.

K: How about if you could go from F to H?

B: Yes, I think I would do that.

K: And where might you be indifferent to F between G and H? How about 125,000 acres?

B: Yes, that's about as close as I can come.
K: Then let's jump back again to Fig. 7.13 and consider F' versus G' with X_3 at 75,000 acres. Which would you prefer?
B: I would prefer the 75,000 acres and the 1.5 trillion gallons.
K: And how about 140,000 acres?
B: That's very close again.
K: Do you see any reason why it should be different than before?
B: I don't see any differences.

To try to promote independent thinking each time, the order used to converge to indifference is varied. To see this, compare the sequence G, H, J in Fig. 7.12 to the sequence G', H', J', K', in Fig. 7.13. With the given responses, we can reasonably assume that $\{X_3, X_4\}$ is preferentially independent of the other attributes.

K: So now let's try the value tradeoffs between X_3 and X_5, the sulfur dioxide emission. My understanding is that the sulfur dioxide is here for effects other than health. Is that correct?
B: Yes, both X_5 and X_6 are meant to include aspects other than health effects. The health effects of chronic air pollution are considered to be part of attribute X_{10}. The fatalities from acute SO_2 exposure are in X_1.
K: So what effects do you wish to pick up here?
B: Visual effects, damage to buildings, odors, more frequent washing of clothes, damage to property, reducing land values, crop damage, etc., ... things like this.
K: Okay, consider the value tradeoffs between X_3 and X_5 illustrated in Fig. 7.14. Note that one advantage of using the same attribute in the tradeoffs is that you get used to thinking in terms of that attribute.

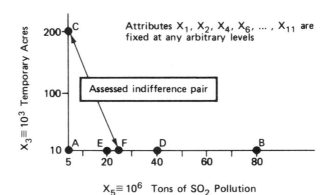

Fig. 7.14. **Value tradeoffs between SO_2 pollution and temporary land use with other attributes at arbitrary levels.**

Suppose you are at the best point A and must move to either B or C.

B: I would prefer to lose the 200,000 acres.

K: And what if we reduce X_5 to 40 million tons? Do you prefer point C or D?

B: I think I would still prefer to go to the 200,000 acres (point C).

K: And if X_5 is 20 (i.e., point E)?

B: I guess at 20, I would take the sulfur dioxide to the 200,000 acres.

K: Where will this break?

B: Oh, 25 million tons of SO_2.

If the client has a hard time answering this last question, I would usually offer a specific level of X_5. In the first case, the client must select an indifference level from an entire range, and in the second case, he or she must decide only on which side of the specific given level indifference lies. The second question is easier to answer.

K: Is it reasonable to assume that your answer above does not depend on where the levels of the other attributes are fixed, since I didn't specify them and you didn't ask me?

B: That's right. I really don't feel any difference on those. As long as they are held fixed, they are not involved in the tradeoff between X_3 and X_5.

K: Let me push a bit farther, because I would personally find it easier to argue against the assumption in this case than in many others. In particular, let's look at the impact due to knowing the particulate pollution level measured by X_6. Suppose I fix the particulate pollution at its worst level—10 million tons—and I again ask if you would be indifferent between C and F in Fig. 7.14.

B: But am I stuck with 10 million tons of particulate in both cases?

K: Yes, you're stuck with the 10 million tons.

B: Then I'd still be indifferent between C and F.

K: Now suppose you have just 0.2 million tons of particulate in all cases, would your answer change?

B: No, I'd still be indifferent.

K: Let me suggest a rationale that would imply that there should be a difference depending on whether particulate pollution was low or high. Of course, remember that there are no right or wrong values. My rationale is that people would view air pollution as a whole. If there were a lot of particulate, a little increase in SO_2 may have serious effects, whereas if particulate pollution were low, this same increase in SO_2 would be relatively unimportant. In such a case, one might give up more in terms of usable land to reduce SO_2 from 40 to 30 million tons if particulate pollution were high than if particulate pollution were low. Such preferences would violate the preferential independence condition.

B: As a matter of fact, there have been some studies which indicate that SO_2 and particulate together cause more health effects than equivalent amounts do separately. However, this is health effects, and these are excluded from attributes X_5 and X_6. In terms of damage costs, it's the acid more than anything from the SO_2, whereas it is the sooting from the particulate. There doesn't seem to be much synergism in this context, so I would remain with my previous responses.

K: Fine, then we can assume that $\{X_3, X_5\}$ is preferentially independent of the other attributes. I'm not sure now if I ought to belabor the point. Is there ... ?

B: I don't think so. If I considered each of these other attributes compared to, say, land use, I don't see why the value tradeoffs would depend on the levels of the additional attributes as long as they are held fixed.

K: Thus, we will assume that each of the pair $\{X_3, X_i\}$ for $i = 1, 2, 4,$ ..., 11 is preferentially independent of the other nine attributes. This satisfies the preferentially independence conditions necessary to invoke either an additive or multiplicative utility function.

The formal result is given by Result 5 in Section 7.2. Strictly speaking, we did not check to see if $\{X_1, X_3\}$ was preferentially independent (PI). However, since $\{X_1, X_2\}$ and $\{X_2, X_3\}$ were each PI, it follows from a result of Gorman [1968a] that $\{X_1, X_3\}$ is PI.

Verification of Utility Independence Conditions

K: Now, we shall begin to look at your preferences for different numbers of fatalities, indicated by attribute X_1, with all of the other attributes fixed. Consider the lottery illustrated in Fig. 7.15. This lottery gives you a one-half chance of 100 fatalities with all other attributes at their best levels and a one-half chance of 700 fatalities with all other attributes fixed at their best levels. The question is, would

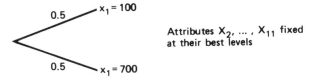

Fig. 7.15.   A lottery involving fatalities with other attributes at their best levels.
(Before assessing the certainty equivalent.)

you prefer the lottery or an option which gives you 600 fatalities for sure with all other attributes at their best levels?

B: I'd take the lottery.

K: Okay, how about if fatalities are 150, with other attributes at the best level?

B: I'd take the 150.

K: How about 200?

B: I'd take the 200.

K: 450?

B: That's pretty close. At 450, I'd take the lottery I guess.

K: 375? The average of the lottery as you know is 400.

B: It would take something slightly under 400 for me to choose it. I'd take 375.

K: And what if it were 400 versus the lottery?

B: At 400 I'd take the lottery, but if it were slightly under 400, I'd be very tempted to take the sure consequence.

K: At 390?

B: Yes, I'd choose around 390 as an indifference point.

K: Now, why is this slightly under 400 as opposed to exactly 400?

B: Well, I feel that as long as the expected value is the same, I'd prefer to accept the risk for the chance that it might come out right. But if there is a little bonus in there, I think it not worth the risk that 700 people may die.

K: With that reasoning, should you perhaps prefer 399 to the lottery? That's a bonus of one expected life.

B: That's right. Maybe that's it. I'd be indifferent at 399. For all practical purposes, I guess it could be 400.

After that process, the lottery in Fig. 7.15 ends up looking like that in Fig. 7.16.

K: Going on, let's ask a similar set of questions concerning levels of fatalities with all other attributes fixed at the worst levels now. Refer to Fig. 7.17, which is a lottery with a one-half chance of 100 fatalities

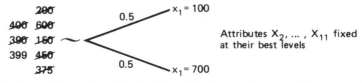

Fig. 7.16. **A lottery involving fatalities with other attributes at their best levels.** (After assessing the certainty equivalent.)

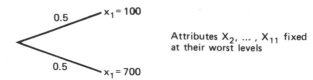

Fig. 7.17.   A lottery involving fatalities with other attributes at their worst levels.

and a one-half chance of 700 fatalities. Would you prefer the lottery or 600 fatalities?

B: I'd take the lottery.

K: How about the lottery versus 250?

B: I'd take the 250.

K: How about 300?

B: I'd take 300.

K: 500?

B: I'd take the lottery.

K: [All of the previous four questions seemed easy to answer so I asked a general question.] And where would you be indifferent?

B: Essentially at the same point, just a shade under 400.

K: The thing to note here is that it appears that your indifference point does not depend on the levels of the other attributes. The relative preferences which you attach to different levels of fatalities seem to be independent of the other attributes as long as their levels are fixed. Is this true in the general case?

B: That's right.

K: This implies that X_1 is utility independent of the other attributes. This assumption, together with the preferential independence assumptions which we already verified, implies that your utility function must be either additive or multiplicative. [The conditions for Result 5 in Section 7.2 have been verified. These conditions also imply that each attribute must be utility independent of all of the others.] But let's try just one more attribute as a check. How about taking X_8, radioactive waste storage, since we haven't said much about that? Let's fix all attributes at their worst levels and examine the lottery in Fig. 7.18. Here you get either 200 or 0 metric tons, each with a proba-

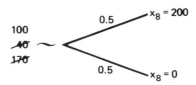

Fig. 7.18.   A lottery over radioactive waste with certainty equivalent shown.

bility of one-half. Would you prefer the lottery or 40 metric tons with all other attributes at their worst levels?

B: I'd take the 40.

K: How about 170 metric tons?

B: I'd take the lottery. In this case, I'd go right to the expected value of 100. Yes, 100 metric tons would be my indifference point.

K: Does your answer to this depend on the other attributes?

B: No.

K: [I felt that the plutonium produced might have some effect on the previous response. Although Buehring's general response implied that it did not, a specific check to see if some aspect was overlooked is sometimes prudent.] For instance, suppose I told you that X_9 was high—that many tons of plutonium were produced. If I told you theft was high, would it change your 100 indifference level in Fig. 7.18?

B: No.

K: Fine. Now let's check X_8 between 0 and 100 metric tons produced. If you had a 50–50 lottery in Fig. 7.19 yielding 0 or 100 metric tons, again with other attributes at their worst levels, where are you indifferent?

B: Right at 50.

K: Does this answer depend on the other attributes?

B: No.

K: To cover the range of X_8, is 150 always indifferent to a lottery yielding 100 or 200 with equal probabilities, as long as other attributes do not vary?

B: Yes.

K: And now the general question. For any such lottery questions involving X_8, regardless of where the other attributes are fixed, are you linear? Would you always be indifferent at the expected value?

B: Yes. Of course, there is one little complication. Since both X_8 and X_9 are nuclear effects, if terrorist activity related to plutonium theft were very high, I suppose there would be some extra resentment of radioactive waste. Is that okay?

K: I'll let you answer that. Suppose there is that high resentment, what is your indifference point to the lottery in Fig. 7.18 yielding 0 or 200 metric tons of waste?

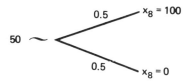

Fig. 7.19. A lottery over radioactive waste with certainty equivalent shown.

B: I'm not sure I feel this myself, but I could see how someone might
      say, "If there were a lot of plutonium blackmailing going on, I am going
      to feel worse about radioactive waste storage. Therefore I'd demand
      something lower than 100 before I'd be indifferent."

K: I think you may be mixing up two things. Suppose you said that 80
      metric tons was indifferent to the lottery.

B: Okay.

K: Well, then I offer you 90 versus the lottery, and you say no, since
      there is so much resentment. Well, there would be a lot more resent-
      ment to that 200 metric tons which you're apparently willing to risk.

B: That's true.

K: From my viewpoint, let me try to state what I think is your concern.
      Let's return to where you had said 100 is indifferent to the lottery in
      Fig. 7.18 regardless of where other attributes were fixed. Take the
      case where there is no theft and no resentment. Then, sloppily
      speaking, you might say that the jump from 0 to 100 isn't too impor-
      tant, but then, the jump from 100 to 200 isn't too important either.
      They are equally important, but neither one is critical. However, if
      the theft is high and there is much resentment, then the jump from 0
      to 100 is very important because of all of the concern about waste,
      but the jump from 100 to 200 is also very important. Again, they are
      equally important. What we are concerned with in finding your indif-
      ference level to the lottery of Fig. 7.18 is whether the jumps from 0
      to 100 and 100 to 200 are equally important, given the other attri-
      butes are fixed, and not, for instance, whether it is more important to
      go from 100 to 200 if theft and resentment are high or low. The quali-
      tative feeling that you're giving to me is that you would be a lot more
      concerned about high levels of radioactive waste storage if there were
      theft than if there were no theft.

B: Yes.

K: That does not imply that your relative preferences for various
      storage levels change depending on the level of theft.

B: Yes, I agree.

K: Now, such an attitude may affect your value tradeoffs between, say,
      X_3 and X_8, given levels of X_9. [That is, it may affect the preferential
      independence condition.] For instance, if theft were high, you might
      be willing to give up more land temporarily used to reduce radioac-
      tive waste from 100 to 50 metric tons than you would be willing to
      give up if theft were low. This is the type of preference indicated by
      your comments. So it has to do with the evaluation of radioactive
      waste versus other attributes as a function of tons of plutonium pro-
      duced, rather than with the relative preferences of various levels of
      waste as a function of plutonium produced.

B: I believe I was thinking of simultaneous changes in the level of theft at the same time as I was changing radioactive storage levels. I can see how the argument says that 100 should be my indifference level for the 50–50 lottery of 0 or 200 metric tons of waste.

K: Okay, then we'll assume that X_8 is utility independent of the other attributes.

Next we went back and explicitly checked whether Buehring did feel that value tradeoffs among X_3, land temporarily unusable, and X_8, radioactive waste, depended on the tons of plutonium produced. For him, they turned out not to. Hence, we continued to assume that $\{X_3, X_8\}$ was preferentially independent of the other attributes.

Ordering the Scaling Constants

K: Now we come to the assessment part. The conditions we have just verified imply that either the additive form or the multiplicative form of the utility function discussed in Section 7.2 must hold. To assess either of these, we need to get the k_is and u_is. [From Result 5 in Section 7.2, k is calculated from the k_is if the multiplicative form holds.] The tough part is probably assessing the k_is.

As a first step, let's try to order the k_is. To do this, refer to Table 7.4 and assume that all attributes are at their worst levels. To get the rankings, we need to know the order in which you would push these attributes up from their worst to their best level if you had the choice. First, if you could push just one of these attributes from the worst to the best level, which attribute would you choose? To help you think about this, let me go through some of them pair-wise.

Take attributes X_1 and X_2. Consider an option leading to 700 fatalities and 2000 acres of land permanently unusable; both attributes are at their worst levels. Would you rather move up to 100 fatalities or 0 acres of land?

B: 100 fatalities.

K: This answer, which implies that k_1 is greater than k_2, seemed clear from the beginning of our discussion. [Had Buehring responded 0 acres, I would have asked him to explain his reasoning.]

B: Right.

K: So now I'll take the better of these two and compare it with temporary land unusable. Would you rather go from 700 to 100 fatalities or from 200,000 to 10,000 acres of land?

B: Change the fatalities. [This implies that k_1 is greater than k_3.]

K: How about water evaporated, 1.5 to 0.5 trillion gallons, or …

B: Fatalities, the 600 fatalities is going to be the most important, I think.

K: Well, let's try radioactive waste: 200 to 0 metric tons stored or 700 to

100 fatalities. Now presumably, some of the thoughts here concern possible genetic impacts of the radioactive wastes.

B: Yes, that's true.

K: So, is that worth the 600 people between now and the year 2000?

B: No, it isn't, the 600 is still worth more. [Thus, k_1 is greater than k_8.]

K: How about the nuclear safeguards? Is it better to go from 50 to 0 tons produced or from 700 to 100 fatalities?

B: Still, I'd prefer to save the people.

K: And the lead produced, measured by X_{10}?

B: Chronic health effects—that's a mysterious one. That could be worth more than 600 actually, but I don't think it is.

An analysis of preferences often indicates questions like this which are important in determining policy, but for which the client needs more information. Often, such information is available. Once the question is clearly articulated, one can begin to look for the answer.

K: So you'd take the 600?

B: Yes.

K: And how about electricity generated, 0.5 to 3?

B: That's an interesting one, preferences go the other way.

K: I think the way to think about this involves what happens to Wisconsin if there is really only 0.5 trillion kWhr produced, etc.

B: It's hard for me to think about X_{11}, electricity generated. The level of 0.5 trillion kWhr might not cause that much suffering. [Then Buehring checked some electricity consumption tables for Wisconsin.] At our current consumption rate, we shall use 0.9 trillion kWhr between now and 2000. That is a cut of almost one-half. But I think I'd still take the choice to save the 600 people.

K: Of course, some of this electricity may run kidney dialysis machines, for example.

B: Yes, that's true, but I'm assuming that the cuts would be selective and that such things as hospitals and schools would stay in operation.

K: However, with a 50% cut in electricity, you would certainly affect life style. But, anyway, you choose the 600 fatalities to be the most important?

B: That's right.

K: This means the largest k_i is k_1.

A common error made in many studies is to ask which of several attributes is most important, independent of their ranges. If the range of fatalities were changed to 700 to 690, rather than 700 to 100, changes from best to worst on several other attributes would have been more important than from 700 to 690 fatalities. See Chapter 5 of Keeney and Raiffa [1976] for details.

K: Now we need to look for the next most important change after fatalities. To be quick—based on your previous answers—let's start with radioactive waste, chronic health effects, and electricity generated, all at their worst levels. Which of these would you rather move up to its best?

B: Chronic health effects, I'd have to say. [This implies that k_{10} is greater than k_8 or k_{11}.]

K: How about chronic effects relative to nuclear safeguards?

B: That's very close, but I think chronic effects?

K: Just glance over the attributes now, energy needed, X_7, for example?

B: That one doesn't bother me so much.

K: So chronic health effects would be number 2. This implies that k_{10} is second largest, next to k_1.

We continued in this manner and found the following order:

$$k_1 > k_{10} > k_{11} > k_9 > k_5 > k_8 > k_2 > k_3 > k_4 > k_6 > k_7.$$

K: Many of your responses above could have been inferred from earlier choices when we were checking for preferential independence. For example, look at the value tradeoffs between temporary unusable land and water evaporated in Fig. 7.12.

B: Okay.

K: There you said you preferred consequence B to consequence C. Thus, if you began at point K in that figure and had to go to either B or C, you would prefer to go to B. This says you would rather move from 200,000 to 10,000 acres used than move from 1.5 to 0.5 trillion gallons of water evaporated, which is exactly what you said in evaluating k_3 versus k_4.

Other information given in checking for preferential independence conditions was also consistent with the ordering of the k_is.

Assessing the Scaling Constants: Value Tradeoffs
between Attributes

K: Now that we have the order of the k_is, let's assess the tradeoffs to get their relative values. Let's start with X_1 and X_{10}. Refer to Fig. 7.20. You said previously that you prefer consequence B to consequence A. Thus, if you were at A, 700 fatalities and 0 tons of lead, you would

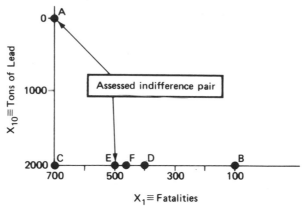

Fig. 7.20.   Value tradeoffs between fatalities and chronic health effects.

be willing to increase lead to 2000 tons in order to decrease fatalities to 100. Is that right?

B: Yes.

K: What if you only got to move to 400 fatalities? That is, would you be willing to move from A to D?

B: I would still rather save the 300 people [i.e., 700–400].

K: What if you can only go to E—500 fatalities—and you are saving 200?

B: I'd say that's pretty close to what I feel is equivalent to the chronic effects, so at that point I might switch.

K: You would switch or be indifferent?

B: Be indifferent.

K: Let's look at what this says. Because of the preferential independence conditions, we can assume that all other attributes are at their worst level—so $u_i = 0$ for $i \neq 1$, 10—and equate the utilities of points A and E since you are indifferent between them. Using either the additive or the multiplicative utility function, we find that the utility of A is k_{10} and that the utility of E is $k_1 u_1(500)$.

B: Okay.

K: Hence, the relationship between k_1 and k_{10} is $k_{10} = k_1 u_1(500)$, where the utility function u_1 is measured on a zero to one scale. Based on what you told me in checking for utility independence, your utility function for fatalities is essentially linear. Since $u_1(100) = 1$ and $u_1(700) = 0$, then $u_1(500)$ must be about 0.333. Thus, we have $k_{10} = 0.333 k_1$. We'll refine this later, but for now let's go on. Look at the tradeoffs between X_{11} and X_1 in Fig. 7.21. We want to find a point on the X_1 axis, when $x_{11} = 0.5$, that is indifferent to point A.

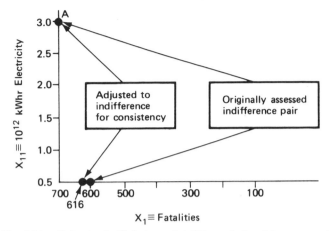

Fig. 7.21. Value tradeoffs between fatalities and electricity generated.

The question is, how many fatalities must you save in order to accept the decrease in electricity from 3.0 to 0.5 trillion kWhr? That's tough, I know, but I'll ask it anyway.

B: That is tough. It is certainly less than the last question, less than the 200 [i.e., the point $x_1 = 500$].

K: That follows in order to have $k_{10} > k_{11}$.

B: All right, well, I'm still confident of that. About 100 at the most.

K: Let's try 50. Suppose you had 650 deaths and 0.5 trillion kWhr or 700 deaths and 3.0 trillion kWhr. Which would you prefer?

B: I might take the 700 and 3.

K: And what if it's 550 and 0.5 or 700 and 3?

B: Okay, I'd take the 550.

K: And where would you be indifferent? How about at 600?

B: That's about it. I'd say that's pretty close.

K: What this implies is that $k_{11} = k_1 u_1(600)$ or $k_{11} = k_1(0.167)$, because that's . . .

B: One-sixth.

K: Now we can run checks on this. Refer to Fig. 7.22 and presume you are at point A, 2000 tons of lead and 0.5 trillion kWhr. Would you rather eliminate the lead (point B) or increase electricity production to 3.0 (point C)?

B: Lose the lead.

K: That's consistent with your previous responses, since, with other attributes at their worst levels, both the additive and the multiplicative utility functions imply that the utility of B is k_{10} and that the utility of C is k_{11}. And you have said that k_{10} is greater than k_{11}. Now increase

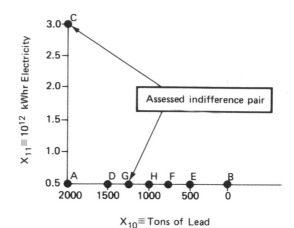

$$X_{10} \equiv \text{Tons of Lead}$$

Fig. 7.22. Value tradeoffs between chronic health effects and electricity generated.

X_{10}. Suppose you could only go to 1500 lead (point D) or 3.0 million kWhr (point C). Which do you prefer? Another way to think of this is, you're at point D with 0.5 electricity and 1500 lead and you are told you can increase electricity to 3.0 if you are willing to accept 500 more tons of lead. Would you do it?

B: Yes, I guess I would.

K: How about if you started at 500 tons of lead (point E), would you accept the additional 1500 to jump up to 3?

B: No.

K: If you started from 750 tons (point F)?

B: No.

K: How about at 1250 (point G)?

B: Okay. That's pretty close. That sounds about where I'd be indifferent.

K: How about if you are at 1000 (point H) and someone says, "For an additional 1000 tons of lead, I can move you to 3." Would you accept that or not?

B: No, I don't think so.

K: You had to tell me that, because if you are indifferent to accepting 750 more tons (i.e., $x_{10} = 1250$), you had better not accept 1000 more. What this says is ...

B: ... is I'm probably confused. This probably isn't consistent.

K: Since C and G are indifferent, we set their utilities equal and find that $k_{11} = k_{10}u_{10}(1250)$. I don't know what $u_{10}(1250)$ is, but we can do a quick assessment of u_{10}. We've got a range of X_{10} from the worst point 2000 tons to the best point 0 tons. Because of our scaling convention, we set $u_{10}(2000) = 0$ and $u_{10}(0) = 1$. Now consider the 50–50 lottery of 0 to 2000 tons shown in Fig. 7.23 and suppose you have this option or $x_{10} = 500$ for sure. Which would you take?

B: 500.

K: The lottery or 1500?

B: I'd take the lottery.

K: How about at 1200?

B: I'd still probably take the lottery.

K: Then we go to 800?

B: I'd take the 800 for sure.

K: 1000?

B: I'd take the 1000, I think.

K: The average in the lottery is 1000, as you know. So, 1100?

B: That's pretty close—you can probably say 1100 is indifferent. I'd take 1050?

K: You'd take 1050 over the lottery?

B: Yes. I'd take 1050 over the lottery.

K: And not 1100?

B: 1100, I don't know, that's pretty close.

K: How about 1200?

B: At 1200, I'd take the lottery.

K: Okay. I'll take 1100 as indifferent.

B: All right.

K: This says that the utility assigned to 1100 must equal the utility of the lottery. It's assigned that way so that we can use expected utilities in evaluating alternatives. Hence, we assign $u_{10}(1100) = 0.5$ and plot it on Fig. 7.24. Would you prefer the 50–50 lottery yielding 0 or 1100 tons of lead or an option of the average 550 tons for sure?

B: I'd take the 550.

K: And which would you choose between the 50–50 lottery yielding 1100 or 2000 and 1550 for sure?

B: Again, I'd take the sure consequence, the 1550.

K: These last answers imply that you are risk averse in the attribute chronic health effects so, as a first approximation, we can sketch in the concave utility function u_{10} in Fig. 7.24.

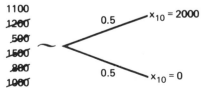

Fig. 7.23.   **Finding the point of indifference to a lottery.**

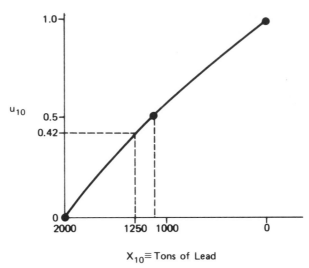

Fig. 7.24.   Utility function for chronic health effects.

Later, a constantly risk averse function will be fit. This degree of preci-
sion on the single-attribute utility functions in a multiattribute problem is
probably sufficient in most cases. Subtle differences in risk attitudes on
the individual attributes are likely to have little effect relative to variation
in the k_i values and the general shape of the u_1 functions.

K: Now we can return to the equation $k_{11} = k_{10}u_{10}(1250)$. Eyeballing it
     from Fig. 7.24, I'd say that $u_{10}(1250) = 0.42$, implying that
     $k_{11} = 0.42k_{10}$. With this, we have assessed three equations relating
     k_1, k_{10}, and k_{11}: namely, $k_{10} = 0.333k_1$, $k_{11} = 0.167k_1$, and
     $k_{11} = 0.42k_{10}$. [They were quickly checked for consistency.] They are
     reasonably consistent, but a slight alteration is required. If just one
     of them is changed, we find that the parameter in the first one must
     be 0.4, so $k_{10} = 0.4k_1$; that the second becomes $k_{11} = 0.14k_1$; or that
     the third becomes $k_{11} = 0.5k_{10}$. Let's see how much your answers
     leading to the original three equations would have to change in order
     to get the new consistent equations. Assuming linear preferences for
     fatalities, $u_1(460) = 0.4$, so points A and F in Fig. 7.20 would have to
     be indifferent to adjust the first equation. Alternatively, in Fig. 7.21,
     since $u_1(616) = 0.14$, you would have to move fatalities from 600 to
     616 to be indifferent to point A to change the second equation. To ad-
     just the third equation, you could either be indifferent between
     $(x_{10} = 1100, x_{11} = 0.5)$ and $(x_{10} = 2000, x_{11} = 3)$ in Fig. 7.22 or adjust
     from 1100 to 1250 the value of X_{10} for which you are indifferent to the

lottery in Fig. 7.23. Of course, there are options of adjusting each of these by a small amount. However, it is easier just to move one to be consistent.

B: Yes, the one I feel the least strongly about is the one in Fig. 7.21. You said you could move the 600 to 616 fatalities, is that right?

K: Right.

B: I think I would notice the 40 additional fatalities in Fig. 7.20; 1100 seems low on the tradeoffs in Fig. 7.22, and 1250 seems high as an indifference amount for the lottery. Yes, I think I'd be happy to change the 600 to 616 in Fig. 7.21.

K: Fine, then for now we are consistent in our value tradeoffs between attributes X_1, X_{10}, and X_{11}.

Next we went through the same procedure for each of the other eight attributes, evaluating value tradeoffs between two attributes at a time. We could have investigated these tradeoffs relative to deaths, attribute X_1, or relative to attribute X_{10} or X_{11}. Since it is often hard to balance deaths, we chose attribute X_{10}, tons of lead, since we already had a rough utility function u_{10}. The first tradeoff was X_{10} versus X_9, nuclear safeguards, because X_9 was the attribute whose scaling factor was the fourth largest. The next pairs were $\{X_5, X_{10}\}$ and $\{X_8, X_{10}\}$. Then, because k_{10} was clearly much larger than the scaling constants k_2, k_3, k_4, k_6, and k_7, of the remaining five attributes, we chose X_8 for the basis of comparison with them. That is, we considered tradeoffs between $\{X_8, X_2\}$, $\{X_8, X_3\}$, etc. As a final result, we had 10 equations with 11 unknowns: k_1, k_2, ..., k_{11}. These are displayed in Table 7.5.

TABLE 7.5

INITIAL TRADEOFFS USED TO
DETERMINE THE SCALING CONSTANTS

Implied equations	Measure (units)
$k_{10} = k_1 u_1 (500)$	x_1 in deaths
$k_{11} = k_1 u_1 (616)$	x_1 in deaths
$k_9 = k_{10} u_{10} (1500)$	x_{10} in tons
$k_5 = k_{10} u_{10} (1600)$	x_{10} in tons
$k_8 = k_{10} u_{10} (1700)$	x_{10} in tons
$k_2 = k_8 u_8 (50)$	x_8 in metric tons
$k_3 = k_8 u_8 (75)$	x_8 in metric tons
$k_4 = k_8 u_8 (100)$	x_8 in metric tons
$k_6 = k_8 u_8 (150)$	x_8 in metric tons
$k_7 = k_8 u_8 (180)$	x_8 in metric tons

Selecting an Additive or Multiplicative
Utility Function

Two separate methods were used to check whether u was additive or multiplicative.

K: Now our 10 equations specify the relative values of k_is. To get their absolute values, I want to ask you one very tough question. It is not necessary to ask such a difficult question; however, it does simplify the calculations which are needed to determine the k_is. It is also illustrative of another method for determining scaling constants, so let's try.

Consider the two options in Fig. 7.25. Option A is a consequence with fatalities at its best level, that is, 100 fatalities, and all other attributes at their worst levels as shown in Table 7.4. Option B is a lottery which gives you all 11 attributes at their best levels with probability p or, otherwise, all attributes at their worst levels with probability $1 - p$. The question is, what is p such that you are indifferent between options A and B?

Let's try out some numbers. Suppose p is 0.8 and $1 - p$ is 0.2, which would you prefer?

B: With $p = 0.8$, I'd have to go with the lottery, I think.

K: One way to look at this is as follows. Suppose you have the consequence in option A and decide to switch it for the lottery with $p = 0.8$. Then, if you are unlucky and move to the worst case, the difference from option A is 600 additional deaths and this occurs with probability 0.2. If you are lucky, which has a 0.8 chance, you maintain the lowest level of 100 fatalities and improve on the other 10 attributes. Does that seem reasonable?

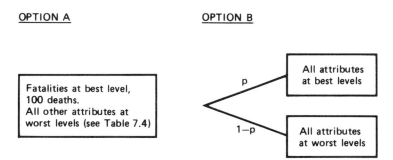

Fig. 7.25.   Assessing the indifference probability.
For the assessed probability $p = 0.6$, options A and B are indifferent.

B: Yes.

K: How about your preference between options A and B when $p = 0.7$?

B: I think I'd still take the lottery at 0.7.

K: How about $p = 0.2$?

B: I'd take option A in that case.

K: And if $p = 0.4$?

B: I'd take option A.

K: How about 0.6?

B: At 0.6, that's pretty close, I think. At $p = 0.5$, I'd still take option A. Yes, I think 0.6 is about the indifference point.

K: What this implies is that the utility of option A must equal the utility of option B when $p = 0.6$. The utility of A is simply k_1 using either the additive or multiplicative utility function, and the utility of B is $p \times 1$, the utility of all attributes at their best levels, plus $(1 - p) \times 0$, the utility of all attributes at their worst levels. Hence, $k_1 = p = 0.6$. Now we can combine this equation with the previous 10 to calculate values for all 11 k_is.

This was done roughly and quickly by hand in a couple of minutes. The sum of the k_is, that is, $k_1 + k_2 + \cdots + k_{11}$, was 1.14. As indicated by Result 5 in Section 7.2, if $\Sigma k_i = 1$, the utility function is additive, and if $\Sigma k_i \neq 1$, it is multiplicative. Because the sum of the k_is is quite near 1.0, an additive utility function may be appropriate. We will now try to find out if this is so.

K: Now let's try to get a qualitative feeling for your preferences in situations involving more than one attribute being varied. Consider the two options I and II in Fig. 7.26. Option I gives you a one-half chance at 100 deaths with 0 lead and a one-half chance at 700 deaths with 2000 tons of lead. Option II is similar, but it gives you one-half chances at either 100 deaths with 2000 tons of lead or 700 deaths with 0 lead. You can consider all attributes other than fatalities X_1 and lead X_{10} to be fixed at any levels, but the same fixed levels for each

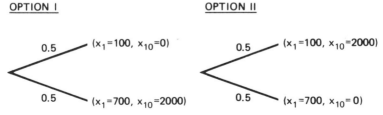

Fig. 7.26.   Preferences for combinations of fatalities and tons of lead.
Options I and II were found to be indifferent (X_1 measured in deaths and X_{10} in tons of lead).

option. Which option, I or II, do you prefer? Before answering, let
me point out that with both options you have an equal chance at
either 100 or 700 fatalities. Also, you have an identical chance at 0 or
2000 tons of lead. So, considering one attribute at a time, the consequences are the same. However, with option I, you get either the
best or the worst of both attributes, whereas with option II, you will
get the best of one but the worst of the other attribute. Do you have a
preference or are you indifferent?

B: I think I am indifferent. Yes, I am indifferent.

K: Let me suggest very rough arguments for preferring one or the other.
You may say that with either consequence in II, the situation will be
"very bad," whereas, at least with option I, there is a one-half
chance to come out okay. This implies I is preferred to II. Alternatively, you may say I can handle either case resulting from II, but the
second possibility in option I is simply untenable, therefore I'd
prefer II. Or these two effects may balance each other and you would
be indifferent.

B: I understand the two positions and I like the idea of having a shot at
both at their best, but it is very close to indifferent.

K: What this implies is that the k_is should sum to 1.0. If you had preferred II to I, the Σk_i should be greater than 1.0, and, if you had preferred I to II, the Σk_i should be less than 1.0.

B: That's interesting, because, if I had selected one, I would have taken
II.

K: [I now repeat the same test for additivity with a pair of attributes
which I feel might indicate nonadditivity. In assessing utility functions, the assessor should play the devil's advocate.] Consider one
more similar question involving the attributes X_5, SO$_2$ pollution, and
X_6, particulate pollution. In Fig. 7.27, with option III, you get either
5 million tons of SO$_2$ pollution with 0.2 million tons of particulate or
80 million tons of SO$_2$ with 10 million particulate. And I think option

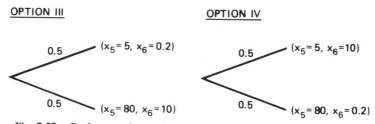

Fig. 7.27. Preferences for combinations of SO₂ and particulate pollution.
Options III and IV were found to be indifferent (X_5 measured in 10^6 tons of SO$_2$ and X_6 in 10^6
tons of particulate).

IV is clear. Which do you prefer or are you indifferent? Here again, the implications are identical, taken one attribute at a time. The difference is in how the attribute levels are combined.

B: Again, I'm reasonably close to indifferent. Although there is perhaps a little synergistic effect with these two attributes, I would still be very close to indifferent.

K: Then we shall assume your utility function is additive, implying again that the k_is sum to 1.0. This relationship together with the 10 equations relating the relative values of the k_is imply [after a little calculation] that $k_1 = 0.526$. Let's return to Fig. 7.25 and examine this implication. It means you should be indifferent between option A and option B when $p = 0.526$. Does this seem reasonable?

B: Yes, it does. I don't think that distorts my feelings.

Assessing the Single-Attribute Utility Functions

K: Good, then the only assessments that remain are the individual utility functions, the u_i. Actually, we have already assessed u_1 and u_{10}, so let's try u_2. We have scaled X_2, the permanent land use, from 2000 acres, the worst point, to 0 acres, the best. Thus, we assign $u_2(2000) = 0$ and $u_2(0) = 1$. Now consider a choice between a 50–50 lottery yielding either $x_2 = 2000$ or $x_2 = 0$ and an option giving you 800 acres used for sure. Which would you prefer?

B: 800.

K: How about 1400 versus the lottery?

B: The lottery.

K: And 900 versus the lottery?

B: At 900, I would take it.

K: 1000?

B: That's going to be the point of indifference.

K: So then the utility function is probably very close to linear as shown in Fig. 7.28.

B: In this case, I think so.

K: Good. Then let's go on to u_3. Temporary land use goes 200,000 to 10,000 acres. What if you had a 50–50 lottery yielding either of these or 80,000 for sure, which would you take?

B: The 80,000.

K: How about 130,000 acres versus the lottery?

B: I'd take the lottery.

K: How about 110,000 acres?

B: I'd stick with the lottery. Again, I think in this case I would be indifferent at the mean for such lotteries.

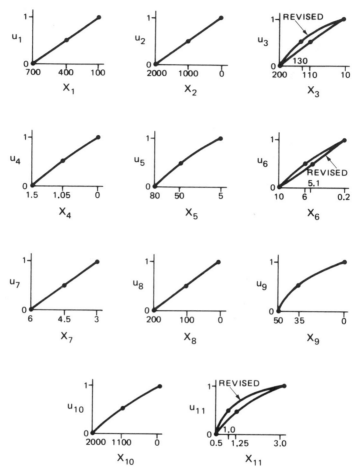

Fig. 7.28.   The single-attribute utility functions.

K: Then u_3 is also linear.

We assessed utility functions for the other attributes in a similar fashion.
The results are shown in Fig. 7.28. For public problems, it seems to be especially true that several utility functions are linear in their respective attributes. This is largely a function of the range of possible consequences.
Let me illustrate this with an excerpt from the assessment of the utility
function for radioactive waste.

K: Let us now assess u_9. You can probably figure out what the question
      will be.

B: Yes. This one I've thought about; it's going to be linear. The max-
     imum is only 200 metric tons. Now if that were 2000 metric tons, my
     answer would be much different. My indifference point to a 50–50
     lottery of 0 or 2000 metric tons would be quite a bit over the mean.

We shall also illustrate part of the assessment of the utility for electricity
generated because it is a proxy attribute.

K: Electricity generated goes from 0.5 to 3.0 trillion kWhr. Because this
     is a proxy attribute, you've got to think about what you would do
     with the various energy amounts if you had them. Consider a 50–50
     lottery of 0.5 or 3.0 versus 2.0 for sure.
B: If I could have 2.0 for sure, I'd take it.
K: How about 1.75?
B: Let's see, the mean of this lottery is 1.75. I'd take 1.75 for sure rather
     than lottery.
K: How about 1.0 versus the lottery?
B: Now 1.0 is about the current level of electricity. I'd come close to
     taking 1.0, but I guess I would take the lottery.
K: How about 1.5?
B: I'd take the 1.5.
K: And 1.25?
B: That's about it, I think. That's where I'm indifferent.
K: That seems reasonable to me, too. This completes a first-cut assess-
     ment of the u_is, and we now have all of the information needed to
     specify your utility function. We have found out that, for you,

$$u(x_1, x_2, \ldots, x_{11}) = \sum_{i=1}^{11} k_i u_i(x_i),$$

where the k_is are found by solving the equations in Table 7.5 together
with $\Sigma k_i = 1$, and the u_is are shown in Fig. 7.28. Let us examine
some implications of your utility function. Refer to Fig. 7.29, where
we have pictured $\{X_1, X_8\}$ consequence space. Now, since $k_1 > k_8$,
there must be some consequence, call it C, between points A and B
which is indifferent to point D. If, for C, the level of X_1 is designated
by x_1', then, equating the utilities of C and D, we find that $k_1 u_1(x_1') = k_8$.
Given the values of k_1 and k_8 which we have calculated, this implies that $u_1(x_1') = 0.0667$. From Fig. 7.28, it follows that $x_1' = 660$.
Put together, this implies you should be indifferent between C:
$(x_1 = 660, x_8 = 200)$ and D: $(x_1 = 700, x_8 = 0)$. Does this seem reasonable, out of the question, a little high, or ... ?
B: It seems quite reasonable.

Several other implications were examined.

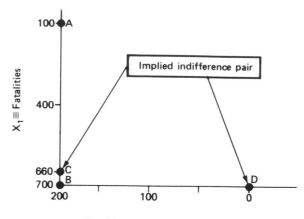

$X_8 \equiv$ Metric Tons of Radioactive Waste

Fig. 7.29.   Implied value tradeoffs between fatalities and chronic health effects.

7.6.3   THE SECOND ASSESSMENT OF BUEHRING'S UTILITY FUNCTION

I did not do all of the curve fitting and calculations necessary to specify the overall utility function, given the assessed information. The reason was that we planned to reassess aspects of Dr. Buehring's utility function in a few days, after enough time for reflection. In the meantime, Buehring assessed Wes Foell's utility function over the same attributes. Dr. Foell was the head of a project on Integrated Energy Systems at IIASA and, as mentioned, was also the leader of the Wisconsin research team which developed the Wisconsin model. This interaction allowed Buehring to "get some feedback on his preferences." For instance, if Foell's preferences were radically different from Buehring's, then Buehring could ask for the reasons and incorporate the reasoning (i.e., the new information) in modifying his own preference structure. What follows is our second-cut assessment of Dr. Buehring's utility function. Because of the work already done, it is obviously much more streamlined.

K: Could you give me the ordering of the k_is, that is, the order in which you would like to move attributes from their worst to best levels in Table 7.4. If we run into inconsistencies later, we can simply revise the list.

B: All right: 10, 1, 9, 5, 8, 11, 3, 2, 4, 7, 6. I think that's it.

K: This means you would prefer going from 2000 to 0 tons of lead to going from 700 to 100 fatalities.

B: Yes.

We continued down the list in this way as a simple check and found that no changes were necessary. This implied that

$$k_{10} > k_1 > k_9 > k_5 > k_8 > k_{11} > k_3 > k_2 > k_4 > k_7 > k_6.$$

K: In Table 7.6 we have the attribute list with the old and new rankings of the k_i scaling constants. If we compare these lists, nothing moved more than one position except k_{11}, which moved three positions. Why do you feel this happened?

B: I think that I overvalued the increase of energy from 1.5 to 3.0 trillion kWhr in my previous answers. The shape of the utility function over the last part of the curve for energy generated will be very close to flat. It is only slightly better to have 3.0 rather than 1.5 trillion kWhr.

K: There are three places where there are single position interchanges among the k_is. The first is between k_{10} and k_1. How did this come about?

B: I've always felt that the trace elements are very important. After interviewing Wes, I decided that the health impact of 2000 tons of lead could be much larger than 600 quantified fatalities. Furthermore, there are the aesthetic impacts due to the lead pollution. The more I thought about it, the worse it became.

We also examined the reversals between k_2 and k_3 and between k_6 and k_7.

K: Okay, let's go ahead and get your relative k_i values. Consider Fig. 7.30. You have said point A is preferred to point B.

TABLE 7.6

ATTRIBUTE LIST WITH OLD
AND NEW RANKINGS OF k_i
SCALING CONSTANTS

Attribute, scaling constant	Ranking	
	Old	New
k_1	1	2
k_{10}	2	1
k_{11}	3	6
k_9	4	3
k_5	5	4
k_8	6	5
k_2	7	8
k_3	8	7
k_4	9	9
k_6	10	11
k_7	11	10

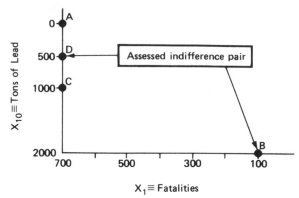

Fig. 7.30.   Reassessed value tradeoffs between fatalities and chronic health effects.
(See Fig. 7.20.)

B: Yes.

K: How about point C versus B? Would you prefer C: ($x_1 = 700$, $x_{10} = 1000$) or B: ($x_1 = 100$, $x_{10} = 2000$)?

B: I'd take B.

K: Point D versus B?

B: Here I'd be indifferent. Yes, that's about it.

K: Okay, this means that $k_1 = k_{10}u_{10}(500)$ since the utility of points B and D must be equal.

We continued in the manner illustrated before and generated the 10 equations in Table 7.7.

TABLE 7.7

REVISED TRADEOFFS USED TO DETERMINE
THE SCALING CONSTANTS

Implied equations	Measure (units)
$k_1 = k_{10}u_{10}(500)$	x_{10} in tons
$k_9 = k_{10}u_{10}(1200)$	x_{10} in tons
$k_5 = k_{10}u_{10}(1700)$	x_{10} in tons
$k_8 = k_5u_5(10)$	x_5 in million tons
$k_{11} = k_5u_5(20)$	x_5 in million tons
$k_3 = k_5u_5(60)$	x_5 in million tons
$k_2 = k_3u_3(50)$	x_3 in thousands of acres
$k_4 = k_3u_3(75)$	x_3 in thousands of acres
$k_7 = k_3u_3(125)$	x_3 in thousands of acres
$k_6 = k_3u_3(150)$	x_3 in thousands of acres

K: Let me now ask you a question on additivity. Maybe since you assessed Wes' utility function you've already thought more about it.

B: Yes, I have and I am additive.

K: Well, let's try one check. Reconsider the two options in Fig. 7.26 involving X_1 and X_{10}, the two attributes whose ranges are most heavily weighted. Do you have preference between them?

B: No, I am indifferent.

This implies again that $\Sigma k_i = 1$, which, together with the equations in Table 7.7, gives us 11 equations with 11 unknowns: $k_1, k_2, ..., k_{11}$. Later on we shall solve for these.

K: Going on, refer to Fig. 7.31. Option I gives you fatalities of 100 for sure and option II gives you a p chance at all of the attributes at their best or a $1 - p$ chance of all at their worst. Which would you choose if $p = 0.5$?

B: That's tough.

K: Well, suppose you have option I. Would you risk a 0.5 chance at fatalities increasing to 700 for a 0.5 chance at all other attributes raised to their best level?

B: At 0.5 I'd take the lottery. [For consistency, this had to be the case, since k_1 will equal the indifference probability and the sum of the k_is equals one, but $k_1 < k_{10}$.]

K: What if $p = 0.4$?

B: I'd still take the lottery, but just barely.

K: At $p = 0.3$?

B: At $p = 0.3$ ..., at $p = 0.35$, I'd essentially be indifferent.

K: Now consider the same type of question, but between options III and IV in Fig. 7.31. If $q = 0.5$ which would you choose?

B: At 0.5 I think I'd take the lottery, but that is close.

K: How about $q = 0.3$?

B: At 0.3, I'd take option III.

K: What if $q = 0.4$?

B: At 0.4, I'm almost indifferent. That's a little low. How about 0.45?

K: This is interesting and quite consistent with earlier responses in this session. Which do you prefer between the two sure options I and III?

B: Well, III, as I've already said.

K: Sure, so if option I is indifferent to II for some value $\bar{p}$ and III is indifferent to IV for some value $\bar{q}$, which should be bigger, $\bar{p}$ or $\bar{q}$?

B: I guess $\bar{q}$, since IV must be preferred to II, given the respective indifference options.

K: Yes, and, in fact, an in-the-head calculation implies the ratio of 0.35 to 0.45 is very consistent also. [These numbers should equal k_1 and

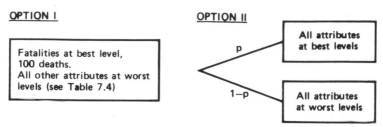

Options I and II were assessed to be indifferent for p = 0.35.
This was adjusted to p = 0.3 to be consistent with the response
for options V and VI below.

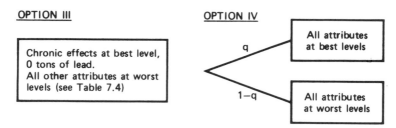

Options III and IV were assessed to be indifferent for q = 0.45.
This was adjusted to q = 0.4 to be consistent with the response
for options V and VI below.

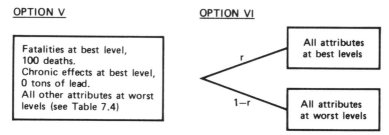

Fig. 7.31. Assessment of indifference probabilities.

Options V and VI were assessed to be indifferent for $r = 0.7$. For consistency with the
additive utility function, this indifference r must equal the indifference p plus the indifference q.

k_{10}, respectively. A later calculation indicates the implied ratio is, in
fact, very consistent.] Now consider option V with lead and fatalities
both at their best levels and all other attributes at their worst levels
or option VI with probability $r = 0.5$.

B: At 0.5, I'd take the sure thing, option V.

K: Suppose $r = 0.9$.

B: At 0.9, I'd take it.

K: You'd take the lottery?

B: Yes, I'd take it.

K: At $r = 0.6$?

B: At 0.6, I don't think I'd take the lottery. I'd go back to option V. What I'm saying is that the other attributes don't mean much here, aren't I?

K: What you are implying is that you are not willing to change from V and take a 0.4 chance at 2000 tons of lead and 600 additional deaths in order to get a 0.6 chance at pushing all the other attributes up to their best level.

B: Well, I guess I do feel this way. It's not that those others are meaningless, though; this bothers me a little bit. If this were an answer consistent with the two previous choices, would the indifference probability $\bar{r}$ be 0.8?

K: Yes, for consistency with the additive utility function, $\bar{r}$ must equal $\bar{q} + \bar{p}$.

B: At 0.8, it seems too high. Maybe the other indifference probabilities should be a little lower.

K: That is exactly the type of thinking we want to promote with utility assessments.

B: That is a good check.

K: Let's return to $r = 0.6$.

B: I'd still take option V at $r = 0.6$, but maybe, at $r = 0.7$, I'd be indifferent. Yes, I can't believe it's as high as 0.8, so there must be something wrong with the other indifference probabilities.

K: Okay, then the sum of the first two indifference probabilities must be 0.7. A simple way to do this is to make $\bar{p} = 0.3$ and $\bar{q} = 0.4$. [I marked these on the respective lotteries for options II and IV.] How does this seem?

B: I'll buy that. Yes, at 0.3 and 0.4, they seem very close.

K: We can do a quick sloppy check to see if these numbers are at all reasonable. We'll assume that all of the utility functions are linear, an approximation that is probably okay for present purposes. Referring to Table 7.7, we see that k_9 would be about $\frac{8}{20}$ of k_{10}, or 0.16, since the decrease from 2000 to 1200 tons of lead is equivalent to the entire range of nuclear safeguards, 50 tons to 0. Similarly, k_5 would be $\frac{3}{20}$, or 0.06. Now k_8 would be almost the same as k_5. Specifically, it would be $\frac{14}{15} \times 0.06$, but we'll assume that it is 0.06. And k_{11} would be $\frac{12}{15}$ of k_8, or about 0.05. Continuing, we find .... Anyway, all summed up, we see that the 11 k_is would equal approximately 1.1. For additivity, as you know, they should sum to 1.0. However, given the roughness of our calculations, the numbers seem to check out reasonably well.

B: That's not bad; it's amazing.
K: The last assessments we'll do here will be to get the single-attribute utility functions. I need only one certainty equivalent on each of these. Let's take the lead first and consider a 50–50 lottery yielding either 0 or 2000 tons. For what sure level are you indifferent to this lottery?
B: It's a little over 1000, 1100 maybe.
K: Okay. The next assessment is for X_1.
B: That's going to be linear, that's fatalities.
K: In our last session, you felt it was very close to linear. Now do you think it *is* linear, or do you have the same feeling as before?
B: Fatalities? I think it is linear, perfectly linear.
K: The third one is X_9, nuclear safeguards. Consider a 50–50 lottery yielding 0 or 50 tons of plutonium produced or 25 tons for sure. Which would you choose?
B: I'd take the sure 25.
K: How about 40 for sure versus the lottery.
B: At 40, I'd take the lottery.
K: 35?
B: I think that's about it.

This procedure was continued for all of the other attributes. The results are shown in Fig. 7.28. If the revised assessment differed from the first, both are indicated. The three points marked by dots in the figure were those used in the assessment. The middle one was the certainty equivalent for a 50–50 lottery of the outer two.

K: We can compare these responses to what you gave before if you are interested.
B: I am, actually.
K: Okay, let's look at Fig. 7.28. On attribute X_{10}, before you gave 1100, and this time you also gave 1100 for the certainty equivalent of a 50–50 lottery yielding 0 or 2000, which seems *rather* consistent. On attribute X_1, you were linear both times. For X_9, you gave 35 as your certainty equivalent both times. Actually, the only three that seem to be much different at all are X_{11}, X_3, and X_6. On X_{11}, you gave 1.0 this time and 1.25 last time. You can reflect on this and change your mind if you want.
B: I think I should stay with the 1.0, because I feel the utility of 2.0 trillion kWhr is very close to one. There is simply not much difference between 2.0 and 3.0 trillion kWhr.
K: Suppose it is 0.9, what does that mean to you?
B: That means that over the range 0.5–3.0, I'll have received 90% of

utility possible by reaching 2.0. Said another way, if I have a 50–50 lottery yielding 0.5 or 2.0 or a 50–50 lottery yielding 0.5 or 3.0, my certainty equivalents probably would not be that different.

K: That's right.

B: And so I want to stick with 1.0 as the certainty equivalent.

K: Good. Let's go on with X_3; you now have the 50–50 lottery yielding 10 or 200 thousand acres indifferent to 130 thousand acres, whereas before you were linear. Do you know anything you were thinking about which seemed different in the two assessments?

B: I'm thinking now of some sort of cumulative effect. By the time we get to 200 thousand acres, it's getting to be a very noticeable impact.

K: Finally, now you are linear in attribute X_6, particulate pollution, whereas, before, you were a little off linear. Any reflection on that?

B: As I think about it, the worst case for particulates is not very noticeable. Therefore, there are no real cumulative effects in this range.

K: Well, now I guess all of the information is here to calculate the utility function.

7.6.4 CALCULATING THE UTILITY FUNCTION

From Result 5 presented in Section 7.2 and from our assessments, we know that the utility function

$$u(x_1, x_2, \ldots, x_{11}) = \sum_{i=1}^{11} k_i u_i(x_i), \tag{7.52}$$

where u and the u_is are scaled zero to one, the k_is are positive, and $\Sigma k_i = 1$. To specify the utility function u, we need to calculate the u_is and k_is.

First, exponential curves were fitted to the nonlinear single-attribute utility functions using the revised data in Fig. 7.28. The final results are given in Table 7.8. Next, given the utility functions in Table 7.8, we could solve the set of 10 equations in Table 7.7 and the equation $\Sigma k_i = 1$ for the 11 unknown k_is. The 11 equations used for the solution are given in Table 7.9, as well as the solution itself in the last column. The final utility function is given by (7.52) and Tables 7.8 and 7.9.

7.6.5 DISCUSSION OF THE ASSESSMENT PROCEDURE

Let me briefly comment on the assessment procedure. First, to better illustrate the techniques of assessment and for convenience in calculation,

TABLE 7.8

THE SINGLE-ATTRIBUTE UTILITY FUNCTIONS

Attribute	u_i	Unit of measurement for x_i	Range Worst	Range Best
X_1, fatalities	$u_1(x_1) = (700 - x_1)/600$	Deaths	700	100
X_2, permanent land use	$u_2(x_2) = (2000 - x_2)/2000$	Acres	2000	0
X_3, temporary land use	$u_3(x_3) = 1.496 - 0.466e^{0.00581x_3}$	10^3 Acres	200	10
X_4, water evaporated	$u_4(x_4) = 1.784 - 0.520e^{0.822x_4}$	10^{12} Gallons	1.5	0.5
X_5, SO$_2$ pollution	$u_5(x_5) = 1.784 - 0.742e^{0.011x_5}$	10^6 Tons	80	5
X_6, particulate pollution	$u_6(x_6) = (10 - x_6)/9.8$	10^6 Tons	10	0.2
X_7, thermal energy needed	$u_7(x_7) = 4.260 - 2.495e^{0.0892x_7}$	10^{12} kWhr (thermal)	6	3
X_8, radioactive waste	$u_8(x_8) = (200 - x_8)/200$	Metric tons	200	0
X_9, nuclear safeguards	$u_9(x_9) = 1.198 - 0.198e^{0.036x_9}$	Tons of plutonium	50	0
X_{10}, chronic effects	$u_{10}(x_{10}) = 3.017 - 2.017e^{0.0002013x_{10}}$	Tons of lead	2000	0
X_{11}, electricity generated	$u_{11}(x_{11}) = 1.039 - 2.003e^{-1.313x_{11}}$	10^{12} kWhr (electric)	0.5	3

TABLE 7.9

SOLVING FOR THE SCALING CONSTANTS[a]

k_i	Relationship to k_{10}	k_i equals	Value of k_i
k_{10}		k_{10}	0.339
k_1	$k_1 = k_{10}u_{10}(500)$	$0.786k_{10}$	0.266
k_9	$k_9 = k_{10}u_{10}(1200)$	$0.449k_{10}$	0.152
k_5	$k_5 = k_{10}u_{10}(1700)$	$0.177k_{10}$	0.060
k_8	$k_8 = k_5u_5(10) = k_{10}u_{10}(1700)\,u_5(10)$	$0.169k_{10}$	0.057
k_{11}	$k_{11} = k_5u_5(20) = k_{10}u_{10}(1700)\,u_5(20)$	$0.152k_{10}$	0.051
k_3	$k_3 = k_5u_5(60) = k_{10}u_{10}(1700)\,u_5(60)$	$0.062k_{10}$	0.021
k_2	$k_2 = k_3u_3(50) = k_{10}u_{10}(1700)\,u_5(60)\,u_3(50)$	$0.054k_{10}$	0.018
k_4	$k_4 = k_3u_3(75) = k_{10}u_{10}(1700)\,u_5(60)\,u_3(75)$	$0.048k_{10}$	0.016
k_7	$k_7 = k_3u_3(125) = k_{10}u_{10}(1700)\,u_5(60)\,u_3(125)$	$0.033k_{10}$	0.011
k_6	$k_6 = k_3u_3(150) = k_{10}u_{10}(1700)\,u_5(60)\,u_3(150)$	$0.023k_{10}$	0.008
		$\Sigma = 2.953k_{10}$	$\Sigma = 1.0$

[a] Solving $\Sigma k_i = 2.953k_{10} = 1$ yields $k_{10} = 0.339$, from which the other k_is are evaluated.

some of the questions asked of Dr. Buehring in specifying his utility function were difficult to consider. These questions are not necessary, especially if one has some computer support. For instance, one never needs to ask for indifference probabilities directly as we did with the options in Fig. 7.31. In this case, we did not actually use them in specifying the utility function.

A second point is that, had there been a preference between the options in Fig. 7.26, then the overall utility function u would have been multiplicative. As is seen from Result 5, this would mean that an additional scaling constant, the k in (7.12), would need to be specified. It is evaluated directly from the values of the 11 k_is. The point is that the multiplicative utility function is only slightly more difficult to specify and use than the additive one. See Chapters 3 and 9 and Keeney and Raiffa [1976] for several applications.

The utility function u should be carefully scrutinized to make sure it does capture Mr. Buehring's preferences. For instance, it would now be easy to draw sets of indifference curves given u. By examining these, one may find aspects of the utility function which are not appropriate. When this is the case, the "errors" should be corrected.

As an example, the values of k_1 and k_{10} in Table 7.9 imply that $\bar{p}$ and $\bar{q}$ in Fig. 7.31 should be 0.266 and 0.339 rather than 0.3 and 0.4, respectively, if the value tradeoffs between the other nine attributes and X_1 and X_{10} remain the same and if the additive utility function is to be used. Alternatively, if the assessed values 0.3 and 0.4 seem to be more reasonable and if the above-mentioned tradeoffs remain fixed, a multiplicative utility function must be employed. Such discrepancies need to be reconciled before using the utility function to evaluate policy.†

The set of assessments discussed here took about eight hours of Dr. Buehring's time. Consequently, one fairly common comment about such assessments is, "This all seems fine, but when is anybody who is a real decision maker going to take all of the time necessary to do this? We need simpler procedures to get the sense of the decision maker's preferences quickly, even if they oversimplify." I agree that this often is a problem. On the other hand, I feel that with the person-years of effort and millions of dollars being spent to model such crucial problems as those concerning energy policy and siting, we should be able to "free" a real decision maker (decision makers) who has a comprehensive knowledge of the problem area for a week or so—at least long enough to structure his or her preferences reasonably. It may even be prudent to have a team of policy makers and analysts work together in a few person-months effort to construct a good preference model. This would then be coupled with the impact model for evaluating policy. The default, of course, is to expect our decision makers to simultaneously consider and balance all of the multidimensional consequences of the impact model, as well as their implications, in their heads and then to arrive at a responsible decision.

I appreciate very much Bill Buehring's comments on this dialogue, as well as his willingness to have his preferences appear in print. The dialogue is clearly altered from the way it was assessed. However, the changes were very minor—mainly correcting grammar, deleting uh's and huh's, removing interruptions, and referring to "point D of Fig. 7.14" rather than to "this point here" as it was pointed to on the page. The complete sense of the discussion is preserved.

Three months after the assessment took place, Dr. Buehring and I went over the results here and informally discussed his preferences. In the interim, he had evaluated selected policies using his utility function, assessed some other individual's preferences, and learned more about some of the consequences of various levels of the proxy attributes (e.g., SO_2 emissions) used in the assessment. These have led to some minor changes in his preferences, as one might expect. In light of this, as well as the fact that the assessments were done with some time pressure, it is inappropriate to interpret the utility function specified by (7.52) with Tables 7.8 and 7.9 as "the final utility function of Bill Buehring." However, had he been required to make a policy decision at the time his preferences were assessed, the expressed utility function could have been of considerable help in examining and choosing among the alternatives.

† See Buehring [1975] for an evaluation of six policy options using such a utility function.

7.7 Evaluation of Impacts with Multiple Clients

Throughout this chapter, we have been mainly concerned with siting problems involving a single client. This situation represents the majority of cases, especially when one allows the client to include the viewpoints of others as was done in Section 7.4. However, there are some siting problems involving multiple clients. This section will indicate the additional information necessary to adapt the results of this chapter to the multiple client problem.

To provide a setting for the problem we are talking about, consider the following. There are proposals to build an oil terminal and marine service base site on Kodiak Island, Alaska to support offshore oil operations. The Division of Community Planning of the Alaska Department of Community and Regional Affairs sponsored a study to evaluate several possible sites. Several groups will be able to influence the siting decision when it is taken. These include the Kodiak fishermen, the natives, the municipality of Kodiak (the major town on Kodiak Island), and the petroleum companies. For the study briefly described later in this section, these four groups were treated as clients for this problem.

7.7.1 STRUCTURING THE VALUES OF THE MULTIPLE CLIENTS

Let us generalize and assume there are C clients for the problem with respective utility functions u_c, $c = 1, ..., C$. Each of these utility functions could be assessed as described in this chapter. The extra complication of the multiple client problem is to combine the separate evaluations in some manner. This may be done informally or formally. If done informally, each client may individually evaluate the various sites using his or her utility function before proceeding with discussions to reach a mutual choice. Such a procedure may be very helpful and lead to a happily agreed upon site. The procedure requires no more formal modeling than that for the single client problem.

If the clients decide to formalize further, this means that it will be necessary to proceed in one of two ways:

1. to combine the individual client utility functions u_c, $c = 1, ..., C$, into a group utility function u and then evaluate the sites with u, or

2. to evaluate the sites with each utility function u_c, $c = 1, ..., C$, to get expected utilities $\bar{u}_c$ for each site, and then combine these into an expected utility Eu_j for each site S_j.

Both procedures require essentially the same new information about the value tradeoffs between clients which imply the relative weights assigned to each.

In notational form, the first approach is to find a function g such that the group utility function u is given by

$$u(x) = g[u_1(x), \ldots, u_C(x)], \qquad (7.53)$$

and then calculate the expected utility for each site S_j from

$$Eu_j = \int_x u(x)p_j(x)\, dx, \qquad j = 1, \ldots, J, \qquad (7.54)$$

where p_j is the probability distribution describing possible impacts at site S_j. The second approach is to calculate

$$\bar{u}_c(S_j) = \int_x u_c(x)p_j(x)\, dx, \qquad c = 1, \ldots, C, \qquad (7.55)$$

and then find a function h to obtain group expected utilities from

$$Eu_j = h[\bar{u}_1(S_j), \ldots, \bar{u}_C(S_j)], \qquad j = 1, \ldots, J. \qquad (7.56)$$

One should note that both (7.53) and (7.56) are very similar in structure to (7.39), where the aim was to aggregate the impacts on several interest groups from the viewpoint of the single client. In determining the forms to use in aggregation, the same concepts—universal indifference, Pareto optimality, equity, and conservatism—are used. The difference is that the group of clients as a whole must decide which assumptions are appropriate. There is always a chance that an agreement cannot be reached, in which case there is no unique group utility function. If there is no agreement on a single combination procedure, perhaps agreement to try alternative procedures will indicate that the results of the site ranking are insensitive to this aspect.

Any combination rule of the form (7.53) or (7.56) will involve utility functions of the separate clients and scaling factors between them. An example is the additive form for (7.53), so

$$u(x) = \sum_{c=1}^{C} \lambda_c u_c(x), \qquad (7.57)$$

where u_c are scaled zero to one and $\Sigma \lambda_c = 1$. The individual u_c can be assessed by the respective clients. The assessment of the λ_c must be done by the group. This assessment is perhaps the heart of the multiple client problem, because the larger λ_1 is relative to the other λ_c, the larger the influence of client 1 on the siting decision. The two issues to be addressed by the group in determining the λ_c are identical to those outlined in Section 7.4.3. The group should consider both the range of impacts possible to each client and the relative "importance" of each client in assigning the λ_c.

There are no formal procedures for arriving at a consensus set of λ_c values. However, it is probably both easier to achieve and more useful for the ensuing analysis to identify a range for each λ_c. The analysis should include all possibilities within the specified ranges. Related to this is the fact that the additive utility function (7.40) and the multiplicative utility function (7.41) for groups can be written as one utility function as shown by Result 5 in Section 7.2. In terms of the notation of (7.53), this is

$$
u(x) = \sum_{c=1}^{C} \lambda_c u_c(x) + \lambda \sum_{c=1}^{C} \sum_{d>c}^{C} \lambda_c \lambda_d u_c(x) u_d(x)
$$

$$
+ \lambda^2 \sum_{c=1}^{C} \sum_{d>c}^{C} \sum_{b>d}^{C} \lambda_c \lambda_d \lambda_b u_c(x) u_d(x) u_b(x)
$$

$$
+ \cdots + \lambda^{C-1} \lambda_1 \cdots \lambda_C u_1(x) \cdots u_C(x). \tag{7.58}
$$

When $\Sigma \lambda_c = 1$, then $\lambda = 0$ and the additive form (7.57) results. When $\Sigma \lambda_c \neq 1$, then λ is the nonzero solution to

$$
(1 + \lambda) = \prod_{c=1}^{C} (1 + \lambda \lambda_c), \tag{7.59}
$$

so by multiplying each side of (7.58) by λ, adding 1 to each side, and factoring, we conclude

$$
1 + \lambda u(x) = \prod_{c=1}^{C} [1 + \lambda \lambda_c u_c(x)], \tag{7.60}
$$

which is the multiplicative utility function. Thus, by allowing each of the λ_c to independently cover their possible ranges as defined by the group, we are not only examining the sensitivity of the results to changes in weighting among the clients, but also varying the assumptions which were used to combine the client's utility functions.

7.7.2 Uses of the Multiple Client Utility Assessments

The uses of the group utility function u in (7.53) or the group expected utilities Eu_j in (7.56) can be categorized into evaluation and examination. Evaluation is rather obvious. One combines the utility function with the outputs of the impact model (i.e., the $p_j, j = 1, \ldots, J$) and calculates expected utilities of the alternatives. Of course, as discussed in Chapter 8, a sophisticated analysis should include sensitivity analyses, but at least in theory, this is straightforward.

By examination, I mean focusing the discussion of the clients on specific problem assumptions and inputs and appraising the implications of different perspectives of the clients. This includes such things as thinking about one attribute at a time, deciding whether the measure is appropriate, and seeing if critical information (e.g., the relationships between pollution levels and health effects) is lacking. It also includes the identification of differences and similarities in perspectives of the multiple clients. In investigating reasons for the disagreements, directions for additional research to reduce the uncertainty may be suggested. The information gained from this research may tend to reduce disagreement. Once such differences are small, analysis may indicate that the same policy options are preferred using the utility functions of any of the clients. Or at least, there may be uniform agreement to eliminate some of the proposed sites from further consideration. This is indeed the situation which arose in the Kodiak Island case described in Section 7.7.3.

Finally, utility analysis may serve as a mechanism for creative and constructive compromise among individual clients in the decision-making unit. It would indicate which client gains what at whose expense (I recognize that this can sometimes be counterproductive). But more importantly, it could indicate mitigation measures or small changes to a proposed site that might make it acceptable, or even desired, by all clients. Such changes may include slight moves of the site, additional pollution abatement measures, front-end funds to prepare a local community for the influx of people, and so on. More details on such uses are found in Chapter 10 of Keeney and Raiffa [1976].

7.7.3   THE KODIAK ISLAND STUDY†

As stated at the beginning of this section, oil terminals and marine service bases may be constructed on Kodiak Island, Alaska to support offshore oil developments. The Division of Community Planning of the Alaska Department of Community and Regional Affairs sponsored a study to evaluate and rank possible sites. The purpose was to assist the State of Alaska, the Kodiak Island Borough, and other interested parties in the planning process for these onshore developments. The study was

† This section is based completely on the work described in Woodward-Clyde Consultants [1977], which is only briefly summarized in this section.

conducted by Woodward-Clyde Consultants. The decision analysis aspects of the study described here are the work of Alan Sicherman.

The Alternatives. The alternatives for this study included the terminal site and in some cases pipelines on land to connect with the marine pipeline from the offshore wells. Alternative sites were proposed by the Department of Community and Regional Affairs, petroleum companies, and a screening model requiring a 72 ft deep, 4000 ft diameter turning basin for oil tankers and at least 100 acres of suitable land for the terminals. This resulted in the identification of eleven potential sites. Depending on the general location of any discovered oil (northeast, east, or southeast of Kodiak Island), a pipeline over land may or may not be used as part of the alternative for a particular site. In several cases, potential sites were evaluated with both land and sea delivery of the oil.

The Clients. The clients whose preferences were formalized in the study included the petroleum companies, the natives, the fishing interests, and the municipality of Kodiak, which is the largest community on the island.

The Objectives. The objectives of the study concerned economic, environmental, and socioeconomic impacts. All together, there were 15 objectives, one to correspond to each category of the concerns listed in Table 7.10. An attribute was either specified or developed for each of these objectives.

The Impacts. The possible impacts for each siting alternative were developed by the study team. These assessments were point estimates on the attribute scale. The assessments were based on existing literature and data, interviews with individuals familiar with the area and the public's attitudes, professional judgments of the team members, and models for predicting certain types of consequences. The uncertainties inherent in these point estimates were treated in the sensitivity analysis.

The Clients' Utility Functions. The utility functions were assessed as follows. Within the environmental area of concern, siting team biologists made the necessary value judgments, since a knowledge of the implications of certain biological disturbances is necessary to do this appropriately. Similarly, a socioeconomist team member provided the value judgments among socioeconomic impacts. These assessments were conducted using the procedure discussed in Section 7.2 and illustrated in Section 7.6.

Value tradeoffs between economic, socioeconomic, and environmental

TABLE 7.10

IMPACT CONCERNS FOR THE KODIAK
ISLAND SITING STUDY

Economic
  Cost of developing the facility at the site
Environmental
  Salmon escapement
  Bay habitat
  Land mammal concentration
  Vegetation removed by pipeline
  Bays crossed by pipeline
  Seabirds
Socioeconomic
  Recreation
  Archaeological/historical factors
  Land use
  Native lifestyle changes
  Harbor use
  Fishing economics
  Induced population increases
  Demands on municipal facilities

impacts are the more crucial ones for this study. It is these tradeoffs that we expect to be quite distinct for the various groups. To get a feeling for these various value tradeoffs, Alan Sicherman interviewed members of each of the client groups. These discussions were not intended to be formal assessments of the value tradeoffs. They did provide the information necessary to construct a utility representation for each group. This level of detail was felt to be appropriate in view of the planning use of the study.

The Evaluation.   The evaluation was conducted for each of the viewpoints separately for oil discoveries in each of the three areas. Evaluations were also conducted for sea–land and all sea pipelines for the northeast and east fields. A southeast field would surely use sea–land pipelines. The results indicated whether sea–land or all sea pipelines were preferred and why for the different cases. They also indicated that in general some sites seemed to be better than others from essentially all viewpoints. Quoting Woodward-Clyde Consultants [1977] from the conclusion, "In the compilation of the various viewpoints, of the sites investigated, several stand out as highly likely candidates: Kalsin Bay, Izhut Bay, Ugak Bay-9, and Kiliuda Bay-13."

Since the study was to be used as a planning tool, no attempt was made to aggregate formally the client's utility functions. Instead, several ways in which the clients can influence the resulting siting decision when it is made were outlined. Early coordination and cooperation among the clients has the potential to result in a better and more satisfying decision for each of them. This study suggested areas, such as the need for some currently nonexistent data, which could provide a focus for that coordination, and it also clearly articulated the concerns of the various clients and parties to the decision.

CHAPTER 8

ANALYZING AND COMPARING
CANDIDATE SITES

At this stage in a decision analysis of a siting problem, the model is developed and the data—probabilities and utilities—collected. A rather simple mathematical analysis is now conducted to determine what should be done, given that the model and data are appropriate. However, since no model is ever perfect and no data are precisely accurate, numerous sensitivity analyses are conducted to help in the comparison of sites. The creative part of this final step in decision analysis is deciding which sensitivity analyses to conduct and how to interpret the implications of the analysis for the real decision of selecting a site.

Section 8.1 discusses the analysis procedure and results for a given set of data. Sections 8.2–8.4 indicate how this is repeatedly used in sensitivity analyses of the model and input data. Presentation and interpretation of the results of a decision analysis siting study are the topics in Section 8.5.

8.1 The Procedure for Analysis

Let us briefly review the information we have from previous steps in the study. We have a number of sites S_j, $j = 1, ..., J$ to be evaluated on attributes X_i, $i = 1, ..., n$. For each site the possible impacts are quantified by the probability distribution $p_j(x)$. The value structure is

328

quantified by the client's utility function $u(x)$. With this, the expected utility Eu_j of each site is calculated from

$$Eu_j = \int p_j(x)u(x) \, dx. \tag{8.1}$$

If the expected utility for one site is higher than that for a second site, the first site should be preferred; thus the sites can be ranked using their expected utilities. This result, which is a major motivating factor for using utility in problems involving significant uncertainties, is derived from the axioms of decision analysis for siting described in the appendix. The axioms, which appear to be very appealing for evaluating sites, are the logical and theoretical foundations of the decision analysis approach to siting.

Expression (8.1) is a quantitative summary description of the decision analysis model. Even though the forms of u and p_j are often complex and involve many attributes in several time periods, it is clear that the degree of mathematical sophistication required by decision analysis is not great. Integration is all that is used. Even so, the computational effort involved in using (8.1) is often greater than one would wish to attempt without help. This is especially true because of the appropriateness of using a wide range of sensitivity analyses. The answer, of course, is simply to use a computer for the computational task. For any modern computer, the actual computer time needed for the analysis is minimal.

An interactive computer program has been developed to utilize (8.1) on siting studies (Woodward-Clyde Consultants [1979]). The model conveniently permits a broad range of information for specifying the utility function and probability distributions for the computer. From this information, the computer determines the utility function and probability distributions. Then the expected utilities of the sites are calculated. The existing routines make it easy to conduct sensitivity analyses of several input parameters, such as are described in Sections 8.2–8.4. Earlier versions of the computer program are described in Keeney and Sicherman [1976] and Seo *et al.* [1978].

8.1.1 INTERPRETING EXPECTED UTILITY

It is not easy to interpret expected utility for at least three reasons. First, utility is a construct used to simplify the analysis of a complex decision problem; it does not have a physical meaning. Second, the scale for the utility function can be arbitrarily chosen, so knowing the utility without knowing the scale is meaningless.† And third, a utility function provides an index of preference appropriate for those who constructed it. Other individuals and groups may have quite different values and, hence, quite different interpretations of various utilities.

As a result, although expected utility is appropriate and useful for evaluating and ranking sites, it is not particularly helpful for reporting or interpreting the ranking. Fortunately, there is a simple solution. We merely convert the expected utilities back into a set of consequences with utilities equal to the expected utilities. That is, for each expected utility Eu_j, we find an x^j such that

$$Eu_j = u(x^j), \tag{8.2}$$

where x^j is then a consequence with impact equivalent to the expected impact at site S_j. The consequences x^j, $j = 1, \ldots, J$, referred to as "equivalent site" consequences, can be used to get an intuitive feeling for the overall implications of constructing the proposed facility at the various sites.

In practice, the equivalent site consequences are much more useful if the levels of all but one of the attributes are fixed at identical levels throughout the set x^j, $j = 1, \ldots, J$. For instance, if attributes $X_2, \ldots, X_n$ are fixed at $\hat{x}_2, \ldots, \hat{x}_n$, then $x^j = (x_1^j, \hat{x}_2, \ldots, \hat{x}_n)$, so from (8.2)

$$Eu_j = u(x_1^j, \hat{x}_2, \ldots, \hat{x}_n). \tag{8.3}$$

If X_1 is cost, for instance, then we refer to x_1^j as the equivalent cost of site S_j, given the other attributes are fixed at $\hat{x}_2, \ldots, \hat{x}_n$. Now the x_1^j, $j = 1, \ldots, J$ serve as a single-dimensional indicator of the value of the various sites. Because x^j does have a physical interpretation, the implication of each site impact and the differences in site impacts are much better understood than was the case with the Eu_j. The levels $\hat{x}_2, \ldots, \hat{x}_n$ in (8.3) should be selected to facilitate the interpretation of the x_1^j. For example, these other impacts might be set at a status quo (i.e., undisturbed) level or a "normal" level.

In both the nuclear siting study and the pumped storage study discussed in Chapters 3 and 9, respectively, equivalent costs were calculated with (8.3) to help interpret the study results. The equivalent costs were reported in addition to the expected utilities in all the sensitivity analyses. In other studies, it may be appropriate to use attributes other than costs for "equivalents." Sometimes it may even be helpful to use more than one set of equivalents based on different attributes.

---

† This is the case with temperature scales. Simply knowing that the temperature is 75° is useless unless it is known whether the scale is Kelvin, Centigrade, or Fahrenheit.

8.1.2 SENSITIVITY ANALYSES

A key to the successful use of decision analysis for siting of energy facilities is a thorough sensitivity analysis. However, for this analysis to be at all meaningful it is necessary that the earlier steps of the decision analysis be conducted well. From (8.1), it is clear that the set of expected utilities from a siting study depends on the following information:

1. the set of sites: S_j, $j = 1, ..., J$,
2. the set of attributes X_i, $i = 1, ..., n$,
3. the set of probability distributions describing possible site impacts, p_j, $j = 1, ..., J$, and
4. the utility function u.

This information is supplied by the first four steps of the decision analysis. The final step puts all the information together to evaluate the sites. For each evaluation, all of the information must be precisely specified. The analysis which uses the basic set of data is referred to as the base case.

Sensitivity analysis, in general, refers to the repeated evaluation of the candidate sites with some of the information changed in each case. The purpose is to examine how the site evaluations depend on the input information of the model. For instance, with a coal-fired plant, if the transportation cost of coal increases at 10% per year, rather than 5%, would the ranking of the sites change. There are many professional judgments, such as this one, that are necessarily made formally or informally in any siting evaluation study.

A sensitivity analysis will allow identification of the following:

1. professional judgments and value judgments which critically influence the implications of the study,
2. a set of sites to be definitely included in any further evaluation and another set of sites to be eliminated from any further consideration, and
3. differences in the opinions and values of different groups which affect the implications of the study.

Because of the importance of sensitivity analysis, Sections 8.2–8.4 will be devoted to it. These sections discuss the impact data, the values, and the problem structure, respectively. The problem structure is meant to include the sites, the attributes, and the basic assumptions upon which the probability distributions are determined. An example of the latter might be whether or not a particular pollution control bill would become operative. The sensitivity analysis of impact data and values concern changes in the probability distributions and utility functions, respectively.

8.2 Sensitivity Analysis of the Impact Data

The probability distributions p_j in (8.1) may have resulted from data analyses, professional judgments, analytical models, simulation models, or any combination of these. Whatever the sources, in general, it will be possible to think of each p_j as a function of some single-attribute (conditional or marginal) probability distributions p_j^1, p_j^2, ..., p_j^n over attributes X_1, ..., X_n and some parameters q_s, $s = 1$, ..., S. These parameters may be, for instance, the means and standard deviations of distributions, the proportion of high-sulfur oil burned, or the annual rainfall. Thus, we can symbolically write

$$p_j = g(p_j^1, ..., p_j^n, q_1, ..., q_S), \qquad (8.4)$$

where g is a function combining the p_j^i and the q_s. A sensitivity analysis of the impact data means individually and simultaneously varying the p_j^i and q_s. Let us illustrate the idea by referring to a few examples in the book.

8.2.1 SENSITIVITY ANALYSIS OF UNCERTAINTIES

In the WPPSS nuclear power siting study discussed in Chapter 3, there was reason to believe that uncertainties about the costs could be significant. Such factors as delays, interest rates, and fuel prices over which the utility has little control can greatly effect costs. In addition, the weight assigned to cost in the utility function indicated that, given the possible range, costs were relatively important compared to the other attributes. In spite of these two facts, the costs were reported in terms of only one figure—an estimated annualized net present value differential cost above base cost for each site. For these reasons, we conducted a sensitivity analysis as follows.

First, evaluations of the sites were run with average costs at 90, 100, and 115% of the estimated cost and with the standard deviation ranging from 25 to 50% of the average. The results indicated that the site ranking was not sensitive to these changes. However, this is not particularly surprising since the costs were varied up and down together for the set of sites. In defense of this, one would expect the cost estimates to be correlated to the ''true'' costs in this fashion since the same procedures were used in estimating costs for each site and since, in many instances, the same components contribute to cost changes from the base case.

One of the major factors which may lead to a different distribution of costs between sites was felt to be liquefaction. Because of this, the model was altered to include the costs of engineering with a liquefaction potential present. This aspect of the WPPSS sensitivity analysis is discussed in Section 8.4.

Risk Parameters. As described in Section 6.8, a model was developed to calculate the public risks from a proposed liquefied natural gas (LNG) facility in Matagorda Bay, Texas. The principal public hazard results from an LNG spill which vaporizes, travels to a populated area, and then is ignited by a source such as a cigarette lighter or pilot light on a stove. For purposes of the study, it was conservatively assumed that individuals within the LNG vapor cloud at the time of ignition would become fatalities.† If a vapor cloud is traveling toward a populated area, two features which can greatly influence the extent of risk produced by the cloud are (1) the probability of ignition per ignition source and (2) the lowest LNG to air ratio at which the cloud is still ignitable. In the base case calculations, the probability of ignition was set at 0.1 and the cloud was assumed to be ignitable down to a level of 5.0% LNG in the air.

If the true level of either of these parameters were lower, then the area which could be covered by a cloud capable of being ignited would increase. Thus, the possible number of fatalities could increase. As reported in Section 6.8, the ignition probability was set at 0.01 and the lower ignition limit at 2.5% in sensitivity analyses. For both cases, calculated sequentially, the risk per person in the area of the proposed terminal increased approximately 10%. Since these risks were approximately two orders of magnitude below what was felt to be an acceptable level for the problem, these changes in risk were not considered significant.

Geological Uncertainties. For the pumped storage siting study discussed in Chapter 9, deterministic estimates of each attribute were used in the original base case analysis. There seemed to be little uncertainty about the estimates assumed with the three environmental attributes. However, the geological conditions at the sites could potentially influence the costs of reservoir construction and drilling to connect the upper and lower reservoirs. This was investigated in sensitivity analyses in the following manner.

Sensitivity analyses of the problem structure and values were conducted given the deterministic estimates. These indicated that the first two ranked sites had an advantage equivalent to about 3 million dollars of first-year equivalent costs over any other sites for the base case. The geological uncertainties for the contending sites were then quantified using

---

† In the only large accident in the U.S. involving the ignition of a vapor cloud, Van Horn and Wilson [1977] report that 128 people died in Cleveland, Ohio in 1944. It is estimated that approximately 1500 people (many in houses) were within the cloud at the time of ignition.

the professional judgment of geologists who had collected data at the sites. Comparing each site, as evaluated deterministically, to its evaluation with the uncertainties, indicated the interesting fact that the two best sites appeared even better than in the base case, while the other main contenders appeared worse. This simply means that the deterministic estimates were a little pessimistic for the first two sites and a little optimistic for the others. As a result, no further analysis of geological uncertainty was conducted at that time.

8.2.2 SENSITIVITY ANALYSIS OF PROBABILISTIC DEPENDENCIES

In many situations, the impacts on different attributes at a site are correlated or probabilistically dependent. This was the case, for example, in the pumped storage study, between costs and transmission line environmental impact. The reason was that longer transmission lines tended to increase both costs and line impacts. It may also be the case that attributes are correlated or probabilistically dependent between sites. For instance, increasing the transportation costs of fuels would increase the costs for all sites, even though it could differentially impact the sites because of proximity to the fuel supply. There is perhaps a more important dependency resulting from what is sometimes referred to as a "common mode" mechanism. For instance, strikes of workers mining fuel could not only lead to an increased price for fuel, but perhaps to shortages resulting in less efficient or curtailed operations. Although each situation separately may be "manageable," the combined effect could be a disaster.

In theory, probabilistic dependencies are not difficult to appraise. Although the complications may be more involved, the computer can easily take care of them and report expected utilities and equivalents as defined in Section 8.1.1. However, it may be quite difficult to formulate all the dependencies for the computer, unless a simulation model of impacts with dependencies included is available. Furthermore, the degree of dependency may not be known and may require professional judgments for quantification.

The best approach in this situation is perhaps to conduct a sensitivity analysis to find out if a probabilistic dependency makes any difference. The idea is to evaluate the sites assuming no dependency of the type being investigated and then to repeat the analysis assuming a strong dependency. If the results are similar enough, further concern is not required. If the dependency does seem important, it must be further appraised either with a more formal effort or informally.

Another way to handle specific dependencies is to bound the possible implications of the dependency. For example, perhaps simple assumptions can be made which will clearly tend to make site S_1 appear worse than it really is and to make site S_2 appear better. Then if site S_1 is better than S_2 under these conditions, no further appraisal of this dependency may be warranted. The idea of bounding is illustrated by the following example.

Natural Occurrences. The Geysers geothermal power project outlined in Section 6.7 involved the evaluation of whether a landslide induced by an earthquake would occur at one site proposed as a geothermal power plant site in California. This potential appeared to be the main negative feature at an otherwise desirable site. If the landslide potential was small enough, it was felt that the proposed site would be the best one.

There were three earthquake faults capable of producing a landslide at the Geysers site. The overall probability of a slide P was calculated from

$$P = 1 - (1 - P_{sa})(1 - P_{ma})(1 - P_{hr}), \qquad (8.5)$$

where P_{sa}, P_{ma}, and P_{hr} are the probabilities that an earthquake on each of three faults might induce a slide during the life of the proposed project. In using (8.5), we have assumed that the probability that one fault has an earthquake large enough to induce a slide is independent of whether the others have one. This independence assumption is conservative in that it will lead to a higher overall probability P than would be the case if the natural probabilistic dependencies were included in the model. Intuitively, this is true for the following reasons.

Consider just the first fault. There is an assessed probability P_{sa} that it will induce a slide. This possibility is clearly included in (8.5). However, suppose it does not induce a slide. This implies that the slide area is a bit more stable than originally thought. The fact the first fault did not induce a slide may be because there were no major earthquakes on it during the project lifetime. However, there may have been some major earthquakes which did not induce slides. This latter consideration should make us decrease the estimate of P_{ma}, which will, in turn, decrease the derived estimate of P. For simplicity, in the analysis, P_{sa}, P_{ma}, and P_{hr} were all determined independently. Since the conservative estimate of P was then in reality an upper bound and since it seemed reasonably low, no formal analysis of the probabilistic dependency was calculated.

8.3 Sensitivity Analysis of Values

The utility function in (8.1) depends on a number of single-attribute utility functions u_i, $i = 1, ..., n$, and scaling constants k_r, $r = 1, ..., R$. Thus we can symbolically write u as some function f of these so

$$u = f(u_1,\ldots, u_n, k_1, \ldots, k_R), \tag{8.6}$$

where the u_i are meant to include utility functions over different attributes, different time periods, and affecting different groups of people, and the scaling constants concern value judgments between each of these. The sensitivity analysis conducted for the value structure in a problem refers to changes in the u_i, the k_r, or the form of f. Some examples will illustrate this.

8.3.1 SENSITIVITY ANALYSIS OF RISK ATTITUDES

As indicated in Section 7.5, risk attitudes are quantified by the single-attribute utility functions. For the pumped storage case discussed in Chapter 9, the first-year cost attribute was most important, given its possible range from 50 to 75 million dollars. The utility function for costs was assessed from the vice-president in charge of finance for the company involved. He felt that a sure first-year cost of 68.75 million dollars was indifferent to a 50–50 lottery of either 50 or 75 million dollars. Since 68.75 is three-fourths of the way between 50 and 75, this indicates quite a risk-averse position, since the expected cost of the lottery is 62.5 million. The reason for this risk aversion was the difficulty that the vice-president expected in raising necessary capital for the project at the expensive end of the scale. The base case analysis used an exponential utility function for costs which incorporated this strong risk-averse attitude.

In the sensitivity analysis reported in Table 9.6, we appraised the effect that this risk attitude had on the site evaluations. We considered a less risk-averse attitude and a risk-neutral attitude. In the former case, we took 65 million dollars to be indifferent to the 50–50 lottery of either 50 or 75 million dollars. The risk-neutral case, 62.5 million dollars, was obviously indifferent. Those changes improved the equivalent cost evaluation of the best and third ranked sites significantly, whereas the cost equivalent of the second best site in the base case improved only slightly. This indicated that the third site may really be a viable contender if a less risk-averse attitude could be justified by changed funding possibilities.

8.3.2 SENSITIVITY ANALYSIS OF VALUE TRADEOFFS

The constants k_1, $\ldots$, k_R in (8.6) represent the value tradeoffs between attributes, impacts in different years, and so on. By changing them, we are changing the relative weight given to each attribute in the evaluation. Because of their importance to the overall analysis and because they must (and should) be assessed using an individual's values, which are necessarily subjective, it is almost imperative to conduct a sensitivity analysis by varying these constants. This was done in both the nuclear siting and pumped storage studies.

For example, the base case for the four attributes in the pumped storage study had $k_1 = 0.716$, $k_2 = 0.382$, $k_3 = 0.014$, and $k_4 = 0.077$, so $\Sigma k_i = 1.189$. The multiplicative utility function from Result 5 in Section 7.2 was used. Notice that k_1 is about twice k_2 and that the other constants are much smaller. In the sensitivity analysis, we first kept the same ratios of the four k_i and ranged the Σk_i from 0.8 to 1.5. This has the effect of changing the manner in which the values of different attributes interact. For instance, when $\Sigma k_i = 1$, the additive utility function from Result 5 is appropriate. Next, since the value tradeoffs between attributes X_1 and X_2 were clearly most important, we first set $k_1 = k_2$ and then $k_1 = 4k_2$, while holding other value tradeoffs fixed. The implications of these changes in values were not significant, as can be seen from Table 9.6.

Attribute X_4 in the study concerned the riparian community. Because it appeared to be unique at one site, the scaling factor k_4 was greatly increased for that site. This significantly influenced the environmental desirability of the site, eventually leading to its elimination as a possible pumped storage site. It was replaced by a nearby site which turned out to be highest ranked, as described in Chapter 9.

8.3.3 ANALYZING A PROBLEM FROM DIFFERENT VIEWPOINTS

Since a client's utility function u quantifies his or her value structure, it is clear that by changing the utility function, a different value structure can be used on the same problem. This simple recognition allows us to analyze a siting problem from a variety of viewpoints. These viewpoints can represent any of the multiple clients, regulators, impacted groups, and concerned parties if this is desirable.

There are two important reasons for a sensitivity analysis of viewpoints. First, it provides an evaluation from each different perspective. This may allow us to eliminate certain sites from consideration because nobody particularly likes them. It may even provide enough information to select a best site. The second reason is to identify conflicts which might exist between clients (or groups) and to indicate actions which may serve to alleviate them. Such actions may involve, for example, pollution mitigation measures or quick approval of access to cross federal land.

If the different viewpoints can be represented by a utility function with the same structural form for each group, then the sensitivity analysis of viewpoints is very similar to that for the value tradeoffs discussed above. However, there are many problems in which different groups will be interested in completely different attributes, and, hence, in completely different value structures. Even in these cases, it may be possible to have the same structure for each group, but with each group weighting those attributes with which they are not concerned as zero. However, it would probably be easier to assess each group's utility function directly.

There is a direct way to do a thorough sensitivity analysis of viewpoints as indicated by the following characterization. Suppose there are three groups with viewpoints expressed by u_1, u_2, and u_3. Define the utility function u as

$$u(x) = \lambda_1 u_1(x) + \lambda_2 u_2(x) + \lambda_3 u_3(x), \tag{8.7}$$

where $\lambda_1 + \lambda_2 + \lambda_3 = 1$ and each λ is positive. We can then use u to evaluate the options. When $\lambda_1 = 1$ and $\lambda_2 = \lambda_3 = 0$, the analysis is completely from the first group's viewpoint. When $\lambda_2 = 1$ and $\lambda_1 = \lambda_3 = 0$, the analysis is from the second group's viewpoint, and so on. By conducting a sensitivity analysis over all possible values of the λs, we have a rather complete analysis of the implications of the different viewpoints. For some studies, such as the Skeena River salmon case described in Section 7.4 and below, a more robust form of the combined utility function may be appropriate. Any of the results in Section 7.4 may be used rather than the additive form (8.7).

Planning.   In the oil terminal and service base siting study for Kodiak Island, described in Section 7.7, several groups which would be impacted by the construction and operation of such a terminal were identified. These included the natives on Kodiak Island, the fishing interests, the municipality of Kodiak, and the petroleum companies. Discussions with members of each of these groups helped to identify their viewpoints, mainly with respect to the value tradeoffs between economic, socioeconomic, and environmental impacts. Analysis from the different viewpoints helped to identify a set of promising candidate sites and reject others. Since the purpose of the study was to assist in planning and preparation for the proposed terminal, the study served its purpose. See Section 7.7 for more details.

Alternative Generation.   A critical problem effecting the livelihood of several groups was discussed and structured in Section 7.4. The Canadian Department of the Environment must specify a salmon fishing policy for the Skeena River in British Columbia, Canada. The policy has major effects on lure fishermen, net fishermen, sports fishermen, native Indians, and regional development groups. Utility functions were elicited from individuals knowledgeable about each of these groups. These were combined into a utility function which both allowed for analysis of each viewpoint and provided for different types of value interactions between the groups. The main purpose of this exercise was to help generate creative alternatives which may be more satisfying to each of the parties. Although this was not a siting study, the sensitivity analysis of viewpoints provides a good example for illustrating the techniques for a thorough analysis of viewpoints.

8.4 Sensitivity Analysis of the Problem Structure

By problem structure we include the work done in the first two steps of decision analysis in identifying sites and attributes and the construction of a model to help calculate the probability distributions. It is often not as easy to do sensitivity analyses of these components as it is to do sensitivity analyses of the probability distributions themselves or the utility function. The latter simply requires changing parameters in the model which has been provided for the computer, whereas the former requires changes in parts of the model. As a result, some of the sensitivity analyses of the problem structure are done less formally than those of the probability distributions and utility function.

In addition to the fact that less formal analyses are simpler, there is another reason for this approach. After completing the basic model, the base case analysis, and the sensitivity analyses of the probabilities and utilities, the siting team has a great amount of information at its disposal, which was not available earlier. In particular, it should be clear which sites (or site) are really in contention to be recommended for construction of the proposed facility. By knowing how good these sites are, the team will have an understanding of the magnitude of the problem structure changes necessary to alter the best sites.

8.4.1 SENSITIVITY ANALYSIS OF CANDIDATE SITES

In Chapter 4, screening models for identifying candidate areas were developed. These necessarily relied on a number of cutoff criteria based on professional judgments. A major question is whether there is a good chance that sites in areas screened from the study at that stage could be real contenders. If the team simply reconsiders each of these judgments, it may be clear whether each still remains appropriate. For instance, if the cutoff on water availability implied a cost of 20 million dollars over base costs and if the best five sites had cost equivalents (which include non-monetary impacts) of less than 15 million over base costs, it would be safe to conclude that the areas rejected because of water availability did not contain any viable contenders.

If on the other hand, an area is found which may have a potential candidate site or sites, members of the team should visit the area. Then, if no sites can be identified as candidates, the team should be able to justify this in qualitative presentations. If sites are found which may be contenders, probability distributions should be collected for the site and an analysis conducted. At this point, we are in the same position as if the site was contained in the initial set of potential sites identified from step 1 of the decision analysis.

The following example illustrates the same idea from a reverse perspective. The selection of candidate sites in a candidate area will result in omitting possible nearby sites felt to be slightly inferior. Subsequent evaluation may indicate that these candidates are, in fact, inferior to the nearby omitted sites. Sensitivity analysis of the site identification process can lead to the inclusion of the good sites originally omitted.

In the pumped storage siting study of Chapter 9, the sensitivity analysis had determined that one of the ten considered sites could be ranked either first or ninth depending on the environmental uniqueness of the riparian community at that site. Further appraisal by biologists indicated that the community was unique for that part of the country. The team reappraised the area and found that, by selecting a site a few thousand yards removed from the former site, the riparian community would not be damaged and the environmental and economic benefits of the original site could almost all be preserved.

The chronology of what occurred is the following. After the original screening, the team identified the initial site as the best in the acceptable candidate area and it was therefore included in the analysis. At this stage the uniqueness of the riparian community was not known. When this became clear, further appraisal indicated that what was initially thought to be a slightly inferior site was, in fact, much better. This alternative site was then evaluated using the original model and it was ranked best. The current status is that this is the prime site for development.

8.4.2 SENSITIVITY ANALYSIS OF ATTRIBUTES

By its very nature, a useful sensitivity analysis of the attributes must be informal. The only question to be addressed here is: would the results change if attributes were deleted or added. The question of deletions can be formally addressed with the model in an easy manner. This can be done by weighting an attribute with a zero weight in the utility function. However, even if the implications of the study are the same, we have gained very little useful information. Since we have already collected the data for the attribute with regard to the alternatives being evaluated and since the computer computation is simple and cheap, eliminating a nonessential attribute serves almost no purpose unless new alternatives are to be added.

On the other hand, it is useful to know if attributes not included could affect the analysis if they were included. To answer this, we need some idea about the weight the attribute would have relative to the weights of the attributes in the analysis already and/or about the range of impacts over the contending sites with respect to the attribute. Byer [1979] discusses a procedure for screening attributes which uses only the weights. If estimates are used, each estimate can be done rather roughly, as illustrated by the following case.

In the WPPSS nuclear siting study, we initially felt that it may be appropriate to include attributes involving the loss of agricultural land, the number of people living near cooling towers and consequently in additional fog, and the impact on road and shipping traffic due to fogging. As discussed in Chapter 3, the professional assessments of a meteorologist were obtained to appraise the impact of fogging. He used his knowledge of the local area and U.S. Weather Service data to construct regions where additional days of fogging due to the cooling towers could be expected. The maximum additional days of fog turned out to be three days. Using other data on the location of residents, vehicle traffic, and boat traffic in the regions near the candidate sites led us to conclude that the formal inclusion of these attributes in the study would not affect the results. The study implications also did not seem sensitive to the inclusion of agricultural land loss. These reappraisals, conducted after the analysis was completed, reaffirmed earlier decisions not to include the attributes.

At times it is appropriate to do an even more informal appraisal of attributes. For instance, suppose no attribute is included to capture the difficulty—not the cost—of acquiring the land needed for the proposed energy facility. After the analysis, the differences in cost equivalents between the sites could be millions of dollars. The professional intuition and experience of the client may strongly suggest that the difficulty of land acquisition is in no way worth more than a few hundred thousand dollars. This is surely sufficient to justify omission of a "land acquisition" attribute.

8.4.3 SENSITIVITY ANALYSIS OF THE IMPACT MODEL

Many important assumptions which may influence the results are necessary in defining the siting problem. Some of these may involve political or regulatory actions or indirectly related matters. For instance, if a state public utility commission allows large rate increases in the next year, the financial status of the utility may be sounder, allowing lower interest rates on loans. This will reduce the overall cost of building a facility. Some assumptions are necessary about what to include in the model. For example, should the costs of transmission lines or negotiations (intercompany or government) be partially or totally included in the study if such lines or agreements may have multiple future purposes? Other assumptions simply concern exogenous variables which affect the attributes, such as the inflation rate and demand growth for the product of the proposed facility. Two examples may help explain these ideas.

In the WPPSS study, it became apparent that there was a possibility of liquefaction at three of the nine alternate sites. A study to indicate whether liquefaction will be a problem costs only about 100 thousand dollars and engineering costs to alleviate this problem are from 10 to 20 million dollars, whereas the implications of neglecting possible liquefaction could result in the loss of use of the facility as well as tremendous mitigation costs. Clearly, before any of these three sites were chosen, a complete liquefaction study would have to be conducted. Thus, in the results reported in Table 3.6, the sites are ranked assuming no liquefaction potential at the three sites and again assuming there is a liquefaction problem. The principal effect of this was an increase in construction costs, which is evident from the cost equivalents in the table.

The pumped storage study included the site impacts and the cost and aesthetic impact of the transmission lines necessary to transmit electricity to the load centers of the utility company. In the base case analysis, it was assumed that all transmission line impacts would be attributed to the pumped storage study. This clearly puts a premium on sites nearer the load centers. However, to the extent that such lines would become part of the utility's main network and be used for multiple purposes, perhaps less than 100% of the cost should be borne by the sites in this analysis. By altering this structural assumption, we found that for even the fourth best site to become equal in desirability to the second best, 32% of the transmission line costs would need to be allocated elsewhere. For the fourth best site to become number one, 62% would need to be allocated to other projects. Both of these implications assumed all the transmission impacts associated with the best and second best sites would remain associated there. In view of these figures, the clients felt the other sites could not replace numbers one and two by a reallocation of transmission line impacts.

8.5 Reporting and Interpreting the Results

The analyses described in this book are designed to address two problems:

1. identification of the best site for a proposed energy facility and
2. documentation of this site and the siting process for the regulatory authorities and the public.

In some cases, the decision analysis, with the extensive sensitivity analyses just described, may provide enough information and resulting insights that all the members of the client who have a hand in the site selection process agree that one site is best. Then the client should simply select that site and proceed.

However, in other instances, such an obvious site may not be immediately apparent. A major advantage of the decision analysis model is that it provides the information and methodology to help the client select a site in these situations. Stated simply, the model provides a means for identifying and resolving internal conflicts. Subject to the premise that everyone feels that a site should be chosen (i.e., each agrees on the need for such a facility), conflicts about the best site must ultimately rest on either disagreements about some aspect of the analysis or on the interpretation of the analysis. It is much more likely that the conflict occur about the analysis rather than about its interpretation. In either case, it should be easy to ascertain which is the situation.

If the disagreement is about the analysis itself, it is appropriate to go through the analysis step by step to identify the points of conflict. Disagreement about a step is much less serious than disagreement about every aspect of the analysis. In fact, experience has shown (Gardiner and Edwards [1975]) that there is usually much more agreement than disagreement about an analysis, even among adversaries, when the analysis is examined piece by piece. This in itself may provide a sound basis on which to reach overall agreement on a site.

If the disagreement is about interpretation, discussion or perhaps an extension of the analysis is appropriate. The issues of a conflict may be resolved by gathering data not originally available, by conducting additional analysis, by generating new alternatives, and so on. However, it may be that the conflict has no analytical solution. For instance, two individual members of the client may maintain, after serious introspection, that different value structures are appropriate for a particular siting problem. The sensitivity analysis may indicate that site A is best using one value structure and site B is best using another. In this case, the analysis may have helped to identify the conflict, but it clearly cannot alone lead to a resolution. Interpretation of the analysis then includes an understanding of the conflict and its implications. All of this is used to aid the client in reaching a decision.†

8.5.1 REPORTING A SITING STUDY

It is important to recognize that conducting a siting study and reporting the results of such a study are distinct activities. The client may conduct a study to decide which site is best and may or may not release the study for outside dissemination. And, of course, there are many in-between strategies. A client may wish to supress certain parts of a report for very valid reasons. For instance, if the client's estimates of delays expected from various agencies in organizing hearings or approving necessary permits are included in the analysis, the client may wish to delete these in the released report. Particularly political and volatile professional judgments, such as the value tradeoffs between potential fatalities and costs, may also be deleted, especially if sensitivity analyses indicate that such judgments are not critical to the siting implications of the study.

A second important point is that different reports should be prepared for different purposes. The siting study reports for the Board of Directors of the client company, its stockholders, its staff, the public, and different regulatory agencies could, and probably should, be quite different. Notwithstanding the caveats referred to above, let us describe a reasonable outline for a complete report to the client company and a word about the style. Such a report would provide the basis for reports to other groups. The company report should be presented on three levels: an executive summary (1–5 pages), an overview of the siting study and its results (15–50 pages), and a complete documentation of all assumptions, models, and data used in the study and a complete presentation of all the analyses. The executive report and overview both cover the scope of the entire study, whereas the other information could be considered as attachments to the overview.

Since the executive summary is simply a concise synopsis of the overview, we shall describe the overview. It should begin with a definition of the siting study, with special care to indicate features such as areas or concerns to be excluded from the study. Since no study can include everything, it is permissible to exclude possibly relevant features as long as these are clearly agreed upon and presented with the study to avoid possible misinterpretations of the results. Next, the overview should

---

† There is a more detailed discussion of aspects of the use of decision analysis for conflict resolution in Chapter 10 of Keeney and Raiffa [1976].

discuss each part of the decision analysis. The assumptions and results at each stage should be clearly indicated. The results should include specification of all the factors found to be important to the site choice in the sensitivity analysis.

The use of maps, graphs, and tables can be a particularly effective mode for presenting study results. To make these as useful as possible, a rule of thumb is that the reader should be able to interpret the material presented in this fashion without referring to the text. This requires complete labeling and the inclusion of the assumptions used in generating the data in tables and graphs on the tables and graphs themselves.

The overview should conclude with the supported recommendations of the siting team. These recommendations may include eliminating specific sites from further consideration, concentrating on another set of sites, conducting additional studies on certain sites, and suggestions for differentiating the desirability of the contenders.

The backup documentation for the overview should be of such detail that any similarly competent siting team could completely follow the entire history of the study. It should contain all the data collected in the field by geologists, seismologists, environmental scientists, socioeconomists, and so on. The analysis of this data, as well as assumptions made in the study and a justification for each, should be included. The documentation should contain a thorough presentation of the sensitivity analysis, including a summary of all the factors which did not seem critical, and why not. In short, if the client or a regulator asks a question of the study which can not be answered from the detailed study documentation, then the answer should be that the question was not addressed in the study.

8.5.2 INTERPRETING A SITING STUDY

The possible interpretations of a study vary depending on its purposes. The main caveat to keep in mind in interpretation is that the study examines the siting problem as described by the decision analysis model. This model, as does every model, deviates from the "real world." Thus, the study results must be informally combined with considerations not accounted for in the model to arrive at the implications. Especially when the overall implications differ from model implications, it is important to justify the reasoning.

The written report of the decision analysis of a siting problem provides documentation both for which site is chosen and for why that site is chosen. That is to say, it documents both the decision and the decision making process. Depending on the findings of the study, additional work may be necessary to complement the existing results. In fact, especially for the host of regulatory agencies, it is unlikely that the initial study will be sufficient to answer every question posed. It is therefore important that the model be robust and flexible so that it can quickly and inexpensively provide responses to legitimate "what if" questions which regulators, intervenors, or members of the public may raise. Questions such as "if low-sulfur coal is no longer available, what will be the health consequences of a proposed coal-fired power plant; and would it still be considered best with high-sulfur coal?" "if an oil tanker has a major spill next to the proposed refinery, what will be the environmental consequences?" and "if the values are altered to weight more heavily local community impacts relative to company economics, what happens to the site ranking?" This is a big advantage of decision analysis, since it is usually not possible to obtain insights into such questions with most of the procedures currently in use to select sites for energy facilities. A complete siting study is not only useful for its results, but is also useful as a device or tool to assist in meeting contingencies which will undoubtedly occur along the path to the realization of any major energy facility.

CHAPTER 9

EVALUATION OF PROPOSED
PUMPED STORAGE SITES

Chapters 4–8 discuss, in detail, the methodology and procedures for siting energy facilities using decision analysis. With this material as a background, we wish to go through the procedure for a particular case.† It should be possible to appraise the study based on a knowledge of the problem features and the methodology.

The case concerns identifying a suitable site for developing a pumped storage facility with a 600 MWe capacity in the southwestern states. Screening models and on-site visits by a team composed of engineers, geologists, and biologists were used to identify 10 candidate sites. These candidates were evaluated by economic, environmental, and technological concerns related to the sites and their required transmission lines. Using sensitivity analyses, factors critical to the ranking were identified.

The organization of this chapter is as follows. Section 9.1 summarizes the screening process used to identify the candidate sites. Section 9.2 discusses the objectives for indicating the desirability of the several sites. Section 9.3 provides a description of each site in terms of the degree to which the objectives are met. The value model which provides the method used to aggregate the various site costs and impacts into an overall measure of desirability is described in Section 9.4. Section 9.5 discusses the base ranking and the sensitivity analyses used to identify the

† This chapter is liberally adapted from Keeney [1979].

347

sites which are contenders. A summary and recommendations of the overall ranking evaluation process are presented. The final section offers an appraisal of the study.

Background for the Problem

Because of the large difference between peak and average electricity use in most sections of the country, utility companies are interested in identifying feasible means of meeting the peak demand without building unnecessary baseload generating units. One such means involves pumped storage. A pumped storage facility consists of two nearby water reservoirs (or other large water sources) with a difference in elevation of several hundred to a few thousand feet. When baseload energy is available, water is pumped to the upper reservoir. Then, when peak load energy is needed, water flows to the lower reservoir, generating electricity.

As part of their ongoing system planning studies, a large utility company† in the Southwest (UCS) determined that pumped storage hydroelectric generation may be a viable alternative for satisfying its peak capacity needs for the mid-1980s. The proposed facility would have a capacity of 600 MWe operating on a weekly cycle with daily generation of 9–11 hours. UCS selected Woodward-Clyde Consultants (WCC) to identify, select, and rank an inventory of potential pumped storage hydroelectric sites. The overall study, of which the work reported here was a part, was directed by Ashok S. Patwardhan. Other WCC staff contributing to the ranking were D. Beaver, J. Beley, S. James, P. Kilburn, D. McCrumb, K. Nair, D. Rabinowitz, and A. Sicherman. Significant contributions by several members of UCS staff were critical in providing data and preferences necessary to conduct the study.

9.1 Identifying Candidate Sites: The Screening Process

Using data from existing sources and information gathered from on-site visits by a team composed of engineers, geologists, and biologists, a series of screening models was developed to identify progressively smaller geographical areas having a high probability of containing sites suitable for the pumped storage facility. The primary screening resulted in

† For reasons of confidentiality, the name of the client is disguised in the study. However, all of the details of the study are accurately reported.

a series of candidate regions. These were identified by eliminating regions of special scenic, cultural, or aesthetic significance, such as national parks, Indian reservations, and sites near archaeological findings. Areas which could pose a major hazard to towns or villages because of possible dam failure were also eliminated.

The secondary screening of the candidate regions was conducted using hydraulic functional criteria. A proposed candidate area needed to have an effective head (vertical distance between the reservoirs) of more than 500 feet, the ratio of the horizontal to vertical distance between the reservoirs had to be less than 15, and the usable area for the upper reservoir had to be at least 300 acres. This screening resulted in a list of approximately 70 candidate areas.

Aerial reconnaissance and geologic conditions available in the literature were used to examine these areas. Extremely rough terrain or any evidence of landslides or unstable slopes were considered grounds for eliminating the area from further consideration. With the additional information of the aerial reconnaissance available, it was felt appropriate to retain only sites with a horizontal to vertical distance ratio of less than 10. Because there were approximately 20 such locations, the siting team believed the additional focus provided on these 20 sites would outweigh the disadvantage of discarding some other sites.

The 20 sites were visited on the ground by a siting team including geologists, biologists, and engineers. Their collective professional judgments were used to eliminate several sites. As a result, 10 candidate sites were identified.

9.2 Specifying Objectives and Attributes

As is the case with potential candidate sites, a series of screenings is also used to identify the main objectives which are important for differentiating the desirability of the candidate sites.

The first phase of identifying objectives requires that one specify general concerns which may be relevant to the choice. These general concerns, discussed in detail in Section 1.3, were then decomposed into the considerations of Table 9.1, which were reviewed on a site-by-site basis by several individuals familiar with the candidate sites. For example, one aspect of socioeconomic effects concerned the potential archaeological features at each site. Although it was recognized that it is possible to encounter undiscovered archaeological features in the course of a major construction project, it appeared that the likelihood of such a finding did

TABLE 9.1

A PRELIMINARY LIST OF GENERAL CONCERNS AND CONSIDERATIONS

General concerns	Considerations
Health and safety	Consequences of dam breach
	Impact on water quality due to reservoir development
Environmental effects	Terrestrial ecological impact
	Aquatic ecological impact
	Equalization of species composition between reservoirs
	Ecological impact from disposing of blowdown waters
	Transmission line impacts
Socioeconomic effects	Recreation potential
	Preemption of resources
	Archaeological features
	Land acquisition
	Sociopolitical system effects
Economics	Cost for adequate safety and operability
	System reliability
	Reserve capacity
Public attitudes	Public acceptance

not vary from site to site. Consequently, such a consideration was not relevant to site ranking.

As a result of these screenings, the considerations felt to be most appropriate for formal inclusion in the site evaluation process concerned environmental effects and system economics. Some considerations concerning legal and political questions as well as public acceptability were thought to be potentially relevant but, at this stage, the evaluation would not include them. Such considerations could either be included formally in the analysis at a later stage or dealt with separately. Consequently, the primary objectives chosen for the evaluation were:

1. minimize costs,
2. minimize detrimental transmission line impacts, and
3. minimize detrimental environmental impacts at sites.

To indicate the degree to which these objectives were achieved by each site, the attributes in Table 9.2 were specified. Attribute X_1, first-year costs, was defined by UCS to include both operating and capital costs. Because first-year capital cost is about 10 times the annual operating cost at every site, very little information is lost by aggregating these, so this was done.

The attribute X_2 was meant to capture the ecological and aesthetic impact associated with the construction and operation of the new transmission lines to each of the candidate sites. The costs of building and

TABLE 9.2

FINAL ATTRIBUTES FOR THE PUMPED STORAGE RANKING

Attribute	Measure	Range Best	Range Worst
$X_1 \equiv$ First-year cost	Millions of 1976 dollars	50	75
$X_2 \equiv$ Transmission line distance	Mile equivalents	0	800
$X_3 \equiv$ Pinyon–juniper forest	Acres	0	800
$X_4 \equiv$ Riparian community	Yards	0	2000

operating the lines are included in attribute X_1. The measure for the transmission line impact is mile equivalents. Line miles running through environmentally or aesthetically sensitive areas are weighted more heavily than line miles running through less sensitive areas. The calculation of mile equivalents is based on the constructed scale of Table 9.3. This type of scale, which was discussed in detail in Section 5.4, clearly indicates the value judgments necessarily made in combining two aspects such as aesthetic and environmental impacts. As can be seen in the table, the more desirable areas are weighted up to 10 times as much in calculating mile

TABLE 9.3

CONSTRUCTED SCALE FOR TRANSMISSION LINE MILE EQUIVALENTS

Raw mileage	Mile equivalents	Level of impact
1	1	Rural route traversing unpopulated rangeland; not visible from primary highways (3+ miles); does not affect any endangered species or unusual habitats; does not intrude on a "pristine" area.
1	1.1	Suburban route traversing sparsely populated area with minimal scenic intrusion; less than 1 mile of route parallel to and visible (within 3 miles) from a primary highway.
1	2	Urban route traversing populated areas; route traversing Bureau of Land Management or Indian owned lands; or aesthetic intrusion on primary highways (more than 1 mile parallel to and within 3 miles of highway).
1	2.5	Route traversing state or federal forest lands; wildlife management areas; ecologically sensitive wetlands.
1	5	Route traversing pristine undisturbed areas; or areas judged to be extraordinarily scenic.
1	10	Route traversing state or national parks; wildlife refuges; historic monument sites; or habitats containing unusual or unique communities or supporting endangered species.

equivalents. To help interpret the scale, note that one mile equivalent denotes one mile of transmission line through a region of very little ecological value or scenic beauty.

Site ecological impacts are indicated by the attributes X_3, pinyon–juniper forest, and X_4, riparian community. These measures refer to the acres of forest and yards of riparian community that will be lost if the pumped storage facility is built at the site.

Two other minor site environmental factors were included in the analysis. At one site, arroyo seeps (i.e., steep gulleys with water seepage which supports biological species) were present, and at another site the topography seemed particularly suitable as a nesting ground for raptors (hawks or eagles). These considerations were included by adjusting the acreage of the pinyon–juniper forest present.

The ranges in Table 9.2 were specified to include the actual ranges found for the set of all sites.

9.3   Describing Possible Site Impacts

Once the attributes were established, it was necessary to describe each of the 10 sites in terms of the attributes. The data given in Table 9.4 were provided by various staff groups within UCS, by an engineering firm retained by UCS to design and price preliminary plans for each site, and by WCC. The data developed from the design plans of the engineering firm and from site visits is referred to as the base data. There were uncertainties about the necessity of a lining at the upper reservoirs of three sites and about the ecological uniqueness of the riparian community of another site. Because these uncertainties could be resolved by site-specific studies, alternative data were specified for four sites in Table 9.4.

9.3.1   COSTS

The cost data included both capital costs and operating costs. Specifically, the measure for attribute X_1 was first-year costs which were defined by UCS as 18% of capital costs plus annual operating costs. The 18% is based on the financial structure of UCS.

The capital costs included direct construction costs and indirect costs for engineering design, construction management, overhead, surveying and licensing, and contingencies. The indirect costs were defined as 33% of direct costs. The direct costs were provided by the engineering firm, based partially on geological data from WCC and performance criteria

TABLE 9.4

DATA FOR EVALUATING UCS PUMPED STORAGE CANDIDATE SITES

Candidate site	First-year cost (millions $)	Transmission line (mile equivalents)	Pinyon– Juniper (acres)	Riparian community (yards)
		Base data		
S1	56.01	97.8	230	0
S2	59.18	140.0	150	0
S3	61.48	163.0	0	0
S4	59.68	342.3	0	0
S5	64.47	91.0	270	0
S6	61.36	152.7	721[a]	2000
S7	58.23	681.0	0	0
S8	59.92	704.0	240	0
S9	49.71	84.2	260	1900[b]
S10	75.42	392.7	419[c]	1600
		Alternative data		
S9*	49.71	84.2	260	1900[d]
S6*	51.64[e]	152.7	721[a]	2000
S8*	52.98[e]	704.0	240	0
S7*	65.19[f]	681.0	0	0

[a] Includes addition of 350 acres for damage to arroyo seeps.
[b] Assumes this 1900 yards is a unique riparian community.
[c] Includes addition of 200 acres for impact on raptors.
[d] Assumes this 1900 yards is a "normal" riparian community.
[e] Assumes upper reservoir is not completely lined.
[f] Assumes upper reservoir is completely lined.

from UCS. The cost of acquiring land, building transmission lines, and an allowance for funds during construction were provided by UCS.

The basic annual costs which differentiated sites were due to transmission losses, taxes, and water requirements. The former two were furnished by UCS and the latter by WCC. The personnel costs for operating the pumped storage facility were felt to be equal for all sites. Hence, they were not included in the analysis since they could not serve to differentiate between sites.

9.3.2 TRANSMISSION LINES

For each of the candidate sites, it was assumed by UCS that two 345 kV lines would need to be constructed from the plant site. One would end in the major load center of UCS and the other could interconnect with the existing bulk system in the most economical manner. Based on these assumptions, corridors for lines were overlaid on maps indicating the type of land being traversed. The miles running through the various categories defined in Table 9.3 were tabulated. Using the mile equivalent conversion factors, the data in Table 9.4 were generated. In this study, none of the proposed transmission lines went through areas weighted 10 mile equivalents per route mile. However, for nine of the candidate sites, 2–12 miles of route did traverse areas weighted 5.

9.3.3   SITE ECOLOGICAL IMPACTS

The site ecological impacts of importance were defined by WCC biologists who had visited all the candidate sites.† The major adverse impacts that could result were primarily due to (1) potential loss of riparian habitat, which is uncommon and valuable in the Southwest, and (2) loss of open pinyon–juniper forest which is relatively good wildlife habitat, primarily for deer.

Pinyon–juniper forest was confined to upper reservoir sites. It varied considerably in density, but the sites could be categorized as high-density (5–25% pinyon–juniper coverage) or low-density (less than 1% coverage). Low-density sites had little ecological value, so they were neglected in measuring impacts. For those sites categorized as high-density forest, the actual acreage of forest destroyed was taken from the design plans for the upper reservoir site.

Similarly, WCC biologists had identified streams with riparian communities in the vicinity of candidate sites. Once the design plans were available, the length of the stream covered by two reservoirs and the length of stream diverted in a third case were measured from maps.

As seen in Table 9.4, site S6 had arroyo seeps and site S10 was thought to be particularly suitable as a raptor habitat. The WCC biologists felt that the ecological values of these were equivalent to 350 and 200 acres of pinyon–juniper forest, respectively. These numbers were added to actual pinyon–juniper destroyed at the two sites.

9.4   Evaluating Site Impacts

Once the data in Table 9.4 had been obtained, they were evaluated to rank the candidate sites and indicate how much better one was than another.

† One site had only been seen from the air. The other nine had all been visited on the ground.

This step required a model of preferences, as discussed in Chapter 7, which aggregates all the possible impacts into a single indicator. Such a model explicitly addresses the value tradeoffs, which define the desirability of specific changes in one attribute relative to specific changes in another, and the attitudes toward risk. This requires that one assess the preferences of the decision maker. In the following, requisite input information about preferences was provided by UCS personnel, except in one case which is noted. In this case, because UCS biologists had not visited one site, they preliminarily accepted judgments of WCC biologists who had seen the site.

The value judgments and preferences of several UCS personnel were used in calibrating the preference model. In general, there was reasonable agreement among the individuals. However, no effort to reach a consensus was attempted. Because of the basic agreement, we chose to take an average in cases of disagreement to provide a basc model of preferences. Variations of this base model, representing individual differences in preferences, were considered in the sensitivity analyses.

9.4.1 ASSUMPTIONS FOR THE GENERAL PREFERENCE STRUCTURE

In the course of assessing the utility function, several assumptions were verified which implied that the general preference structure could be represented by the utility function

$$u(x_1, x_2, x_3, x_4) = \frac{1}{k} \left(\prod_{i=1}^{4} [1 + kk_i u_i(x_i)] - 1 \right), \tag{9.1}$$

where u and the u_i, $i - 1, 2, 3, 4$ are utility functions scaled from zero to one, the k_i, $i = 1, 2, 3, 4$ are scaling factors between zero and one, and k is another scaling factor calculated from the k_i. The conditions implying (9.1) are preferential independence, utility independence, and the absence of additive independence (see Section 7.2).

In examining value tradeoffs between attributes X_1 and X_2, it was found that they did not depend on the levels of X_3 and X_4. This implied that the pair $\{X_1, X_2\}$ was preferentially independent of $\{X_3, X_4\}$. Similarly, the pairs $\{X_2, X_4\}$ and $\{X_3, X_4\}$ were preferentially independent of their complements. Specific indifference points are illustrated in Fig. 9.1.

In assessing the utility function for costs, we found a first-year cost of $68.75 million for sure was indifferent to a risky option yielding either $50 million first-year cost with a one-half chance or $75 million first-year

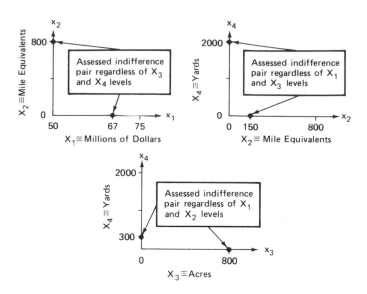

Fig. 9.1.   Invariant value tradeoffs indicating preferential independence.

cost. This was found to be the case regardless of the levels of the other attributes as long as they were the same for all possibilities. Hence, we could assume that the cost attribute X_1 was utility independent of the other three attributes. This condition is illustrated in Fig. 9.2a.

Finally, we considered the choice between the two risky options A and B in Fig. 9.2b. Option A, for instance, yields either cost equal to $50 million and a 100 mile equivalent transmission line or a cost of $75 million with an 800 mile equivalent transmission line, each with a probability of one-half. Option B also yields either a $50 or $75 million cost and a 100 or 800 mile equivalent transmission line. However, in option B, these possibilities are combined differently. Since option B was preferred to option A, additive independence was not appropriate. The reason for preferring option B was to avoid the risk of *both* costs and transmission line impacts being very high.

9.4.2   UTILITY FUNCTIONS

The individual utility functions u_i, $i = 1, 2, 3, 4$, were assessed with indifference information, as illustrated in Fig. 9.2a. Since u_1 is scaled from zero to one and since $x_1 = \$50$ million is the best cost and $x_1 = \$75$ million is the worst, the origin and unit of measure of u_1 are set by

$$u_1(50) = 1, \qquad u_1(75) = 0. \tag{9.2}$$

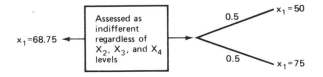

(a) Invariant Indifference Indicates X_1 Utility Independent of $\{X_2, X_3, X_4\}$

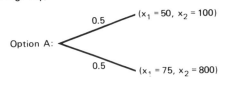

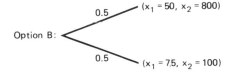

(b) A Preference for Option B Violates Additive Independence

Fig. 9.2. Examining potential independence conditions.

Then since $68.75 million was assessed as indifferent to the risky option in Fig. 9.2a, we set

$$u_1(68.75) = 0.5u_1(50) + 0.5u_1(75) = 0.5. \qquad (9.3)$$

For the sensitivity analysis, we asked for bounds on the indifference level 68.75 and found that the risky option would certainly be preferred to a sure $70 million and that $65 million would surely be preferred to the risky option. These values bound the true indifference point assumed to be $68.75 million in the base preference model.

Next we checked to see if, in fact, preferences for costs were always risk averse. That is the case when any sure option is preferred to any risky option which has an average impact equal to the sure option. In other words, a sure 60 would be preferred to a one-half chance at either 55 or 65 (which has an average impact of 60) by a risk averse individual. It did seem appropriate to assume risk aversion. The information in (9.2) and (9.3) is plotted in Fig. 9.3a and the exponential utility function fitted to the three points. Such a utility function is consistent with a risk averse attitude.

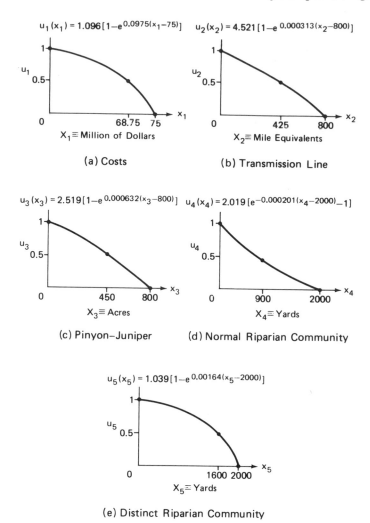

Fig. 9.3.   The assessed single-attribute utility functions.

The other utility functions u_2, u_3, and u_4 illustrated in Fig. 9.3 were assessed in a similar way. The utility function u_5 in Fig. 9.3e, unlike the others which are based on preferences of UCS personnel, is based on the professional judgment of a WCC biologist. The measure represents the amount of riparian community at site S9 which would be destroyed. It is based partially on the observation that the existing community is only 2000 yards long. The strong risk aversion results from the judgment that this riparian community is quite distinct for the Southwest and much different from "normal" riparian communities.

9.4.3 SCALING FACTORS

The first step in calibrating the scaling factors k_1, k_2, k_3, and k_4 in the model (9.1) was to determine their relative magnitude. Referring to Table 9.2, we asked the UCS siting group which attribute they would prefer to move from its worst to best level, given all attributes were at their worst levels. The response was X_1, indicating that the scaling factor k_1 for costs must be the largest. The attribute which they would next prefer to move from worst to best was X_2, indicating that k_2 was second largest. Then k_4 and k_3 followed.

To obtain specific values for the scaling factors, the two largest were assessed as follows for the base preference model. The probabilities p and q in Fig. 9.4 were varied until the compared options were indifferent. This resulted in the indifference probabilities of $p = 0.75$ and $q = 0.4$. Equating expected utilities of these indifferent options with X_3 and X_4 at their least desirable levels using (9.1) yields

$$k_1 = 0.75(k_1 + k_2 + kk_1k_2) \qquad (9.4)$$

and

$$k_2 = 0.4(k_1 + k_2 + kk_1k_2). \qquad (9.5)$$

These equations are consistent with the first value tradeoff in Fig. 9.1. The other two value tradeoffs in Fig. 9.1 imply

$$k_2 = k_4 + k_2u_2(150) + kk_4k_2u_2(150) \qquad (9.6)$$

and

$$k_4 = k_3 + k_4u_4(300) + kk_3k_4u_4(300). \qquad (9.7)$$

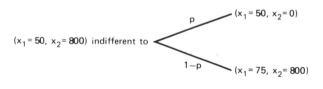

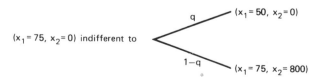

Fig. 9.4. Information assessed to evaluate scaling factors.
Indifference probabilities $p = 0.75$ and $q = 0.4$ were assessed for these two situations.

And evaluating (9.1) with all attributes at their best levels, we find

$$1 + k = \prod_{i=1}^{4} (1 + k k_i). \tag{9.8}$$

Since u_2 and u_4 in (9.6) and (9.7) can be directly evaluated using the functional forms in Fig. 9.3, Eqs. (9.4)–(9.8) were solved to yield

$$k_1 = 0.716, \qquad k_2 = 0.382, \qquad k_3 = 0.014,$$

$$k_4 = 0.077, \qquad k = -0.534. \tag{9.9}$$

The basic utility function used in this problem is defined by (9.1) with the individual utility functions given in Fig. 9.3 and the parameters in (9.9). For site S9, the utility function u_5 must be used in place of u_4. However, the scaling must be adjusted. The WCC biologist felt a loss of 500 yards of the S9 riparian community would be equivalent to a loss of 750 yards of normal riparian community. Also, since a loss of 0 yards for either is equivalent, we define b and c by

$$u_4(0) = b + c u_5(0) \tag{9.10}$$

and

$$u_4(750) = b + c u_5(500), \tag{9.11}$$

so that using the utility functions in Fig. 9.3, we find $b = -7.52$ and $c = 8.52$. For the S9 site, we must therefore substitute $-7.52 + 8.52 u_5$ for u_4 in (9.1) for evaluation.

9.5   Analyzing and Comparing the Candidate Sites

Using the preference model (9.1), the base data of Table 9.4 were evaluated. The results are presented in Table 9.5. The candidate sites are ranked from best to worst by the utility calculated using (9.1). To simplify interpretation of these results, we have computed an equivalent first-year cost for each site. From Table 9.5, one can see that the equivalent first-year cost for site S1 is $58.7 million. A site with this $58.7 million cost, but with absolutely no detrimental effects of transmission lines and no impact on pinyon–juniper or riparian communities, would be indifferent to the actual S1 site. In other words, using the preference model to calibrate, we would pay the difference between the forecast cost of S1, which is $56.01 million in Table 9.4, and $58.7 million to eliminate all the potential transmission line and site ecological impacts.

TABLE 9.5

EVALUATION OF UCS PUMPED STORAGE CANDIDATE SITES

Alternative	Rank	Utility	Equivalent first-year cost[a] (millions $)
		Base data	
S1	1	0.931	58.7
S2	2	0.885	62.0
S3	3	0.846	64.1
S4	4	0.820	65.3
S5	5	0.809	65.8
S6	6	0.799	66.2
S7	7	0.732	68.6
S8	8	0.697	69.7
S9	9	0.694	69.8
S10	10	0.196	78.7
		Alternative data	
S9*		0.941	57.8
S6*		0.905	60.7
S8*		0.780	66.9
S7*		0.596	72.2

[a] A site with this equivalent first-year cost and no transmission line, pinyon–juniper, or riparian community impacts would be found indifferent to the alternative on the respective row. This figure essentially aggregates all impacts into a cost figure defined as "equivalent first-year cost."

Site S2 is ranked second in the base evaluation with an equivalent first-year cost of $62.0 million. Currently site S3 is third, but it is $5.4 million more expensive in equivalent first-year cost and almost dominated by S2. Candidate site S4, which is fourth in the base evaluation, is $6.6 million more expensive than S1 in first-year equivalent cost. Sites S5 and S7 are almost dominated by S1. Except for site S9, which is best on cost and transmission line impacts, the other three alternatives are all dominated. This means they are worse on each of the attributes relative to some other candidate site.

9.5.1 SENSITIVITY ANALYSES

Several sensitivity analyses were conducted to identify critical inputs to the study which may affect the ranking of the sites. Specifically, we examined the implications of changes in the uniqueness of the riparian community at S9, uncertainties, the value tradeoffs, risk attitudes, and the problem definition. In the sensitivity analysis, we focused only on those sites which were not dominated by others and, hence, were contenders for being appropriate for site-specific studies.

Ecological Uniqueness

One of the site impacts critical to the overall evaluation concerned the riparian community at S9. As seen in Table 9.5, this site was ninth in the base evaluation, due mainly to the large consideration being given to the riparian community. Destruction of 2000 yards of riparian community at S9, which is essentially its entire length, was valued approximately eight times as much as the destruction of 2000 yards of "normal" riparian community. This was justified in the opinion of the WCC biologists, who had visited the sites, because the riparian community at S9 is thought to be unique for the Southwest.

If one treats the riparian community at S9 as "normal," then the site is best, with an equivalent first-year cost of $57.8 million. This case is illustrated by S9* in Tables 9.4 and 9.5. As one begins to weight the riparian community more than normal, the site progressively appears worse. When the weighting factor is approximately 1.25 times normal, S9 and S1 are equally desirable. Evaluation, with weighting factors of approximately 2.5 and 3.3 times "normal" will make S9 indifferent to sites S2 and S3, respectively. Hence, if the riparian community at S9 is roughly 3.0 times as valuable as the normal riparian community, then S9 no longer appears to be a contender.

Uncertainties

Some of the major uncertainties of this study concern the geological conditions at the sites, which can have a major impact on the costs. For instance, site S6 ranked sixth in the base evaluation using the cost data provided by the engineering firm, which assumed that the upper reservoir would be completely lined at a cost of approximately $48 million. Based on site visits by WCC geologists, it was thought that the complete lining might not be needed. Assuming a $35 million savings on the lining, this alternative, designated as S6*, would rank second in the base ranking with an equivalent first-year cost of $60.7 million.

Additional analysis of uncertainties found that, while there were major geological uncertainties at sites, it was not likely that the ranking would be significantly impacted until field verification tests were conducted at contending sites. It was determined that if current uncertainties were included in the base evaluation, then S1 and S9* would decrease in equivalent first-year cost by approximately $0.4 and $0.1 million, respectively, and that S2 and S4 would increase slightly.

Because water costs and land costs were such a small part of overall costs, uncertainties in these figures could not have a major impact on the overall evaluation.† At the major contending sites (S1, S6, S2, and S9), transmission line construction costs contribute about 8% to first-year cost. At S4, transmission costs contribute approximately 16%. Hence, it is unlikely that transmission line cost uncertainties would significantly affect the current ranking.

To examine the possible implications of a systematic underestimating or overestimating of costs, we evaluated all the sites using (9.1) with costs equal to 90% and then 115% of those in Table 9.4. The results are given for all the possible contending sites in Table 9.6 using equivalent first-year costs. Although the ranking of sites does vary slightly, there are no major changes.

Value Tradeoffs

The relative weights k_1, k_2, k_3, and k_4 in the preference model (9.1) were varied over reasonable ranges and this did not seem to have much impact on the ranking. First we scaled the weights, which now sum to 1.19, to sum to 0.8, 1.0, and 1.5 and found that the four best sites according to the base evaluation, S9*, S1, S6*, and S2, remained the four highest ranked sites in all cases. Allowing the relative weight of cost to be reduced to that of the transmission line impact also did not have a major influence on ranking. If the weight of cost is four times the transmission line weight, there is still no change in implication. These results shown in Table 9.6 account for the views of different UCS personnel as indicated by assessing different value tradeoffs. However, for the range of preferences exhibited, the implications for evaluation are the same.

Risk Attitudes

The impact of less risk averse preferences was examined. In particular, it was first assumed that a sure $65 million first-year cost was indifferent to a risky option yielding either a $75 or $50 million first-year cost, each with a probability of one-half. In this case, the equivalent first-year cost of S9* improved to $53.8, site S6* moved to second with $57.0 equivalent first-year cost, S1 had $57.7, and S2 had $61.4 million.

---

† There may, of course, be legal difficulties or delays in obtaining rights to the water needed. After discussions with UCS, it was decided to exclude this factor from the formal analysis. It should be considered in addition to the formal evaluation.

TABLE 9.6

IMPLICATIONS OF THE SENSITIVITY ANALYSES

Equivalent first-year costs (in millions of 1976 dollars) under various conditions[a]

Site	Base evaluation	90% costs	115% costs	$\Sigma k_i = 0.8$	$\Sigma k_i = 1.0$	$\Sigma k_i = 1.5$	$k_1 = k_2 = 0.51$	$k_1 = 4k_2$	Less risk averse	Risk neutral
S9*	(1) 57.8	(2) 55.6	(1) 62.1	(2) 62.0	(2) 59.9	(1) 53.2	(2) 63.6	(1) 56.5	(1) 53.8	(1) 52.5
S1	(2) 58.7	(1) 54.5	(3) 65.8	(1) 59.7	(1) 59.2	(3) 57.5	(1) 61.4	(2) 57.3	(3) 57.7	(3) 57.4
S6*	(3) 60.7	(4) 58.6	(2) 64.9	(4) 64.5	(4) 62.7	(2) 56.1	(4) 66.7	(3) 58.8	(2) 57.0	(2) 55.7
S2	(4) 62.0	(3) 57.6	(4) 69.5	(3) 62.8	(3) 62.4	(4) 61.0	(3) 64.7	(4) 60.5	(4) 61.4	(4) 61.2
S3	(5) 64.1	(5) 59.4	(6) 72.1	(5) 64.7	(5) 64.4	(5) 63.4	(4) 66.7	(6) 62.7	(5) 63.9	(5) 63.9
S4	(6) 65.3	(7) 62.0	(5) 71.8	(7) 66.7	(7) 66.1	(6) 63.6	(7) 69.8	(5) 63.4	(6) 64.7	(6) 64.8
S5	(7) 65.8	(6) 60.2	(7) 74.8	(6) 66.0	(6) 65.9	(7) 65.5	(6) 67.2	(7) 65.1	(7) 65.9	(7) 66.0

[a] The ranking of the possible contending sites is given in parentheses for each of the cases examined.

If one goes all the way to a risk neutral attitude where \$62.5 million is indifferent to a one-half chance at \$50 of \$75 million, the top four remain as in the preceding paragraph. Thus, as seen in Table 9.6, the risk attitude has essentially no impact on the overall evaluation.

Problem Definition

For the base evaluation, UCS had assumed that all the impacts and costs of new transmission lines would be allocated to the pumped storage project. It may be that some of these lines would be put to several uses, and consequently only a proportion of the line costs and impacts should be allocated to the pumped storage siting. The implications of this were examined.

For either site S8 or S7* to become a contender, over 50% of the transmission line impacts must be allocated to other projects.† For S8 or S7* to be equivalent to site S2, for example, the percents are 72 and 84%, respectively.

If 32% of transmission line impacts are allocated away from this project, site S4 becomes as desirable as S2; and if 62% is allocated away, it is as desirable as S9*. This is assuming, however, that all the transmission line impacts are still attributed to sites other than S4.

9.5.2 SUMMARY

As a result of this study, it was suggested that more detailed site-specific studies be conducted at sites S1, S6, and S9. These studies should especially focus on appraising the environmental value of the riparian community at site S9 and the need for lining the upper reservoir at site S6. It was also suggested that the uncertainties affecting the costs of those three sites be identified, quantified, and formally incorporated into the ensuing evaluation of the three sites.

The base evaluation and sensitivity analyses were done to differentiate the desirability of candidate sites using data from several sources. The evaluation is obviously contingent on that data. The ranking provided by this evaluation is intended to be one of the major inputs to the decision process. Some other factors important to the overall ranking of sites were explicitly not considered here. Such factors include legal considerations in acquiring land and water rights, political factors, and the interests of groups living in the vicinity of proposed sites. This overall decision process is currently underway.

---

† Sites S8 and S7* were used (instead of S8* and S7) since they assume the upper reservoirs are lined and the engineering firm and WCC feel this is the most likely case.

9.6   Appraisal of the Analysis

Although, as stated throughout the book, the process of identifying sites for energy facilities cannot be divided into clear-cut steps to be followed in order, the discussion of the process can be categorized into the steps listed in Sections 9.1–9.5. These are

1. identifying candidate sites,
2. specifying objectives and attributes,
3. describing possible site impacts,
4. evaluating site impacts, and
5. analyzing and comparing the candidate sites.

For this siting study, I feel that the weakest step was describing possible site impacts, mainly because the uncertainties associated with possible construction costs were simply not considered, yet these could be substantial. The strongest parts of the study were steps 4 and 5. The sensitivity analysis in step 5 provided some rationale for neglecting uncertainties in step 3. These steps will be considered in sequence.

9.6.1   IDENTIFYING CANDIDATE SITES

As with almost all screenings, the major shortcoming in this aspect of the study was that the screening criteria were considered separately. For instance, both the criteria of more than 500 feet of vertical head and less than 10 to 1 ratio of the horizontal to vertical separation of the reservoirs are important because of the economic implications these have for the pumped storage facility. Yet there was no attempt to relate these two criteria to the facility cost and screen on the combined criterion of cost. As discussed in Chapter 4, such a procedure is more accurate in identifying good sites and provides for a meaningful way to examine the appropriateness of the screening assumptions in sensitivity analyses once the site evaluation is completed with cost as an attribute.

9.6.2   SPECIFYING OBJECTIVES AND ATTRIBUTES

There are perhaps two omissions with respect to the objectives. First, there is no attribute concerned with health and safety. In this regard, the main concern is with the danger posed by a possible dam breach. Because such a possibility is very small and all the alternative sites are located away from populated areas, the difference in possible consequences between sites was considered insignificant relative to costs and transmission line impact. However, because of the obvious interest of the public, regulatory bodies, and the client in safety, it might have been better to include explicitly an attribute capturing this concern.

The second omission concerns the impact on local communities. This would include individuals in the vicinity of the proposed facilities and individuals owning land necessary for the reservoirs or transmission lines. Clearly, some of these aspects are included in the existing attributes such as land costs and aesthetic impact of the transmission line. Still, there might have been virtue in making this concern into a separate attribute.

9.6.3 DESCRIBING POSSIBLE SITE IMPACTS

Because of the nature of the problem and the attributes, the assumption that the aesthetic and environmental impacts of the facility and transmission lines are known with certainty seems very reasonable. However, the assumption that facility costs are known with certainty seems inappropriate. There are major uncertainties about the geology of each site. This can greatly influence construction costs. The source of the water required for the sites is not known and the costs of obtaining it could vary greatly. Furthermore, transmission line impacts are subject to major uncertainties. Any of these uncertainties could have a great impact on the evaluation of sites.

The value of the study could have been improved had these uncertainties been included. The existing UCS cost model could still have been used. The only difference, for instance, would have been that various possible descriptions of the geological factors would have been used with their associated probabilities. These could have been specified using the existing geological data, the information collected from site visits, and the professional judgments of the geologists. The effect of these different conditions on facility cost would then need to be identified by the engineering design firm. These results would be input to the UCS cost model. The output would provide a possible first-year cost (i.e., attribute X_1) with a probability corresponding to that set of geological conditions.

9.6.4 EVALUATING SITE IMPACTS

In general, this step was carried out well. The assessments of the component utility functions were conducted with the individuals at UCS most familiar with the attributes concerned. The value tradeoffs were conducted with the group responsible for the siting study at UCS. The aspect of the value structure which should probably be improved concerns preferences over time for the costs. The current UCS cost model uses discounting at a fixed rate. Because of fluctuating interest rates, inflation rates, and uncertainties in the financial markets, it may be better to address explicitly the value structure over time using some of the concepts discussed in Sections 7.2 and 7.3.

9.6.5 ANALYZING AND COMPARING THE CANDIDATE SITES

The main weakness in this part of the study was that some assumptions made in steps 1 and 2 were not examined in the sensitivity analysis. Specifically, the screening assumptions and judgments concerning the elimination of possible attributes (i.e., with respect to safety) were not verified for appropriateness after the completion of the study. The reason for this omission was that some of the necessary information for such sensitivity analyses was not available. With a little more effort and time, such information could have been generated.

This study was presented to the UCS in late 1976. As a result, percolation tests were begun to determine the degree of lining required at sites S6 and S9. This knowledge would considerably reduce uncertainties in the facility construction costs. A thorough appraisal of the riparian community at site S9 did indicate that it was indeed unique. As a result, a more thorough ground investigation in the immediate region of site S9 suggested an alternative site slightly removed from the original location. The characteristics of this new site were such that it could be accurately referred to as S9*. Because S9* would not harm the riparian community, it was selected for the prime site. Detailed investigations concerning funding, acquisition of the land and water required, and legal issues related to the site were then conducted.

The following news item referring to site S9* appeared in the *IEEE Spectrum* of September 1979:

> A large pumped storage power project is planned in a desert region 45 miles west of Albuquerque, New Mexico. The project calls for a 600-MWe generating station that will use reversible turbines and two water reservoirs at 3000 and 1950 meters above sea level. Pumps will raise the water to the upper reservoir during off-peak times, and during the hours of peak demand water from the upper reservoir will drive the turbines. The Stone & Webster Engineering Corporation in Denver is the designer of the project. Geotechnical studies to determine the location of the first 154-MWe reversible-turbine unit are under way. The first unit is scheduled to become operational in 1989. The water for the pumping operation will come from an underground source in a nearby uranium mine.

CHAPTER 10

SITING AND RELATED PROBLEMS

In the beginning of this book, we carefully outlined what is referred to as "the siting problem." This served to exclude a host of related problems which are extremely important in the energy field. The excluded problems might be categorized as follows:

licensability of energy facilities,
design of energy facilities,
selection of an energy source,
capacity of an energy facility,
timing the introduction of a new energy facility,
sequencing of energy facilities, and
setting standards for energy facilities.

The siting problem is intertwined in various ways with each of these problems. The purpose of this chapter is to discuss the relationships and the manner in which siting analyses, as described in this book, are useful in aiding decision making with regard to each of these related problems.

It is probably safe to conclude that no analysis ever addresses every aspect of a complex problem. This is certainly true of the analyses for siting facilities described in earlier chapters. The purpose of the analysis is to model several crucial features and, through the analysis, gain insights into the overall implications of the choice of each alternative site. With the additional information and insights provided by the analysis, the client should be better prepared to make a responsible and justifiable decision, one which he or she feels comfortable with and can defend to others. It is in this same sense that decision analysis of siting alternatives is relevant to the related problems discussed in the chapter. As will become apparent, the siting analysis is relevant in two senses. First, siting is a component part of each of the related problems. For instance, questions about licensability, design, energy source, and so on, may be very dependent on the site to be chosen. Second, each of these related problems can be characterized by the same features as siting problems (e.g., multiple objectives, uncertainties, impacts over long time horizons), as outlined in Section 1.4. Since the decision analysis methodology described in this book addresses these features head-on, the same type of analysis is appropriate for the related problems.

10.1 Licensability of Energy Facilities

Any energy facility requires several licenses for construction and operation. These licenses are issued by various agencies and organizations of the federal, state, and local governments. Licensability, as used here, refers both to whether a proposed facility will ultimately receive the necessary licenses and to what length of time is required for the process. Many individuals (e.g., Denton [1977], Murphy [1978], Nagel [1978], and Greenwood [1979]) in all phases of the energy industry have cited licensing delays and uncertainties about licensability, as critical components of energy siting problems in the United States. Delays have major implications for the economic consequences of any energy facility both because the facility will cost more in the future and because the productive capacity is not ready as soon as was planned. Construction delays of major power plants are estimated to cost in the neighborhood of one million dollars per day. These costs are necessarily passed on to the consumers (i.e., the public) either directly or indirectly.

Licensability is clearly intertwined with siting. A site judged to be best using a decision analysis as described in earlier chapters may be passed over in the actual site selection for a site judged to be slightly inferior in the analysis but which appeared to be much easier to license. However, interpreting whether or not this might be the case would require a careful appraisal of the analysis, since it is possible to implicitly include questions of licensability in the decision analysis procedure as outlined. The explanation is as follows.

Licenses are required to insure that appropriate sites are selected for energy facilities. It should be that, the better a proposed site is, the more likely it is that the necessary licenses will be obtained in the shortest possible time. If the values used in the siting study are chosen to be consistent with those of the license granting authorities, the ranking of sites should provide a ranking for licensability. Thus, as part of the sensitivity analysis of any siting study, it may be appropriate to develop a model of the regulators' values and appraise the alternatives from their viewpoint as well as others. Because different regulatory authorities have different values, it may help to repeat this process more than once with different values.

Given the information provided by the above sensitivity analysis, it will still be necessary to choose a site. Even if licensability were all that mattered, the choice might not be simple because it may be judged that different licenses would be easier to obtain for each site. It is possible to do an analysis of licensability to complement the basic siting study. This analysis would appraise the different strategies (e.g., which license to apply for first, which in conjunction with others) for proceeding through licensing. The results would indicate the probability that each site would obtain all the necessary licenses and a probability distribution of time this would take. These results would be used in conjunction with the siting study to select a proposed site. Greenwood [1979] suggests the necessity of such analyses for domestic pipeline projects.

10.2 Design of an Energy Facility

Design alternatives for the facilities at a particular site can readily be considered in the decision analysis framework for siting (see Tribus [1969]). Here, design alternatives means two distinct but similar facilities of the same size, at the same site, for the same ultimate purpose. For example, a 650 MWe coal-fired power plant may utilize coal of high or low sulfur content, water towers or once-through cooling, a 50 or 150 m emissions stack, scrubbers or no scrubbers. Each of these variations will affect the subsequent impacts of the particular facility. For example, once-through cooling would probably result in more environmental damage in the water source and also in a more economical facility. One could simply treat these two design alternatives as separate sites, with a great amount of similar data, in the analysis. However, because of this similarity of data, it may be that the best design at a particular location can be chosen by considering only a subset of the siting concerns. Crawford, Huntzinger and Kirkwood [1978] discuss such an analysis for a proposed 765 kV transmission line and Smith, Miles, and Goldsmith [1978] compare underground and aboveground nuclear power plants. After such an analysis, of course, only the best design at each site location should be compared.

It is appropriate to note that mitigation measures taken to avoid possible undesirable consequences are, in fact, just another design alternative. As an example, at one possible location for a power plant, it may be necessary to either excavate and transfer archaeological structures or build over or around them. Another example involved the inclusion or exclusion of caribou crossings along the Alaska pipeline. Recognized as design alternatives, mitigation measures can be treated exactly as any other design alternative in a siting study.

10.3   Selection of an Energy Source

Selection of an energy source means, for example, choosing coal or nuclear as the fuel for baseload electrical generation facilities. Thus the selection of an energy source is more relevant to the electrical utility industries than to other energy organizations, such as petroleum companies. The main distinction between the energy source problem and the design problem discussed in Section 10.2 is simply that the alternatives involve different fuels. This additional feature can be addressed in the decision analysis framework by expanding the analysis. Both Barrager et al. [1976] and Garribba and Ovi [1977] have conducted parts of such an analysis.

First, in addition to the characteristics examined in siting studies, one must carefully consider whether it is appropriate to exclude possible benefits of the electricity produced in the analysis. A 1000 MWe coal-fired plant and a 1000 MWe nuclear plant do not necessarily produce the same amount of electricity. This is partially due to possible interruptions of service for different lengths of time (see Lapides [1978]). This can be accounted for in an analysis by designing the alternatives to produce an equal output over time. Of course, even in this case, one must be careful to make sure that the uncertainties with regard to the output are also comparable for the possible energy sources.

Once the benefit aspect is accounted for, the main additional difficulty in the energy source problem concerns the fact that the impacts result from different means. For instance, the major health impacts of a coal plant may result from sulfur dioxide pollution, whereas those of a nuclear plant may result from radiation. If proxy attributes are used in the evaluation, the attributes for the coal siting problem and the nuclear siting problem would probably be different. If more fundamental attributes, such as number of deaths, are used in the problem, then the relationships between fatalities and both sulfur dioxide exposure and radiation must be specified. Supplying two dose–response relationships is obviously a more difficult task than supplying one, which would be the case when selecting a site for a single energy source.

Perhaps the most appropriate way to consider the energy source problem is as follows. First, an analysis to identify the best site for a facility using the first energy source would be conducted as discussed in this book. Next, a similar analysis would be conducted for the second energy source. Then the best sites for each of the energy sources would be compared. Because several of the possible sites for each of the sources would have been eliminated in the two siting analyses involving like sources, the energy source comparison may be reduced in complexity. It is much easier to compare two alternatives—the best sites with each source—than 20. However, an additional complexity in this latter comparison of energy sources may result from the inherent relationships between licensability and the energy source. Because of different processes and difficulties in licensing facilities of different energy sources, it may be necessary to combine the energy source selection with the licensability problem discussed in Section 10.2. Regardless of whether that is the case, the fact that siting analyses have been done for both of the energy sources can be a major step in selecting which source is appropriate.

10.4 Capacity of an Energy Facility

Let us once again consider only one energy source or one type of energy facility. However, let us introduce a new complication to the problem, the elimination of the requirement that the facilities have the same capacity. For instance, we may wish to consider whether an oil refinery which refines 0.1 or 0.2 million barrels daily is better or, in another situation, whether a 400 or a 650 MWe coal-fired power plant is better. In another situation, one may compare a large power plant to several smaller ones with the same total capacity. See Ford and Flaim [1979] for an analysis of one such case. Wyzga [1979] summarizes some current work on this problem.

With facilities of different size, it is definitely necessary to address the benefits of the various facilities in order to select the best option. With the introduction of benefits, one must also be concerned with the demand for the product. Of course, the negative impacts of each facility must also be included in the analysis, using procedures discussed thoroughly in the context of siting in this book.

As was the case with an energy source, it may be possible to greatly reduce the complexity of the capacity problem if separate siting analyses are first carried out for facilities of each size. For instance, with the oil refinery, one could first conduct an analysis to identify the best site for a facility to refine 0.1 million barrels per day. Then a similar analysis would be conducted to identify the best site for a 0.2 million barrel per day facility. The final comparison to help select the capacity would compare the best site of each of the two previous studies. This final comparison may be simplified for two reasons. First, it might be that the same site is identified as being the best for each size facility. This might imply that some of the environmental effects on the local habitat and communities might be the same for both options. If this were the case, such considerations would probably cancel out in the comparison of the two alternatives of different capacity. This would clearly limit the scope of analysis. Second, even if different sites are used, the same mechanism may occur. The best sites of each capacity may be such that there is essentially no impact on local communities or that those impacts are judged to be equivalent. Then of course, these considerations would probably again cancel out in the ensuing analysis of the capacity problem.

10.5 Timing the Introduction of a New Energy Facility

The essence of the timing problem is to balance the demand and supply for the product. The purpose is to find the optimal time for introduction of service. As a result, the timing problem is intertwined with the siting and licensability problems.

For some situations, it may be appropriate to separate siting from the licensability and timing problems. In this case one would conduct a siting study to identify the best site for a particular facility. Then one would conduct a supply and demand study for the project to identify the optimal time to initiate operation of the new facility. This analysis should include information about the possible penalties for early and late introduction. It really sets up the objective function to be used in the timing problem. Finally, one analyzes various procedures for conducting the licensing and construction processes of the new proposed facility to best meet the objectives of the timing problem.

It is perhaps interesting to note that this problem is essentially a more complex version of the licensing problem discussed in Section 10.2. In the original licensing problem, the objective was to license the facility as soon as possible to bring it on-line as soon as possible. In the timing problem, we have simply changed the objective. It may not be appropriate to bring the facility on-line as soon as possible, but rather to bring it on-line at a particular time. It still may be appropriate to obtain the licenses as soon as possible, and then schedule the construction process to best meet the overall objectives of the timing problem.

10.6 Sequencing of Energy Facilities

Unlike all of the previous problems which were designed to identify only one energy facility as best, the sequencing of energy facilities deals with multiple energy facilities. For simplicity of discussion, we shall define the sequencing problem to include where and when to locate a series of similar facilities. For instance, a utility may wish to locate six 600 MWe coal-fired power plants in the next 30 years in its general service area. This would be a sequencing problem as we have defined it. In reality, we would expect sequencing problems to also involve facilities which are similar in function but not similar in characteristics. For instance, typically an electric utility would be locating power plants of different capacities using different energy sources over a period of time (see Eagles *et al.* [1979]). Because we have discussed the problems of capacity and energy sources separately, the relevance of our siting work to the sequencing problem can be discussed in the context of similar facilities.

In the simplest possible case, this sequencing of facilities simply involves a series of timing problems for individual facilities as described in Section 10.6. This would be the case if the timing of the introduction of each new facility, as well as its site, did not depend in any way on the location of previously sited facilities and depended on their timing only with regard to how long it had been since they went into service. Although it may be the case that this assumption about timing dependence of future facilities is reasonable, the siting of facilities certainly depends, in most cases, on the previous sites. To see this in a simple context, suppose that the service area of an electrical utility has two major load centers. If existing plants are located such that they are much closer to the first load center, it would clearly be better to locate a new facility closer to the second load center, given that all other factors are equal. For this reason, if one conducts a siting study to identify the best site for a particular facility and then one builds the facility at that site, it is not appropriate to assume that the second best site in the original ranking is the best location for the next power plant to be built. The obvious interdependency must be addressed in the sequencing of the facilities problem.

As input to the sequencing problem, we may wish to conduct a series of siting studies to identify the best sites under a variety of conditions and, more importantly, to identify the impacts associated with each of a number of possible sites. These studies would be conducted exactly as described earlier in this book. The site evaluations of the environmental, socioeconomic, health and safety, and public attitude impacts should be included as input information into the overall sequencing problem. The alternatives in the sequencing problem involve a series of plants introduced at specific locations and times in the future. The main interaction in sequencing is likely to be the economic impacts, so a separate composite model utilizing the individual siting economic models may be appropriate.

One additional point about the sequencing problem is that the overall objectives for the sequencing of facilities may be very similar to the overall objectives for the siting of one facility. They certainly both relate to the same set of concerns. Although the sequencing problem is clearly of much larger scope than an individual siting problem, it is also just as obvious that the evaluations of each of the possible sites, as described in this book, are critical to the conduct of a successful sequencing analysis.

10.7   Setting Standards for Energy Facilities

The problems of identifying sites for energy facilities and specifying standards relevant to the construction and operation of those facilities are intrinsically interrelated. That is to say, the decisions with respect to one of these problems affects the alternatives and the decisions which are best for the other. To illustrate this, let us assume for our purposes that a standard is defined as a rule, regulation, or guideline which affects the selection of a site for an energy facility. Such standards may result from a law or from accepted professional practice. They may have been established by the legislative authorities, regulatory authorities, professional societies, or the client wishing to select a site. Finally, they may deal with the selection of sites, the measuring of impacts of those sites, or the evaluation of those impacts.

To indicate how standards affect siting and evaluation of sites, let us consider the identification of a site for a large nuclear power plant. Standards of the Nuclear Regulatory Commission require that any site be more than five miles from a capable fault and more than three miles from any center of population of more than 2500 people. Such standards obviously affect the possible sites which can be considered for the facility. Other standards affect what information must be measured to identify impacts and in some cases how it should be measured. Perhaps the best example of this is the National Environmental Policy Act which requires that a broad range of environmental considerations be included in evaluating alternative energy sites. Another example of a standard, in this case in the form of a guideline, for evaluating sites is the following. It is generally accepted by many individuals in the nuclear industry to consider the release of one rem of radioactivity equivalent to a cost of $1000. Such a standard defines the value tradeoff between person-rems and costs.

Standards also affect the design of a facility and, hence, the potential desirability of a particular site. For example, standards require that nuclear facilities be designed to withstand specified potential earthquake hazards in a local area. Sites in more seismically active areas must be constructed to withstand greater shaking due to an earthquake. This obviously affects the design of the specific facility and, therefore, the cost.

The siting of power plants also affects the standards set for their siting, although the relationship is a little less direct than the impacts of standards on siting. To illustrate this, let us consider the case of solar power plants. With the passing of time, knowledge about potentially harmful effects of solar power plants (or any type of energy facility) increases. Simultaneously, public values and priorities also change. This often results in concern for different factors now in comparison with ten years ago. One need only recognize the increased concern for environmental impacts over the last decade to realize that this process can occur rather rapidly. With this increased knowledge and changing values, we may realize that we are now displeased with some of the negative impacts of solar power plant sites. These feelings influence the political process and eventually lead to new or revised standards which will affect siting. These effects may manifest themselves in any of several ways such as described above. They may affect the alternatives available, the measuring of impacts, the evaluation of impacts, and so on.

Finally, it is worth noting that the concerns and complexities of the standard setting problem are the same as those of the siting problem. This case was recently stated clearly by von Winterfeldt [1978] as follows:

> "Regulators and scientists who have been involved in a standard setting problem will readily testify that the task is exceedingly complex and difficult. There usually exists a vast uncertainty about the affects of pollutants [or other impacts of siting] on human well-being. Crucial [value] tradeoffs have to be made among multiple objectives that are often conflicting like engineering, cost, environmental, and social objectives. Some of these objectives are hard to measure or even to express precisely. Conflicting interest groups are involved and affected by standards, each of them having their own experts to back up their case. Finally, the uneven distribution of effects of standards and regulations over time and people complicate the standard setting problem."†

As I hope has been demonstrated throughout this book, the decision anal-

---

† On second thought, maybe it was not stated so clearly, but at least the idea is conveyed.

ysis procedures described and illustrated here are designed to address these complexities directly and explicitly.

10.8   Concluding Remarks

The siting of a major energy facility is a very complex problem. It might simply be said that the complexity is due to multiples. There are multiple objectives to be considered, multiple time periods, and multiple groups impacted by any particular site. There are multiple interest groups, each with its own issues and concerns, who wish to impact the choice of a site. And there are multiple disciplines whose professional knowledge is relevant to the selection of a site. This complexity, along with the importance of the siting problem, renders it worthwhile and appropriate to conduct an analysis of sites to help identify a best site. The decision analysis approach described in this book provides a framework to address this complexity and to lend insights helpful in the decision making process. The analysis is prescriptive. That is, it indicates which site should be chosen to be consistent with the information available and the values of the client (i.e., the decision maker in our problem) and impacted and interested parties.

The analysis of siting alternatives is relevant not only to identifying a best site, but also to a host of related problems. However, no analysis addresses every aspect of the problem. Interpretation and insights of the analysis must be combined with other relevant information not formally analyzed to make the choice of a best alternative. Because of this, one should not be disappointed if he or she feels that the analysis procedures here do not address everything. In addition, it is likely that the client is only one member or one unit in the entire decision making process and that the decision analyzed is only a part of the overall decision to be made. These oversimplifications, as they may be referred to, still leave significant complexity to be addressed and provide an opportunity for gathering the insights which are so critical and useful in the overall decision making process.

Decision analysis for siting is problem oriented. The approach taken begins with the characteristics of the problem. The question then asked is how best to address these characteristics to gain as much from the analysis as possible. The answer to this question has led to the methodology described and illustrated throughout this book.

If one compares decision analysis to what one might think of as the "perfect analysis," decision analysis will have important defects. It is certainly not perfect, but the correct comparison is not to the perfect analysis. Rather, decision analysis should be compared with other procedures that are claimed to be useful in identifying sites for energy facilities. The comparison should not be on whether an answer results, because surely any procedure can identify a best site, but should involve the basic assumptions and the logic and rationality of these. It should also involve the procedures used and the difficulties in implementing them. Finally, the comparison should involve the quality and quantity of information and insights which one can gain from the analysis. With decision analysis, the logic is explicit and sound, the procedures are operational and thought provoking, and the insights are relevant and of broader scope than those provided by alternative procedures. These are the characteristics that the client, the regulatory authorities, and the public desire, demand, and deserve from analyses of energy siting problems.

THE COMPLETE SET OF ASSUMPTIONS FOR THE DECISION ANALYSIS APPROACH TO THE SITING OF ENERGY FACILITIES

The evaluation of sites for a proposed energy facility is clearly a complex task. Regardless of the procedure by which one attempts to accomplish this task, a number of assumptions will necessarily be made. The appropriateness of the procedure, and hence the appropriateness of the implications of the site evaluations, depends very strongly on the appropriateness of the assumptions. If the fundamental assumptions are logically unsound or unjustifiable, the analysis will unlikely provide either insight for selecting a good site or documentation that can withstand the scrutiny of the regulatory process.

The assumptions of the decision analysis siting approach are discussed in this appendix. No additional assumptions are required. If the client believes the assumptions are reasonable for the siting problem, then he or she or the organization must accept the procedures of decision analysis for evaluating the problem. Furthermore, if the client chooses a different procedure, one or more of the basic decision analysis assumptions will necessarily be violated.

In specifying any set of assumptions, there is always the question of where to begin and what to include. Typically, sets of assumptions for decision methodologies specify only what is to be done and how it can theoretically be done. We shall also include assumptions about why it should be done and how to do it in practice. In doing this, we hope to improve the ability of potential clients to appraise the appropriateness of decision analysis for their siting problems.

Section A.1 presents the tenets of decision analysis for siting. These can be thought of as the basic assumptions for conducting a decision analysis of siting. In Section A.2, the fundamental assumptions for what is done in the decision analysis approach and the operational assumptions for how it is done are discussed. The implications of accepting the assumptions are discussed in Section A.3. Section A.4 contains a perspective on the assumptions and their implications as well as an overall appraisal of the decision analysis approach to siting.

A.1 Basic Tenets of Decision Analysis

The following four tenets of siting decision analysis essentially define the siting problem. They are the assumptions for "why do a decision analysis of a siting problem."

Tenet 1. A client exists with an interest in a particular siting problem.

Tenet 2. The client believes that analysis of the siting problem can be useful for deciding which alternative site is the best for a particular energy facility and/or for documenting and justifying the site to regulatory agencies and the public.

Tenet 3. To be useful, the siting analysis should explicitly address the general concerns and complicating features of the siting problem.

Tenet 4. The attractiveness of potential sites should depend on both the likelihoods of the possible consequences of siting at each potential site and the client's preferences for those possible consequences.

Because of the general nature of these tenets, it is difficult to find fault with them. For instance, Tenet 1 simply states that the problem exists. Tenet 2 says that analysis can be useful in siting, not that it will necessarily be useful. A poor analysis might be of no use whatsoever, and we are not arguing that it would be. A premise of Tenet 3 is that the general concerns and features of siting decisions outlined in Sections 1.3 and 1.4 characterize the complexity of the problem. Since the analysis is to provide insight and documentation, the tenet simply states that the characteristics of the problem should be addressed in order to do this. In the jump to Tenet 4, there is the implication that the possible consequences of the alternatives and the client's preferences for those consequences address the general concerns and features important to the siting problem. This tenet is clearly more specific than the first three and therefore requires more elaboration. This is provided by the detailed fundamental assumptions of decision analysis discussed in Section A.2.

A.2   Fundamental Assumptions and Operational Assumptions of Siting Decision Analysis†

The fundamental assumptions of siting decision analysis are the foundations on which the entire approach rests. The operational assumptions specify a procedure for carrying out an analysis consistent with these fundamental assumptions. These operational assumptions are clearly important but they may vary over time as procedures to implement siting decision analysis are improved.

The assumptions discussed in this section will be arranged to correspond to the five steps of siting decision analysis discussed throughout this book. For reference these steps are the following:

1. identifying candidate sites,
2. specifying objectives and attributes of the siting study,
3. describing possible site impacts,
4. evaluating site impacts, and
5. analyzing and comparing candidate sites.

These will be discussed in sequence.

A.2.1   IDENTIFYING CANDIDATE SITES

In order for a siting problem to exist, there must be at least two candidate sites from which to choose. This is formalized as follows.

Assumption 1 (Identification of Alternatives).   The client can identify at least two candidate sites for the proposed facility.

This assumption is not particularly strong. As indicated throughout this book, the difficulty is usually not with finding at least two candidate sites, but rather the fact that a very large number of potential candidate sites exists. For time and cost considerations, it is not practical to analyze an extremely large number of candidate sites in great detail. Hence, operational assumptions are required to rapidly eliminate many of the potential candidate sites (one hopes, the poorer ones) from further consideration. To implement this we have the following operational assumptions.

Assumption 1a.   A region of interest can be defined in which to limit the search for candidate sites.

---

† The fundamental assumptions discussed in this section are only slightly modified from the assumptions of decision analysis found in Pratt, Raiffa, and Schlaifer [1964].

Assumption 1b. The region of interest can be efficiently screened to identify a manageable number of candidate sites.

The number of candidate sites which would be considered manageable would, of course, depend on the particular situation under consideration. For the siting of major energy facilities, a manageable number of sites would probably be less than 20.

A.2.2 SPECIFYING OBJECTIVES AND ATTRIBUTES OF THE SITING STUDY

The selection of any candidate site will eventually result in a consequence. This consequence, which will be multifaceted, is meant to capture the impacts of siting with regard to the general siting concerns discussed in Section 1.3 and the complicating features of siting studies discussed in Section 1.4. The following fundamental assumption is made with regard to site impacts.

Assumption 2 (Bounding of Consequences). The client can identify two meaningful consequences such that one is at least as attractive as any possible consequence resulting from a choice of any of the candidate sites and the other is at least as unattractive as any possible consequence resulting from the choice of any of the candidate sites.

We shall designate the more preferred of these two consequences as c^* and the less preferred as c^0. In order to help the client identify the impacts c^* and c^0, two operational assumptions are made concerning the objectives and attributes.

Assumption 2a. An objectives hierarchy can be generated to indicate the dimensions of potential consequences of interest to the client.

Assumption 2b. Attributes can be specified to indicate the degree to which the client's objectives are achieved.

At the top level of the objectives hierarchy are the general siting concerns. These are meant to be collectively exhaustive to include any possible impact of interest to the client. In practice, when attributes and their associated scales are developed for the lower-level objectives of the hierarchy, there is a chance that some of the impacts are not adequately characterized. This may be particularly apparent if quantitative numerical scales are required for each of the attributes. However, there is no need to rely only on numerical scales. Qualitative constructed scales are certainly adequate and, in fact, more appropriate in many situations. Such constructed scales are discussed in Chapter 5 and illustrated in the nuclear power siting study and the pumped storage siting study covered in Chapters 3 and 9, respectively.

A.2.3 DESCRIBING POSSIBLE SITE IMPACTS

There is a great deal of uncertainty about the eventual consequence that will result from the selection of a particular site. This uncertainty is quantified in decision analysis by probabilities.

Assumption 3 (Quantification of Judgment).   The client can specify the relative likelihoods (i.e., probabilities) for each possible consequence that could occur as the result of selecting each candidate site.

As discussed in detail in Chapter 6, there are a number of procedures to assist the client in specifying these probabilities. The probabilistic estimates are based on available data, information collected in site studies or from experiments, the output of analytical or simulation models, and the results of the assessment of experts' judgments. In order to state the operational assumption behind all of these procedures, we need to introduce the following concept. Define E_i as the occurrence of the consequence c_i as a result of selecting a particular candidate site.

Assumption 3a.   For any possible event E_i, the client can specify a probability $p(E_i)$ such that the following are indifferent: (1) an option that results in c^* if E_i occurs or c^0 if E_i does not occur and (2) an option yielding a probability $p(E_i)$ at c^* and a complementary chance $1 - p(E_i)$ at c^0.

It should be noted that the c^* and c^0 in Assumption 3a are chosen for convenience of presentation only. The essential requirement for Assumption 3a is that there be a difference in the desirability of the two consequences c^* and c^0. Since c^* and c^0 were clearly defined in conjunction with Assumption 2, they were used in Assumption 3a.

A.2.4 EVALUATING SITE IMPACTS

It is necessary to quantify the client's value structure for evaluating the candidate sites. This is provided for formally by the following assumption.

Assumption 4 (Quantification of Preferences).   The client can specify the relative desirability (i.e., utilities) for each possible consequence that could occur as the result of selecting each candidate site.

In order to make Assumption 4 operational, we need the following:

Assumption 4a.   For each possible consequence c_i, the client can specify a utility $u(c_i)$, such that the following are indifferent: (1) c_i for certain and (2) an option yielding a $u(c_i)$ chance at c^* and a complementary $1 - u(c_i)$ chance at c^0.

By using Assumption 4a, the utilities of the various possible consequences are scaled from zero to one, where $u(c^0) = 0$ and $u(c^*) = 1$. As discussed in Chapter 7, this utility scale can be transformed by taking any positive linear transformation and the resulting analysis will have the same implications.

In theory, one could attempt to directly assign a utility to each of the possible consequences. However, because of the large (infinite in the continuous case) number of possible consequences, systematic procedures which employ consistency criteria are important in the assessment of utilities. In Chapter 7, a larger number of conditions helpful for structuring the client's utility function are discussed. We do not make the assumption that any particular set of conditions is appropriate for a particular client in a particular case. However, it is helpful (and very likely if the attributes have been properly defined) that some of the conditions are appropriate in any particular problem. It is not that they must precisely characterize the client's value structure, but rather that the conditions are a reasonable representation of the client's preferences. Here reasonable is a term which is defined in the context of a particular problem. If the conditions represent the client's real value structure closely enough that the implications of the analysis are insightful, then we would conclude that the conditions are reasonable.

A.2.5 ANALYZING AND COMPARING CANDIDATE SITES

The preceding assumptions are necessary to analyze the siting problem using the full power of decision analysis. Actually conducting the analysis requires three further assumptions. These assumptions provide the consistency criteria of the decision analysis siting approach and the criterion for comparing alternative sites.

Assumption 5 (Comparison of Options).   If two options could each result in the same two possible consequences, the option yielding the higher chance of the preferred consequence is preferred.

Assumption 6 (Transitivity of Preferences).   If a first option is preferred to a second option and the second option is preferred to a third option, then the first option is preferred to the third option.

Assumption 7 (Substitution of Consequences).   If an option is modified by replacing some of its consequences with another set of consequences indifferent to those being replaced, then the client is indifferent between the original and modified options.

Assumption 5 is necessary to indicate how various candidate sites should be compared. Assumptions 6 and 7 are often referred to as consistency assumptions which allow one to reduce very complex options, such as that describing the variety of possible consequences of choosing a particular site, to a simple option of the form referred to in Assumption 5. Once it is in this form, then it is clearly easy to compare with others.

In order to discuss these assumptions operationally, it is necessary to define the term lottery. A lottery, which is a formalization of the term option used in the above assumptions, is defined by specifying a number of possible consequences and the probability that each will occur. The following three operational assumptions correspond to the fundamental assumptions above.

Assumption 5a. Define lottery L1 as a p_1 chance at c^* and a $1 - p_1$ chance at c^0, and lottery L2 as a p_2 chance at c^* and a $1 - p_2$ chance at c^0. Then L1 is preferred to L2 if p_1 is greater than p_2 and L1 is indifferent to L2 if p_1 is equal to p_2.

Assumption 6a. If L1, L2, and L3 are three lotteries, then preferences are transitive in the sense that if L1 is indifferent to L2 and L2 is indifferent to L3, then L1 is indifferent to L3, and if L1 is indifferent to L2 and L2 is preferred to L3, then L1 is preferred to L3, and so on.

Assumption 7a. If a lottery is modified by replacing one of its possible consequences with another (possibly a lottery) and if the client is indifferent between the original consequence and the replacement, then the client should be indifferent between the original and modified lotteries.

A.3 Fundamental Implication of the Assumptions

There is one fundamental result which follows from the assumptions stated in the preceding section. If a client accepts these assumptions, then the expected utility of candidate sites is the indicator to use in comparing and evaluating these sites. Specifically, sites with higher expected utilities should be preferred. How one should use the expected utility of each candidate site in comparing and evaluating the sites is discussed in detail in Chapter 8. In particular, as we have so often stated, because the purpose of the analysis is to provide insights about the best sites and documentation for regulatory authorities, it is appropriate to conduct a very thorough sensitivity analysis. The ranking provided by "best estimates of inputs, judgments, and values" should not be considered as the basic result of an analysis.

If a client does not accept the implications of the siting analysis conducted using decision analysis, there are two possible explanations. The client must either disagree with one or more of the basic assumptions or feel that the input information to the analysis was not adequate. Acceptance of the assumptions implies that the input information is inadequate. This helps to identify and focus on the weak part of the analysis, which should lead to improved insights and improved decision making.

A.4   Misconceptions about the Decision Analysis Siting Methodology

In order to appraise the appropriateness of decision analysis for siting, the client needs to consider each of the assumptions in Sections A.1 and A.2. The client should also be aware of those assumptions which are not made and which are not necessary for the procedure. Some of these are discussed in Section A.4.1. Possible misinterpretations of the result of any specific decision analysis siting study follows in Section A.4.2. A brief overall appraisal of the decision analysis siting methodology is found in Section A.4.3.

A.4.1   ASSUMPTIONS NOT REQUIRED
FOR THE DECISION ANALYSIS APPROACH

Although decision analysis sometimes appears to be simple, it is subtly complex both in theory and in practice. Consequently, there can be a tendency to believe that many nonstated assumptions are necessarily implied and utilized by decision analysis. This section is meant to dispel some of the main misconceptions.

The decision analysis approach is not a normative approach; it is a prescriptive approach. With a normative approach, a group of "wise old sages" specifies a set of assumptions (such as those in Section A.2) which they feel are rational. Then the sages conclude that any client who wishes to behave rationally should behave in accord with their assumptions. Anyone not acting in this way is then by definition irrational.

With the prescriptive approach, these same sages may specify a set of assumptions which they think may have appeal for particular problems. However, the assumptions must be discussed and verified with the client before utilizing the technique on the problem. Only when the client feels that the fundamental assumptions are appropriate does it follow that the technique is the one to be used in the siting problem.

Another misconception about the decision analysis approach for siting is that it is a substitute for the traditional approaches which have sometimes been used. This is not necessarily the case. In many circumstances the decision analysis approach can be thought of as a complement to traditional approaches and as a procedure which provides a framework for integrating the information gained from the utilization of those traditional approaches on specific parts of the siting problem. In particular, as is demonstrated throughout the book, decision analysis provides a framework for integrating the economic, environmental, socioeconomic, health and safety, and public attitude aspects of a siting problem. Decision analysis is a competitor of other procedures which claim to provide this integrative requirement for analysis.

There is no assumption that decision analysis of siting problems is easy. Siting problems are very complex. Hence, any analysis which attempts to address the complexity in the problem will necessarily be complex in practice. If the client accepts the assumptions in Section A.2 for a siting problem, then decision analysis is the correct methodology to use in that analysis. However, this in no way implies that the analysis itself is correct or accurate. The analysis may be done following the theory and procedures of decision analysis, but it can be done poorly. More to the point, many judgments, about both impacts and values, will be required for any analysis of siting problems. These judgments should be made responsibly and honestly, but different clients may have very different judgments and values. Consequently, different clients could each conduct a "highly professional" decision analysis of a siting problem and end up with different insights and implications. Each may be appropriate for the individual client but neither would be correct in an overall sense since no result would necessarily be correct for everyone.

A.4.2 MISINTERPRETATION OF THE FUNDAMENTAL RESULT OF DECISION ANALYSIS

It is a misinterpretation to assume that the site associated with the highest expected utility is the best site for a particular energy facility and therefore that it should be the chosen site. There are several reasons for this statement. The first ties in with the definitions of a best site discussed in the very first paragraphs of this book. Basically, a site may be the best site among the candidate sites in the decision problem, but it need not be the best possible site for an energy facility. The "best possible site" may not have been included as a candidate site. If you have a very good site and it is felt that a slightly better site may be found with a tremendous amount of effort, it is possible to conclude that the additional effort is not worth the additional benefit. Hence, the already identified site is the best to choose but not the overall best for the facility.

The second reason why the site with the highest expected utility may not be the best site is that the analysis may be poor. This needs no elaboration; if poor information is used in an analysis, the implications are worthless.

A third reason why the best site may not be identified is incompleteness of the analysis, particularly of the objectives and attributes. No siting study can include every single nuance that is important. For instance, the client may wish to exclude certain legal considerations from the analysis of potential candidate sites. Then, of course, the results of the study would indicate which sites are better with regard to aspects other than legal considerations. The site associated with the highest expected utility may have very negative legal implications whereas the site with the second highest expected utility may have positive legal implications. If the difference in legal implications is in some sense greater than the difference in the considerations captured by the expected utility, then the site with the second highest expected utility should be preferred in some overall sense. The judgment about the appropriate balance of the considerations formally included in the analysis and those excluded from the analysis must be done externally (either formally or informally) by the client.

A.4.3   An Overall Appraisal of the Decision Analysis Siting Methodology

There are two general lines of reasoning which support the decision analysis approach to siting. Both rely on the decision analysis assumptions. The first examines the assumptions to see if they are appropriate. The second compares the assumptions to those of other approaches to see if they, whether appropriate or not, are the best set of assumptions currently available. Rather than attempt to elaborate on every detail, it is probably more reasonable simply to outline each.

The first line of reasoning requires the acceptance of the tenets of decision analysis discussed in Section A.1. These are basically that the problem exists, that it is complex, and that the analysis should address the complexity of the problem. The fundamental assumptions of decision analysis in Section A.2 then indicate how one should address the complexity. The operational assumptions indicate how one can address it. Although it is hard to obtain the information needed for the decision analysis, if one assumes that the information is relevant to the problem, then one can also assume that other procedures are going to entail just as much difficulty in obtaining this information. Many of the alternative procedures do not attempt to obtain this information and therefore address a much smaller part of the siting problem. Consequently, they do not provide the number of insights desired for a complex decision problem and they are not suitable for supporting a decision to regulatory authorities.

The second line of reasoning in support of decision analysis is that its assumptions are weaker (i.e., less restrictive) than the assumptions of alternative siting methodologies. The comparison of siting methodologies is complicated because, aside from decision analysis, no siting procedure is based on a clearly articulated set of logically consistent behavioral assumptions. However, all siting methodologies require assumptions and the implicit assumptions necessary for the use of other techniques are identifiable.

Essentially all siting methodologies make the same fundamental assumptions as decision analysis does (outlined in Section A.1). Then most make oversimplifing implicit assumptions to address (or neglect) each of the concerns and features of siting problems. For instance, some approaches rapidly but inadequately screen a region of interest to identify candidate sites. Others use greatly oversimplified sets of objectives, often concerning mainly economics. In many instances, siting procedures do not formally address the uncertainties or the values in the siting problem. These oversimplifying assumptions reduce the task of implementing their methodology to a very routine level. The problem is that siting is not a routine exercise and routine procedures are inadequate for such a purpose.

REFERENCES

Arrow, K. J. (1951). "Social Choice and Individual Values." Wiley, New York (2nd ed., 1963).

Barrager, S. M., Judd, B. R., and North, D. W. (1976). The economic and social costs of coal and nuclear electric generation. A framework for assessment and illustrative calculations for the coal and nuclear fuel cycles. National Science Foundation, Office of the Science Adviser, Washington, D.C.

Bell, D. E. (1977a). A utility function for time streams having interperiod dependencies. *Oper. Res.* **25,** 448–458.

Bell, D. E. (1977b). A decision analysis of objectives for a forest pest problem. *In* "Conflicting Objectives in Decisions" (D. E. Bell, R. L. Keeney, and H. Raiffa, eds.). Wiley (Interscience), New York.

Bell, D. E. (1979a). Consistent assessment procedures using conditional utility functions. *Oper. Res.* **27,** 1054–1066.

Bell, D. E. (1979b). Multiattribute utility functions: decompositions using interpolation. *Management Sci.* **25,** 744–753.

Bell, D. E., and Raiffa, H. (1979). Marginal value and intrinsic risk aversion. HBS 79-65, Graduate School of Business Administration, Harvard Univ., Boston, Massachusetts.

Bell, D. E., Keeney, R. L., and Raiffa, H. (eds.) (1977). "Conflicting Objectives in Decisions." Wiley (Interscience), New York.

Bell, M. C. (1973). Fisheries Handbook of Engineering Requirements and Biological Criteria. North Pacific Division, Army Corps of Engineers, Portland, Oregon.

Black, S., Niehaus, F., and Simpson, D. (1978). Benefits and risks of alternative energy supply systems. Paper presented at the Haus der Technik-Tagung "Reaktortechnik und Kernenergieversorgung," February 27–28, Essen, Germany.

Brown, R. V., Kahr, A. S., and Peterson, C. (1974). "Decision Analysis for the Manager." Holt, New York.

Buehring, W. A. (1975). A model of environmental impacts from electrical generation in Wisconsin. Unpublished doctoral dissertation, Dept. of Nuclear Engineering and Institute of Environmental Studies, Univ. of Wisconsin, Madison, Wisconsin.

Buehring, W. A., and Foell, W. K. (1974). A model of environmental impact of Wisconsin electricity use. IES Rep. 32, Institute for Environmental Studies, Univ. of Wisconsin, Madison, Wisconsin.

Buehring, W. A., Foell, W. K., and Dennis, R. L. (1974). Environmental impact of regional energy use: a unified approach. Paper presented at the *French-Amer. Conf. Energy Syst. Forecasting, Planning, and Pricing, September, Madison, Wisconsin.*

Business Week. (1979a). A dark future for utilities, May 28, pp. 108–124.

Business Week. (1979b). Washington Public Power: going all-nuclear was the wrong reaction, June 4, p. 86.

Byer, P. H. (1979). Screening objectives and uncertainties in water resources planning. *Water Resources Res.* **15**, 768–773.

California Energy Resources Conservation and Development Commission. (1977). California Energy Trends and Choices. Power Plant Siting, Volume 7. 1977 Biennial Report of the State Energy Commission. Sacramento, California.

Carter, L. J. (1977). Auburn Dam: earthquake hazards imperil $1-billion project. *Science* **197**, 643–649.

Carter, L. J. (1978). Virginia refinery battle: another dilemma in energy facility siting. *Science* **199**, 668–671.

Comar, C. L., and Sagan, L. A. (1976). Health effects of energy production and conversion. *Ann. Rev. Energy* **1**, 581–600.

Council on Environmental Quality (1978). Regulations for Implementing the Procedural Provisions of the National Environmental Policy Act. Reprint 43 FR 55978-5607, 40 CFR Parts 1500-1508, U.S. Government Printing Office, Washington, D.C.

Craik, K. H., and Zube, E. H. (eds.) (1976). "Perceiving Environmental Quality." Plenum Press, New York.

Crawford, D. M., Huntzinger, B. C., and Kirkwood, C. W. (1978). Multiobjective decision analysis for transmission conductor selection. *Management Sci.* **24**, 1700–1709.

Dawes, R. M., and Corrigan, B. (1974). Linear models in decision making. *Psycholog. Bull.* **81**, 95–106.

Debreu, G. (1960). Topological methods in cardinal utility theory. *In* "Mathematical Methods in the Social Sciences, 1959" (K. J. Arrow, S. Karlin, and P. Suppes, eds.). Stanford Univ. Press, Stanford, California.

Denton, H. R. (1977). Nuclear power licensing: opportunities for improvement. NUREG-0292, U.S. Nuclear Regulatory Commission, Washington, D.C.

Doane, J. W., O'Toole, R. P., Chamberlain, R. G., Bos, P. B., and Maycock, P. D. (1976). The cost of energy from utility-owned solar electric systems. JPL 5040-29, Jet Propulsion Laboratory, California Institute of Technology, Pasadena, California.

Dobry, R., *et al.* (1970). Influence of magnitude, site conditions, and distance on significant duration of earthquakes. *World Conf. Earthquake Eng., January, New Dehli, India.*

Drake, A. W. (1967). *Fundamentals of Applied Probability Theory.* McGraw-Hill, New York.

Duke, K. M., Dee, N., Fahringer, D. C., Maiden, B. G., Moody, C. W., Pomeroy, S. E., and Watkins, G. A. (1977). Environmental quality assessment in multiobjective planning. Battelle, Columbus, Ohio.

Dyer, J. S., and Sarin, R. K. (1977). An axiomatization of cardinal additive conjoint measurement theory. Working paper No. 265. Western Management Science Institute, Univ. of California, Los Angeles, California.

Dyer, J. S., and Sarin, R. K. (1979). Measurable multiattribute value functions. *Oper. Res.* **27**, 810–822.

Eagles, T. W., Cohon, J. L., Revelle, C. S., and Current, J. R. (1979). Multiobjective programming in power plant location planning. FWS/OBS-78/55, U.S. Department of the Interior, Washington, D.C.

Edwards, W. (1977). Use of multiattribute utility measurement for social decision making. *In* "Conflicting Objectives in Decisions" (D. E. Bell, R. L. Keeney, and H. Raiffa, eds.). Wiley (Interscience), New York.

Electric Light and Power. (1977). Florida P&L gets realistic project cost estimates via computer science, June, p. 54.

El Paso Atlantic Company, El Paso Eastern Company, El Paso LNG Terminal Company. (1977). Joint LNG safety report respecting the proposed Algeria II project before the Federal Power Commission. Docket Nos. CP 73-258 *et al.,* Volume II of II.

Fairfax, S. K. (1978). A disaster in the environmental movement. *Science* **199,** 743–748.

Farquhar, P. H. (1977). A survey of multiattribute utility theory and applications. *In* "Multiple Criteria Decision Making" (M. K. Starr and M. Zeleny, eds.) Vol. 6, pp. 59–89. North-Holland/TIMS Studies in the Management Sciences.

Feller, W. (1950). "*An Introduction to Probability Theory and Its Applications,*" Vol. 1. Wiley, New York.

Finch, W. C., Postula, F. D., and Perry, L. W. (1978). Probabilistic cost estimating of nuclear power plant construction projects. *AACE,* 209–213.

Fischer, G. W. (1976). Multidimensional utility models for risky and riskless choice. *Organizational Behavior and Human Performance* **17,** 127–146.

Fischer, G. W. (1979). Utility models for multiple objective decisions: do they accurately represent human preferences. *Decision Sci.* **10,** 451–479.

Fishburn, P. C. (1964). "Decision and Value Theory," Wiley, New York.

Fishburn, P. C. (1965). Independence in utility theory with whole product sets. *Oper. Res.* **13,** 28–45.

Fishburn, P. C. (1970). "Utility Theory for Decision Making." Wiley, New York.

Fishburn, P. C. (1973). "The Theory of Social Choice." Princeton Univ. Press, Princeton, New Jersey.

Fishburn, P. C. (1977). Multiattribute utilities in expected utility theory. *In* "Conflicting Objectives in Decisions" (D. E. Bell, R. L. Keeney, and H. Raiffa, eds.). Wiley (Interscience), New York.

Foell, W. K. (ed.) (1979). "Management of Energy/Environment Systems. Methods and Case Studies." Wiley (Interscience), New York.

Ford, A. (1977). Summary description of the BOOM1 model. *Dynamica* **4,** 3–16.

Ford, A. (1979). Breaking the stalemate: an analysis of boom town mitigation policies. *J. Interdisciplinary Modeling and Simulat.* **2,** 25–39.

Ford, A., and Flaim, T. (1979). An economic and environmental analysis of large and small electric power stations in the Rocky Mountain West: summary of preliminary results. Los Alamos Scientific Laboratory, Los Alamos, New Mexico.

Ford, A., and Gardiner, P. C. (1979). A new measure of sensitivity for social system simulation models. *IEEE Trans. Syst., Man, and Cybernetics* **SMC-9,** 105–114.

Ford, C. K., Keeney, R. L., and Kirkwood, C. W. (1979). Evaluating methodologies: a procedure and application to nuclear power plant siting methodologies. *Management Sci.* **25,** 1–10.

Fulton, L. A. (1970). Spawning areas and abundance of steelhead trout and coho, sockeye, and chum salmon in the Columbia River Basin—past and present. National Marine Fisheries Service, Special Scientific Rep. No. 618.

Gándara, A. (1977). Electric utility decisionmaking and the nuclear option. R-2148-NSF, The Rand Corporation, Santa Monica, California.

Gardiner, P. C., and Edwards, W. (1975). Public values: multi-attribute utility measurement for social decision making. *In* "Human Judgment and Decision Processes" (M. F. Kaplan and S. Swartz, eds.). Academic Press, New York.

Garribba, S., and Ovi, A. (1977). Statistical utility theory for comparison of nuclear versus fossil power plant alternatives. *Nucl. Technol.* **34**, 18–37.

Glenn, A. H., and Associates (1976). Meteorological-oceanographic conditions affecting design and operations of proposed LNG tanker terminal facilities, La Salle, Terminal, Southwest Matagorda, Texas. New Orleans, Louisiana.

Gorman, W. M. (1968a). The structure of utility functions. *Rev. Econom. Stud.* **35**, 367–390.

Gorman, W. M. (1968b). Conditions for additive separability. *Econometrica* **36**, 605–609.

Gotchy, R. L. (1977). Health effects attributable to coal and nuclear fuel cycle alternatives. NUREG-0332, U.S. Nuclear Regulatory Commission, Washington, D.C.

Greensfelder, R. W. (1977). Seismicity, ground shaking, and liquefaction potential. Unpublished Special Rep. No. 120 by M. C. Huffman and C. F. Armstrong, California Division of Mines and Geology in cooperation with the Sonoma County Planning Department.

Greenwood, M. (1979). How to cope with increasing government/public involvement. *Pipe Line Industry* January, 41–45.

Gros, J. (1975). Power plant siting: a paretian environmental approach. *Nucl. Eng. Design* **34**, 281–292.

Gruhl, J. (1978). Alternative electric generation impact simulator—aegis, description and examples. M.I.T. Energy Laboratory, MIT-EL 79-0, Cambridge, Massachusetts.

Gutenberg, B., and Richter, C. F. (1954). "Seismicity of the Earth." Princeton Univ. Press, Princeton, New Jersey.

Harbridge House, Inc. (1974). The social and economic impact of a nuclear power plant upon Montague, Massachusetts and the surrounding area. Boston, Massachusetts.

Harsanyi, J. C. (1955). Cardinal welfare, individualistic ethics, and interpersonal comparisons of utility. *J. Political Economy* **63**, 309–321.

Hilborn, R., and Walters, C. J. (1977). Differing goals of salmon management on the Skeena River. *J. Fisheries Res. Board of Canada* **34**, 64–72.

Hobbs, B. F. (1979). Analytical multiobjective decision methods for power plant siting: a review of theory and applications. Brookhaven National Laboratory, Upton, New York.

Hobbs, B. F., and Voelker, A. H. (1978). Analytical multiobjective decision-making techniques and power plant siting: a survey and critique. ORNL-5288, Oak Ridge National Laboratory, Oak Ridge, Tennessee.

Holling, C. S. (ed.) (1978). "Adaptive Environmental Assessment and Management." Wiley (Interscience), New York.

Hub, K. A., and Schlenker, R. A. (1974). Health effects of alternative means of electrical generation. *In* "Population Dose Evaluation and Standards for Man and His Environment." International Atomic Energy Agency, Vienna, Austria.

Inhaber, H. (1978). Risk of energy production. AECB-1119/REV-2, Atomic Energy Control Board, Ottawa, Ontario, Canada.

Inhaber, H. (1979). Risk with energy from conventional and nonconventional sources. *Science* **203**, 718–723.

Kalelkar, A. S., Partridge, L. J., and Brooks, R. E. (1974). Decision analysis in hazardous material transportation. *Proc. 1974 Nat. Conf. Control of Hazardous Mater. Spills.* American Institute of Chemical Engineers, San Francisco, California.

Keeney, R. L. (1968). Quasi-separable utility functions. *Naval Res. Logistics Quart.* **15**, 551–565.

Keeney, R. L. (1972). Utility functions for multiattributed consequences. *Management Sci.* **18**, 276–287.

Keeney, R. L. (1974). Multiplicative utility functions. *Oper. Res.* **22**, 22–34.

Keeney, R. L. (1977a). A utility function for examining policy affecting salmon on the Skeena River. *J. Fisheries Res. Board of Canada* **34**, 49–63.

Keeney, R. L. (1977b). The art of assessing multiattribute utility functions. *Organizational Behavior and Human Performance* **19**, 267–310.

Keeney, R. L. (1979). Evaluation of proposed storage sites. *Oper. Res.* **27**, 48–64.

Keeney, R. L. (1980). Equity and public risk. *Oper. Res.* **28**, 527–534.

Keeney, R. L., and Kirkwood, C. W. (1975). Group decision making using cardinal social welfare functions. *Management Sci.* **22**, 430–437.

Keeney, R. L., and Lamont, A. (1979). A probabilistic analysis of landslide potential. *Proc. U.S. National Conf. Earthquake Eng., August 22–24,* Stanford Univ., Stanford, California.

Keeney, R. L., and Nair, K. (1977). Selecting nuclear power plant sites in the Pacific Northwest using decision analysis. *In* "Conflicting Objectives in Decisions" (D. E. Bell, R. L. Keeney, and H. Raiffa, eds.) Wiley (Interscience), New York.

Keeney, R. L., and Raiffa, H. (1976). "Decisions with Multiple Objectives." Wiley, New York.

Keeney, R. L., and Robilliard, G. A. (1977). Assessing and evaluating environmental impacts at proposed nuclear power plant sites. *J. Environmental Econom. and Management* **4**, 153–166.

Keeney, R. L., and Sicherman, A. (1976). An interactive computer program for assessing and analyzing preferences concerning multiple objectives. *Behavioral Sci.* **21**, 173–182.

Keeney, R. L., Kulkarni, R. B., and Nair, K. (1978). Assessing the risk of an LNG terminal. *Technol. Rev.* **81**, 64–72.

Keeney, R. L., Kirkwood, C. W., Ford, C. K., Robinson, J. A., and Gottlieb, P. (1979a). An evaluation and comparison of nuclear powerplant siting methodologies. NUREG/CR-0407, U.S. Nuclear Regulatory Commission, Washington, D.C.

Keeney, R. L., Kulkarni, R. B., and Nair, K. (1979b). A risk analysis of an LNG terminal. *Omega* **7**, 191–205.

Kemeny, J. G., et al. (1979). The accident at Three Mile Island. Report of the President's Commission, Washington, D.C.

Kemp, H. T., Little, R. L., Holoman, V. L., and Darby, R. L. (1973). "Water Quality Criteria Data Book, Volume 5, Effects of Chemicals on Aquatic Life." Water Pollution Control Research Series18050 HLA 09/73. Environmental Protection Agency, Washington, D.C.

Kirkwood, C. W. (1972). Decision analysis incorporating preferences of groups. Technical rep. No. 74, Operations Research Center, M.I.T., Cambridge, Massachusetts.

Kirkwood, C. W. (1979). Pareto optimality and equity in social decision analysis. *IEEE Trans. Syst., Man, and Cybernetics* **SMC-9**, 89–91.

Kohler, J. E., Kenneke, A. P., and Grimes, B. K. (1974). The site population factor. A technique for consideration of population in site comparison. WASH-1235, U.S. Atomic Energy Commission, Washington, D.C.

Koopmans, T. C. (1960). Stationary ordinal utility and impatience. *Econometrica* **28**, 287–309.

Koopmans, T. C. (1972). Representation of preference orderings over time. *In* "Decision and Organization" (C. B. McGuire and R. Radner, eds.). North-Holland Publ., Amsterdam.

Krantz, D. H. (1964). Conjoint measurement: the Luce-Tukey axiomatization and some extensions. *J. Math. Psychol.* **1**, 248–277.

Krantz, D. H., Luce, R. D., Suppes, P., and Tversky, A. (1971). "Foundations of Measurement," Vol. 1. Academic Press, New York.

Lapides, M. (1978). Predicting generating unit performance. *EPRI J.* April, 26–30.

Lee, W. W. L. (1979). "Decisions in Marine Mining: The Role of Preferences and Trade-offs." Ballinger Publ., Cambridge, Massachusetts.

Leistritz, F. L., Murdock, S. H., and Jones, L. L. (1978). Integrating environmental dimensions into economic-demographic impact project models. Presented at the Advisory Workshop, East Sound, Washington.

Lewis, H. W., Budnitz, R. J., Kouts, H. J. C., Loewenstein, W. B., Rowe, W. D., von Hippel, F., and Zachariasen, F. (1978). Risk assessment review group report. NUREG/CR-0400, U.S. Nuclear Regulatory Commission, Washington, D.C.

Lincoln, D. R., and Rubin, E. S. (1979). Cross-media environmental impacts of coal-fired power plants: an approach using multi-attribute utility theory. *IEEE Trans. Syst., Man, and Cybernetics* **SMC-9**, 285–290.

Luce, R. D., and Raiffa, H. (1957). "Games and Decisions." Wiley, New York.

Luce, R. D., and Tukey, J. W. (1964). Simultaneous conjoint measurement: a new type of fundamental measurement. *J. Math. Psychol.* **1**, 1–27.

MacCrimmon, K. R., and Toda, M. (1969). The experimental determination of indifference curves. *Rev. Econom. Stud.* **36**, 433–451.

MacCrimmon, K. R., and Wehrung, A. (1977). Trade-off analysis: the indifferent and preferred proportions approaches. *In* "Conflicting Objectives in Decisions" (D. E. Bell, R. L. Keeney, and H. Raiffa, eds.). Wiley (Interscience), New York.

Mark, R. K., and Stuart-Alexander, D. E. (1977). Disasters as a necessary part of benefit-cost analyses. *Science* **197**, 1160–1162.

McBride, J. P., Moore, R. E., Witherspoon, J. P., and Blanco, R. E. (1978). Radiological impact of airborne effluents of coal and nuclear plants. *Science* **202**, 1045–1050.

Meier, P. M. (1975). Energy facility location: a regional viewpoint. BNL 20435, Brookhaven National Laboratory, Upton, New York.

Meyer, R. F. (1970). On the relationship among the utility of assets, the utility of consumption, and investment strategy in an uncertain, but time invariant world. *In* "OR 69: Proceedings of the Fifth International Conference on Operational Research" (J. Lawrence, ed.). Tavistock Publ., London.

Meyer, R. F. (1977). State-dependent time preference. *In* "Conflicting Objectives in Decisions" (D. E. Bell, R. L. Keeney, and H. Raiffa, eds.). Wiley (Interscience), New York.

Miller, S. (1976). "The Economics of Nuclear and Coal Power." Praeger Publ., New York.

Morgan, M. G., Morris, S. C., Meier, A. K., and Shenk, D. L. (1978). A probabilistic methodology for estimating air pollution health effects from coal-fired power plants. *Energy Syst. and Policy* **2**, 287–310.

Murphy, A. W. (1978). The licensing of power plants in the United States. Seven Springs Center, Yale Univ., New Haven, Connecticut.

Myhra, D. (1975). One nuke gets a warm welcome. *Planning* **41**, 13–18.

Myhra, D. (1976). Montana Power builds new town for miners. *Practicing Planner* **6**, 13–15.

Nagel, T. J. (1978). Operating a major electric utility today. *Science* **201**, 985–998.

Nash, J. F. (1950). The bargaining problem. *Econometrica* **18**, 155–162.

Nash, J. F. (1953). Two-person cooperative games. *Econometrica* **21**, 128–140.

National Oceanic and Atmospheric Administration (1971). Wind distribution by Pasquill stability class (star program) seasonal and annual. National Climatic Center Federal Building, Asheville, North Carolina.

National Safety Council (1975). "Accident Facts, 1975 Edition." Chicago, Illinois.

Neyman, J. (1977). Public health hazards from electricity-producing plants. *Science* **195**, 754–758.

North, D. W., and Merkhofer, M. W. (1976). A methodology for analyzing emission control strategies. *Comput. Oper. Res.* **3**, 185–207.

Nuclear News. (1978). Defeat of referendum stalls plant. **21**, 50.

Okrent, D. (1980). Comment on societal risk. *Science* **208**, 372–375.

Oksman, W. (1974). Markov decision processes with utility independent objective functions. Unpublished doctoral dissertation, Harvard Univ., Cambridge, Massachusetts.

Olds, F. C. (1974). Power plant capital costs going out of sight. *Power Eng.* August, 36–43.

Organisation for Economic Co-operation and Development (1979). "The Siting of Major Energy Facilities." OECD Publ., Paris, France.

Otway, H. J., and Edwards, W. (1977). Application of a simple multi-attribute rating technique to evaluation of nuclear waste disposal sites: a demonstration. RM-77-31, International Institute for Applied Systems Analysis, Laxenburg, Austria.

Panel on Earthquake Problems Related to the Siting of Critical Facilities (1980). Earthquake research for the safer siting of critical facilities. National Academy of Sciences, Washington, D.C.

Papetti, R. A., Dole, S. H., and Hammer, M. (1973). Air pollution and power plant siting in California. R-1128-RF/CSA, The Rand Corporation, Santa Monica, California.

Parzen, E. (1960). "Modern Probability Theory and Its Application." Wiley, New York.

Pattanaik, P. K. (1971). "Voting and Collective Choice." Cambridge Univ. Press, London and New York.

Pollak, R. A. (1967). Additive von Neumann–Morgenstern utility functions. *Econometrica* **35**, 485–494.

Pratt, J. W. (1964). Risk aversion in the small and in the large. *Econometrica* **32**, 122–136.

Pratt, J. W., Raiffa, H., and Schlaifer, R. O. (1964). The foundations of decision under uncertainty: an elementary exposition. *Amer. Statist. Assoc. J.* **59**, 353–375.

Raiffa, H. (1968). "Decision Analysis." Addison-Wesley, Reading, Massachusetts.

Raiffa, H. (1969). Preferences for multi-attributed alternatives. RM-5868-DOT/RC, The Rand Corporation, Santa Monica, California.

Reichle, L. F. C. (1977). The economics of nuclear power. *Public Utilities Fortnightly*, February 3, pp. 24–32.

Richard, S. F. (1972). Optimal life insurance decisions for a rational economic man. Unpublished doctoral dissertation, Graduate School of Business Administration, Harvard Univ., Boston, Massachusetts.

Rosenbaum, N. (1977). The evolution of citizen involvement in governmental decision-making. Working paper 1226-09, The Urban Institute, Washington, D.C.

Ryall, A., Slemmons, D. B., and Gedney, L. (1966). Seismicity, tectonics, and surface faulting in the Eastern United States during historic times. *Seismolog. Soc. of Amer. Bull.* **56**, 1105–1135.

Sarin, R. K. (1980). Ranking of multiattribute alternatives with an application to coal power plant siting. *In* "Multiple Criteria Decision Making-Theory and Application" (G. Fandel and T. Gal, eds.). Springer-Verlag, Berlin.

Savage, L. J. (1954). "The Foundations of Statistics." Wiley, New York.

Savage, L. J. (1971). Elicitation of personal probabilities and expectations. *J. Amer. Statist. Assoc.* **66**, 783–801.

Schlaifer, R. O. (1969). "Analysis of Decisions under Uncertainty." McGraw-Hill, New York.

Schnabel, P. B., and Seed, H. B. (1972). Accelerations in rock for earthquakes in the Western United States. Report No. EERC 72-2, Earthquake Research Center, Univ. of California, Berkeley, California.

Seaver, D. A., von Winterfeldt, D., and Edwards, W. (1978). Eliciting subjective probability distributions on continuous variables. *Organizational Behavior and Human Performance* **21**, 379–391.

Sen, A. K. (1970). "Collective Choice and Social Welfare." Holden-Day, San Francisco, California.

Seo, F., Sakaya, M., Takahashi, H., Nakagami, K., and Horiyami, H. (1978). An interactive computer program for multiattribute utility analysis. GE18-1890-0, Tokyo Scientific Center, IBM, Tokyo, Japan.

Shaw, K. R. (1979). Capital cost escalation and the choice of power stations. *Energy Policy* **7**, 321–328.

Smith, J. H., Miles, R. P., Jr., and Goldsmith, M. (1978). An application of multi-attribute decision theory to the underground siting of nuclear power plants. Rep. 5030-224, Jet Propulsion Laboratory, Pasadena, California.

Spetzler, C. S., and Stael von Holstein, C-A. S. (1975). Probability encoding in decision analysis. *Management Sci.* **22**, 340–358.

Starr, C. (1969). Social benefit versus technological risk. *Science* **165**, 1232–1238.

Starr, C. (1979). "Current Issues in Energy." Pergamon, New York.

Tribus, M. (1969). "Rational Descriptions, Decisions, and Designs." Pergamon, New York.

Tversky, A., and Kahneman, D. (1974). Judgment under uncertainty: heuristics and biases. *Science* **185**, 1124–1131.

U.S. Federal Energy Administration (1976). National Energy Outlook. FEA-N-75/713, February. Washington, D.C.

U.S. Nuclear Regulatory Commission (1975a). Draft environmental statement related to construction of Montague nuclear power stations units 1 and 2. NUREG-75/109, Docket Nos. 50-496 and 50-497. Washington, D.C.

U.S. Nuclear Regulatory Commission (1975b). Reactor safety study. WASH-1400 (NUREG/74/104), October. Washington, D.C.

U. S. Nuclear Regulatory Commission (1976). Preparation of environmental reports for nuclear power stations. Regulatory Guide 4.2, Revision 2. NUREG-0099. Washington, D.C.

U.S. Nuclear Regulatory Commission (1977). Reactor site criteria. Title 10, Part 100, "Code of Federal Regulations-Energy," pp. 409–420. U.S. Government Printing Office. Washington, D.C.

U.S. Nuclear Regulatory Commission (1979a). NRC staff recommends against New York nuclear power plant site. News Release. Office of Public Affairs, February 13. Washington, D.C.

U.S. Nuclear Regulatory Commission (1979b). Report of the Siting Policy Task Force. NUREG-0625, Office of Nuclear Reactor Regulation. Washington, D.C.

Van Horn, A. J., and Wilson, R. (1976). Liquefied natural gas: safety issues, public concerns, and decision making. Energy and Environmental Policy Center, Harvard Univ., Cambridge, Massachusetts.

Van Horn, A. J., and Wilson, R. (1977). The potential risks of liquefied natural gas. *Energy* **2**, 375–389.

von Neumann, J., and Morgenstern, O. (1947). "Theory of Games and Economic Behavior," 2nd ed. Princeton Univ. Press, Princeton, New Jersey.

von Winterfeldt, D. (1978). A Decision Theoretic Model for Environmental Standard Setting and Regulations, Social Science Research Institute, Univ. of Southern California, Los Angeles, California.

von Winterfeldt, D., and Edwards, W. (1973a). Evaluation of complex stimuli using multiattribute utility procedures. Technical Report, Engineering Psychology Laboratory, Univ. of Michigan, Ann Arbor, Michigan.

von Winterfeldt, D., and Edwards, W. (1973b). Flat maxima in linear optimization models. Technical Report, Engineering Psychology Laboratory, Univ. of Michigan, Ann Arbor, Michigan.

Winkler, R. L. (1967a). The assessment of prior distributions in Bayesian analysis. *J. Amer. Statist. Assoc.* **62**, 776–800.

Winkler, R. L. (1967b). The quantification of judgment: some methodological suggestions. *J. Amer. Statist. Assoc.* **62**, 1105–1120.

Winkler, R. L. (1969). Scoring rules and the evaluation of probability assessors. *J. Amer. Statist. Assoc.* **64**, 1073–1078.

Winter, J. V., and Conner, D. A. (1978). "Power Plant Siting." Van Nostrand-Reinhold, Princeton, New Jersey.

Woodward-Clyde Consultants (1977). Oil terminal and marine service base sites in the Kodiak Island Borough. Alaska Depatment of Community and Regional Affairs, Division of Community Planning, Juneau, Alaska.

Woodward-Clyde Consultants (1978). Ranking of eight sites for coal-fired power plant development. San Francisco, California.

Woodward-Clyde Consultants (1979). Environmental assessment methodology: solar power plant applications, decision analysis computer program. ER-1070, Volume 4. Electric Power Research Institute, Palo Alto, California.

Wright, S. G. (1974). Columbia River fish runs and commercial fisheries, 1938–1970: status report and addendums. Joint Investigational Report of Oregon Fish Commission and Washington Department Fisheries, Vol. 1.

Wyzga, R. E. (1979). Siting of coal-burning power plants. *EPRI J.* October, 56–57.

INDEX